MATERIALS SCIENCE RESEARCH
Volume 7

SURFACES AND INTERFACES OF GLASS AND CERAMICS

MATERIALS SCIENCE RESEARCH

MATERIALS SCIENCE RESEARCH • Volume 7

SURFACES AND INTERFACES OF GLASS AND CERAMICS

Edited by

V. D. Fréchette, W. C. LaCourse, and V. L. Burdick

Division of Engineering and Science
New York State College of Ceramics
Alfred University
Alfred, New York

PLENUM PRESS • NEW YORK AND LONDON

Library of Congress Cataloging in Publication Data

International Symposium on Special Topics in Ceramics, Alfred University, 1973.
Surfaces and interfaces of glass and ceramics.

(Materials science research, v. 7)
"Tenth in the university series in ceramic science."
Includes bibliographical references.
1. Ceramic materials—Congresses. 2. Glass—Congresses. 3. Surfaces (Technology)—Congresses. I. Frechette, Van Derck, 1916-   ed. II. LaCourse, W. C., ed. III. Burdick, Vernon L., ed. IV. Title. V. Series.
TA430.I57   1973   620.1'4   74-17371
ISBN 978-1-4684-3146-9   ISBN 978-1-4684-3144-5 (eBook)
DOI 10.1007/978-1-4684-3144-5

Proceedings of an International Symposium on Special Topics in Ceramics
held August 27-29, 1973 at Alfred University, Alfred, New York

©1974 Plenum Press, New York
Softcover reprint of the hardcover 1st edition 1974

A Division of Plenum Publishing Corporation
227 West 17th Street, New York, N.Y. 10011

United Kingdom edition published by Plenum Press, London
A Division of Plenum Publishing Company, Ltd.
4a Lower John Street, London W1R 3PD, England

Preface

Awareness of the great significance of surface constitution in understanding the behavior and performance of
materials has been growing in proportion to the means
which have become available for surface study. Recent
years have seen important advances in analytical tools and
methods; their applications to date will certainly suggest
many other fruitful lines of investigation.

The Conference "Surfaces and Interfaces of Glass and
Ceramics" held at the New York State College of Ceramics
at Alfred University under the sponsorship of the U.S. Army
Research Office, Durham, and the National Aeronautics and
Space Administration, in August 1973, was tenth in the University Series in Ceramic Science, held in rotation among
North Carolina State University, the University of California
at Berkeley, the University of Notre Dame and Alfred University.

The chapters are arranged in order of their particular
emphasis beginning with those principally concerned with
analytical methods. Chapters dealing with friction and wear
follow, highly topical in the present-day concern with efficient use of energy in finishing processes, on the one hand,
and the avoidance of premature failure by frictional damage
to moving parts on the other. Surface reactions are then
considered, including the important questions of physiological
interactions with ceramic candidates for prosthetic applications. Material-material interfaces and transition zones
are discussed through examples which include grain boundaries in ceramics as well as interfaces among various solid,
liquid and gaseous phases. Finally, surfaces generated by
fracture are treated with respect to mechanisms and energetics of formation.

It is a pleasure to acknowledge the generous support
of our faculty and staff colleagues in arranging the Conference, particularly to Mrs. Coral Link who acted as its
secretary, and to Mrs. Jean Wassel who typed the photo-
ready copy. Special thanks are due Dr. H. M. Davis and
Dr. J. C. Hurt of AROD and Dr. James J. Gangler of
NASA for the encouragement of their support and for helpful advice.

Alfred, N. Y.   V. D. Frechette
May, 1974   W. C. LaCourse
                                       V. L. Burdick

Contents

CHARACTERIZATION OF SURFACES AND INTERFACES

Jhan M. Khan

Lawrence Livermore Laboratory
Livermore, California

The purpose of this paper[*] is to develop an overview of the techniques that are currently available for the characterization of surfaces and interfaces. The approach is to concentrate upon external agents (for example, electrons) that may be brought to bear upon a given specimen and to indicate the kinds of observables that are available for measurement. The specific probes considered here are: (1) electrons, (2) ions, (3) photons, and (4) electric fields. Of the observables, three are chosen as externally observable. These are: (1) electrons, (2) ions/atoms, and (3) photons. Also included among the observables is material response, that is, changes in the specimen brought about by exposure to the probing agent. For the incident and observed electrons, ions and photons, energy ranges have been established which are characteristic of either fundamental phenomena or commercially available instrument development.

The basic presentation is designed to give a rapid summary of the generally observed phenomena associated with a given probe and an energy range associated with that probe. In general, a specific tool will monitor only one or perhaps two of the many quantities available for observation.

*Work performed under the auspices of the U.S. Atomic Energy Commission.

1

There are two reasons for having an association of the types of interactions taking place. The first is that it may be possible to extend the capabilities of a given instrument to monitor other quantities of interest. The second is associated with possible adverse effects which may either affect the interpretation of the results or produce some irreversible change in the specimen under observation.

The table summarizes the incident probes, the observables, and the characteristic phenomena; where there are commercially available instruments that measure a specific observable, these instruments are shown in parentheses. Even in cases where an instrument is shown, not all aspects of the associated specific phenomena may be measured by the instrument noted. Electrical fields have been included as a probe purely for the sake of completeness; in general, this probing method cannot be considered to be a general purpose tool, but rather is associated with in situ study of some surface property.

BASIC PROCESSES

Classes of Information

There are four general classes of information that one may seek using the interactions noted in the table.[1]

Composition. It is frequently desirable to know the chemical composition at the surface of the specimen. In addition, knowledge of the variation in composition as one probes to greater depths of the specimen may be necessary to isolate purely surface effects from near-surface bulk composition variations.[2]

Structure. Given the chemical composition, it next becomes necessary to establish the surface topography, surface discontinuities which may be associated with grain boundaries, or impurity clusters. Again, this must be contrasted with bulk microscopic or macroscopic structure.

TABLE I.

OBSERVED → / INCIDENT ↓	ELECTRONS 0–5 keV	ELECTRONS 5 keV–100 keV	IONS/ATOMS 0–20 keV	IONS/ATOMS 20 keV–2 MeV	PHOTONS 0–5 keV	PHOTONS 5 keV–100 keV	MATERIAL RESPONSE
ELECTRONS 0–5 keV	1. Characteristic (Auger) 2. Secondary (SEM) 3. Scattered (LEED)	✕	1. Desorbed atoms 2. Ionized atoms	✕	1. Optical fluorescence 2. X ray fluorescence (microprobe) (Appearance potential spectroscopy)	✕	1. Broken chemical bonds 2. Compositional changes on surface 3. Color center formation 4. Thermal effects 5. Surface charging
ELECTRONS 5 keV–100 keV	1. Characteristic (Auger) 2. Secondary (SEM) 3. Scattered (RHEED) (SEM)	1. Characteristic (Auger) 2. Secondary (SEM) 3. Scattered (RHEED)	1. Desorbed atoms 2. Ionized atoms		1. Optical fluorescence 2. X ray fluorescence (microprobe)	1. X ray fluorescence (microprobe)	1. Atom displacement (radiation damage) 2. Broken chemical bonds 3. Color center formation 4. Stress/strain field formation 5. Thermal effects near surface 6. Surface charging 7. Compositional change at surface
IONS 0–20 keV	1. Secondary 2. Characteristic (Auger)	✕	1. Scattered (LEBS) 2. Secondary ions (SIS) 3. Sputtering 4. Implantation 5. Desorbed atoms	✕	1. Optical fluorescence 2. X ray fluorescence	✕	
IONS 20 keV–2 MeV	1. Secondary 2. Characteristic (Auger)	✕	1. Secondary ions 2. Sputtering 3. Implantation	1. Scattered (HEBS) 2. Secondary ions 3. Implantation	1. Optical fluorescence 2. X ray fluorescence	1. Optical fluorescence 2. X ray fluorescence 3. Nuclear gammas	
PHOTONS 0–5 keV	1. Photo electrons (ESCA) 2. Characteristic (Auger)	✕	1. Desorbed atoms 2. Ionized atoms	✕	1. Optical fluorescence 2. X ray fluorescence 3. Reflected photons 4. Scattered photons	✕	1. Color center formation 2. Broken chemical bonds 3. Composition changes near surface.
PHOTONS 5 keV–100 keV	1. Photo electrons 2. Secondary 3. Characteristic (Auger)	1. Photo electrons 2. Secondary 3. Characteristic (Auger)	1. Desorbed atoms 2. Ionized atoms	1. Ionized atoms	1. Optical fluorescence 2. X ray fluorescence 3. Scattered photons	1. X ray fluorescence 2. Scattered photons	
ELECTRIC FIELDS	1. Electron emission from surface (FEM) 2. Spark generation in high stress gradient situations.	✕	1. Ion emission from surface (FIM)	✕	1. Optical fluorescence		

Chemical State. For chemical interactions between the specimen and the outside world it is extremely desirable to know the chemical binding state of the atoms at the surface, again as contrasted with the bulk.

Electronic State. For special kinds of interactions, the distribution of electrons between valence states, impurity levels, and the conduction band may greatly influence the nature of the chemical or physical interactions at the surface of the specimen.

Interactions

In attempting to generate an overview of useful techniques for characterizing surfaces and interfaces, it is necessary to realize that the incident probe and the observable associated with that probe are connected generally by a complex sequence of events. By appreciating the basic nature of these events, it is possible to develop an intuitive feeling for the techniques that are capable of being applied in a quantitative manner as opposed to those that are inherently qualitative.

Dominant Interactions. Within the various combinations of probes and observables, it is possible to identify those where the dominant interaction is basically binary in nature. An example is the ionization of an atom by an electron resulting in the emission of an x-ray. In other interactions, more complex in nature, a collective response on the part of the atoms in the specimen give rise to the observable. An example of this class of interaction would be the emission of low energy secondary electrons resulting from exposure of the sample to the beam of a scanning electron microscope.

Binary	Collective
Auger Electron Production	Electron Diffraction
X-ray Production	Secondary Electron Emission
Ion Scattering	Secondary Ion Emission
Photo-Electron Production	Desorption of Ions or Atoms

Nuclear Gamma Ray Pro- Sputtering Process
  duction   Optical Photon Production
          Optical Photon Reflection

General Characteristics of Probes

Electrons. Electron beams are potentially the most useful general tool for material characterization. One reason is that electron beams having energies above several keV may be focused to dimensions smaller than typical grain sizes. At higher energies, beams may be focused to dimensions of the order of 10 nm or less. The dominant interactions of beam electrons are with the bound electrons within the specimen. Binary events can, in principle, be quantitatively predicted. Collective interactions can result in a wide variety of scattered primary electrons and secondary electrons. The scattered primary electrons that satisfy the conditions for constructive interference can give much useful information regarding short- and long-range order. In addition, these primary electrons may break chemical bonds, deposit a large amount of thermal energy on the surface of the specimen, and by charge collection on the surface of the specimen, produce very high surface potentials.

Ions. The dominant interaction associated with energetic incident ions is momentum transfer to the specimen atoms. This can result in sputtering, secondary ion production, and scattered primary ions. For high energy ions of lower mass, ionization and excitation of atomic electron shells may take place. The diameter of ion beams is generally larger than one micrometer and is frequently on the order of millimeters. Only in a few demonstrated practical cases have ion probe techniques been used to obtain surface structural information. The detrimental effects associated with ion beams are generally direct consequences of atom displacement. These displacements may result in breaking chemical bonds, production of thermal spikes, general heating near the surface, and charging of the surface to high potentials.

<u>Photons</u>. Photons include optical photons, through high energy x-rays. Depending upon the energy and wavelength of the photons, and the density of free and bound electrons at the surface of a specimen, many qualitatively different phenomena take place. The dominant interactions of photons of energy less than 100 keV are with electrons.

Many optical techniques are being employed for the analysis of surfaces. This general field will not be reviewed here. IR spectroscopic techniques have proved to be of great value in the analysis of chemical bonds. One specific technique has evolved in the recent years due to the ready availability of high-speed computers. This technique is referred to as ellipsometry and requires complex calculations to reduce the data to the fundamental quantities of interest - namely, the real and imaginary indices of refraction at the surface.

With photons of higher energy, absorption processes start to dominate. This results in absorption of the photons in the near-surface region. When absorption is associated with the ejection of electrons from atomic levels, two characteristic processes result. Observation of the photoelectron is employed in Electron Spectrometry for Chemical Analysis (ESCA). When characteristic electrons result from the radiationless reorganization of the atom, the Auger Electron Spectrometry (AES) technique is invoked. In view of the basic interaction of photons with electrons, effects associated with the breaking of chemical bonds may be expected. One effect may be change of composition of the specimen near the surface.

CURRENT TECHNIQUES AND METHODS

The system characteristics of a practical instrument reflect the characteristics of both the exciting agency and the system for detecting the observable. In general, the major improvements have been associated with greater sophistication in detection and with expansion of the number of phenomena monitored for a given specific exciting agent. This is clearly the present trend in surface characterization.

One question of considerable importance to the materials scientist is spatial resolution. It is convenient to quote a probe diameter as a characteristic of the tool. While this may set a lower limit upon the source region of the observed quantity, it by no means fully characterizes the resolution in the plane of the surface. The reason is that there may be a large sequence of events that must occur before the emitted electron, ion, or photon is emitted. The result is that the effective source region may appear much larger than the probe diameter. A good example of this effect is in the production of characteristic x-rays from a high energy (50 keV) electron beam. The source area characteristic of the x-rays may well be micrometers in diameter, even though the electron beam is only 100 nm.

Good depth resolution is not readily achieved in surface characterization tools. The only "non-destructive" technique is limited to cases where the atomic specie of interest is greater in atomic weight than the bulk. There is a family of destructive techniques. In general, they employ sputtering to remove surface layers. This may be alternated with surface characterization or in some cases they may be simultaneous. This removal process may produce changes in the specimen that are detrimental to the study. Care should be exercised in applying such tools.

The final question in tool selection and application is sensitivity to surface roughness. The same arguments apply as were developed in the discussion of resolution. When there are secondary processes that increase the effective source volume over the probe diameter, then one may expect surface roughness effects. These effects may also be expected when either the probe or the detection system views the system at a glancing angle.

Techniques Based Upon Scattered Primary Electrons

An incident beam of electrons may interact with a solid in two ways, reflecting the wave-particle duality of quantum mechanics. As a wave the electron may be diffracted by an orderly array of atoms to give information related to atomic arrangement. In addition, the electron may lose energy to the solid by collisions. The energy loss may be associated with excitation of specific levels within the atoms or excitation of collective motion of electrons in the conduction band of metals (plasmons). A special case of the collision case is elastic scattering where the electron loses no significant portion of its initial energy.

Low Energy Electron Diffraction (LEED). This technique[3-5] observes the two-dimensional structure on the surface of monocrystalline solids. It employs a primary beam of monoenergetic electrons of energy less than several hundred electron volts. LEED is used to study nucleation and growth phenomena for adsorbed or deposited species on crystal surfaces.

It is limited in applicability by the need for preparing clean, well-ordered surfaces. Many materials undergo phase transitions during the cleaning process that leave the surface in a disordered state when averaged over beam dimensions of the order of a millimeter.

Reflection High Energy Electron Diffraction (RHEED). The RHEED technique[6] employs an electron beam with energy in the range from 10-100 keV. The beam is normally incident at a grazing angle (1° - 3°). The electron wave is diffracted as it scatters from the surface. In general, the electron is actually transmitted through small surface irregularities. The sensitivity of RHEED is comparable to LEED, both sampling the very near surface region. RHEED is not limited to monocrystalline samples, however, and can give information concerning crystallite size, lattice spacing, and crystallite orientation.

Techniques Based on Secondary Electron Emission

For purposes of surface analysis, attention is focused upon the characteristic portion of the secondary electron spectra. These characteristic electrons emitted are produced by specific transitions within the atoms of the solid. They are somewhat analagous to x-ray fluorescence. The characteristic electrons studied are generally quite low in energy and originate in the first 10 nm of the surface or less. Surface contamination is a very serious problem for these measurements.

Auger Electron Spectroscopy (AES). The AES technique[7-10] employs an electron beam to excite the atoms that emit Auger electrons. Under most circumstances AES gives an elemental analysis of the first several atom layers of the surface with a sensitivity of 1/10 of an atom layer or about 10^{18} atoms/M^2 or better. In general, the results are qualitative or at most semiquantitative. In spite of some limitations in this area, it is proving to be a versatile tool which may be applied to metals or to non-conductors. It may be combined with sputter removal of surface layers to give a depth profile.

Recent developments indicate that scanning has been added to commercial instrument capabilities, which adds the ability of identification of the region of analysis. Scanning in an SEM mode (5μm spot size) or in an AES mode is now possible.

Electron Spectroscopy for Chemical Analysis (ESCA). ESCA[11,12] also looks at characteristic electrons, but these are photo electrons. An intense source of characteristic x-rays is employed. The principal difference between AES and ESCA lies in the sharpness of the electron spectrum observed. For AES the full width at half maximum is of the order of 1 eV, while for ESCA it may be as low as 0.1 eV. This latter value allows ESCA to observe chemical shifts in a number of cases. With information of this sort, it is possible to identify the binding state of an

atom on the surface, thereby shedding light upon the
surface chemistry of the system.

Scanning Electron Microscopy (SEM). The SEM con-
ventionally displays the intensity of the low energy secondary electron spectrum and the high energy elastically
scattered electrons. In the former case surface topography is observed due to changes in efficiency of emission.
The latter case utilizes the increase in cross section or
probability scattering, with charge on the scattered atom.
Thus a display is presented of a relative distribution of
"light" versus "heavy" atoms on the surface.

A common accessory to the SEM is the nondispersive
x-ray detector. In this mode elemental analysis is possible.
Due to secondary x-ray generation processes, the spatial
resolution is much larger than the beam spot size in many
cases.

The SEM may be upgraded to allow incorporation of
Auger Analysis (SEM-AES).[13] The changes are most
associated with improvement in the vacuum environment.
The partial pressures of condensible, reactive species
should be very much below 10^{-6} Pa (10^{-8} torr). Under
these conditions, the AES-SEM system allows the study
of surface composition with a spatial resolution which may
be comparable to the beam dimensions, i.e., 10 nm,
although secondary production processes may yield effective
sources much larger than these dimensions.

Techniques Based on Ions

Ions incident upon a solid sample lose energy by interactions such as electronic excitation and ionization and
nuclear scattering. The latter may be either elastic or
inelastic. Elastic scattering implies a conservation of
momentum and kinetic energy. Three quantities are of
interest: (1) energy of recoil of the incident particle,
(2) cross section or probability of scattering the incident

particle into a specific element of solid angle, and (3) probability that the incident particle is charged after the scattering event. We will wish to look at two regimes of scattering, i.e., high energy ($E_o \gg 10$ kev) or low energy ($E_o \leq 10$ kev). In both cases the recoil energy, E_1, from a two-particle event is given by:

$$\frac{E_1}{E_o} = \frac{M_1^2}{(M_1+M_2)^2} \left[\cos \theta + \left(\frac{M_2^2}{M_1^2} - \sin^2 \theta \right)^{1/2} \right]^2$$

where M_1 and M_2 are the masses of the primary ion and the target atom, respectively, and θ is the scattering angle in laboratory coordinates.

High Energy Scattering. Typically the incident energies are of the order of 1 MeV and the incident ions are of low mass (H^+, He^+, C^+, N^+, O^+). The detection systems are usually only energy sensitive; the charge state is irrelevant. The application of this technique yields:
    1.   Mass analysis.
    2.   Depth distribution to the order of 0.5 μm with a resolution of 30 nm (300 a.u.) or greater.
    3.   Spatial distribution with typical resolution of 1 mm.
    4.   Sensitivity: $M_2 > M_s$ (bulk masses): 10^{18} atoms/M^2
                          $M_2 < M_s$: 10% of M_s
From the above it is clear that the mass analysis is averaged over a minimum depth of 30 nm or so.

Low Energy Ion Scattering.[19-22] Typical incident energies are of the order of several keV. Under these conditions only incident ions that are scattered from the outermost atoms of the solid remain charged. If the detection system uses electrostatic analysis (as is usual) then only these charged species are detected. Therefore, this technique gives a mass distribution of those atoms in the first atom layer of the solid. Inspection of the recoiling equation given above indicates that the energy separation between adjacent masses decreases as the mass increases.

This results in relatively low mass resolution at high
masses. In addition, multiple scattering processes have
not been fully quantified yet. Their general effect is
broadening of lines. In spite of this qualification to the
technique, it has been employed to study a variety of
surface corrosion effects.

Secondary Ion Spectroscopy.[23] When the inelastic
scattering processes are considered, one result is
sputtering. This treats the ejection of atomic species
from the surface of the specimen. A low fraction (10^{-3})
of the sputtered atoms may be charged, either positively
or negatively, depending upon the respective electron
affinity of the atom and the surface.

The secondary ion spectroscopy technique is based
upon the mass analysis of the ejected ions. The energy
of the incident ions is usually around 20 keV. At these
energies beam spot sizes of the order of 10 µm have been
obtained in commercial instruments. Mass analysis can
be achieved by either quadrupole mass analyzers or by
more conventional analyzer systems.

The secondary ion technique has one element that may
cause some difficulty in obtaining quantitative results. The
fraction of the atoms ejected that are charged is very
dependent upon the composition of the surface. Some
success has been achieved by employing incident ions of
oxygen or cesium in many cases.

Technique Based on Optical Reflection (Ellipsometry).
This technique[24-29] is based upon the measurement of the
reflectivity of the components of light with planes of polarization perpendicular and parallel to the plane of incidence.
The method has been employed to characterize surfaces
subjected to a variety of treatments: mechanical polishing,
chemical etching, sputtering by argon ions, cleavage and
annealing. The resulting reflectivities, and indices of
refraction obtained were very sensitive to treatment. This
sensitivity indicates that ellipsometry is a technique that

may have a valuable place in the laboratory.

SURFACE PREPARATION METHODS

As a general statement, all surfaces must be considered active. That is, the surface is prone to the accumulation of impurity atoms, whether they be chemisorbed or physisorbed. The origin of the impurities may be internal or external to the bulk. Two classes of impurities exist: surface and bulk. There is a liberal interchange between these, in practice. Burton[2] modeled the equilibrium concentration of impurities on crystal surfaces. In his model, he generated the following expression:

$$S = B \exp(-\Delta F / kT)$$

where S and B are the surface and bulk impurity concentrations, respectively, and ΔF is the change in free energy of the crystal when the impurity atom is moved to the surface. By considering surface sites, and substitutional and interstitial sites, he concluded that at temperatures below the melting point, impurity atoms that are strongly bound to the solvent atoms will move to the bulk, while impurity atoms for which the impurity-solvent bond is weaker than the solvent-solvent bond will concentrate on the surface. Rate constants will, of course, be determined by diffusion constants. For the surface-concentrating impurity, the bulk may represent a semi-infinite reservoir just waiting for conditions to become favorable to mass transport, such as temperature rise or radiation damage (radiation-enhanced diffusion). This temperature rise could be due to the exposure of the surface to an analyzing probe.

Bulk impurities may supply low-lying donor electron levels in the case of dielectrics or semiconductors. This phenomenon is exploited in the solid state electronics industry. These impurities may also influence mechanical properties as they can affect defect motion.

At surfaces of grains or on the gas-solid interface,
impurities may significantly influence physical, chemical,
and mechanical properties in what may appear to be
unpredictable ways.

As a result of the foregoing, all operations involving
the external surface should be accompanied with some
experimental test of the surface condition. In the extreme,
as many of the operations as possible should be performed
in a controlled environment so that the full history of the
surface is known. In practical circumstances, this may
not be possible on a routine basis. Under these conditions
the systems employed should be fully understood prior to
these uses so that limits may be placed on the contributions
of different sources of imperfections.

Preparation of Solid Surfaces

The starting point for preparation of a surface is the
bulk material. The chemical composition and structural
characteristics of the bulk should be ascertained prior to
any surface preparation. In addition, the completeness of
the chemical analysis must also be assured. Given confidence in the bulk composition, then a variety of surface
preparation methods may be applied.

Each method or sequence of methods of surface preparation has its own set of advantages and disadvantages,
so that the techniques to be used for a given application
should, in principle, be the result of systematic study of
the different preparation methods on the desired physical
and chemical properties of the finished surface.

Mechanical Operations. The act of cutting, grinding,
or polishing will inherently cause disorder in mono or
polycrystalline solids. In addition, the polishing agents
will be present in or on the surface. The depth of these
effects will vary but may extend microns in general.

The principal purpose of mechanical operation is to give the sample an overall shape. Typically, after cutting and grinding of hard material, polishing is employed. The last stages of polishing usually employ a succession of finer and finer abrasive powders in a suspension. The last particle size may be a fraction of a micron. Disruption and contamination may be expected to be found to depths of the order of the abrasive particle diameter. It will be necessary to remove this portion of the surface to remove such effects, if they are found to be detrimental.

Chemical Etching. To remove surface material, a chemical etch is frequently employed. The particular etch employed will vary with the solid. Again, care must be taken. Some of the problems encountered are:

1. Concentration of less soluble impurities at the surface.
2. Impurities in the etchant remaining bound to the surface.
3. The etchant may partially decompose on the surface, leaving reactive elements behind (such as fluorine).
4. Selective action by the etchant that will partially destroy the smoothness left by the final stage of polishing. As in all cases of surface preparation, the surface should be characterized following the etching operation.

Ion Bombardment.[30] A recent technique that has been used to clean surfaces is inert ion bombardment at low energies (<10 keV). Typical ions are Ar^+, Ne^+, Xe^+. Material removal may be effected with reasonable efficiency. Two potential problems arise: (1) implantation of the ion into the surface, and (2) production of a highly reactive surface in the case of dielectrics. Disorder produced in the surface by low energy ion bombardment is well documented. Frequently, annealing may be used to remove disorder and drive off residual inert gas atoms from the surface region.

Annealing. Annealing to remove disorder and impurities is well known. However, impurities may diffuse
to the surface as a result of this operation.[2] In cases such
as the preparation of atomically clean, ordered silicon
surfaces, many cycles of ion bombardment and annealing
in an ultra-high vacuum environment may be required to
reduce the impurity level to an acceptable value.

Cleaving/Fracturing. The cleaving technique is deemed
best for obtaining an atomically clean surface. Recent
experiments show that stress-strain fields may exist in the
cleaved surfaces up to microns in depth. If it is necessary
to obtain a strain-free surface, annealing may be employed.
The final surface should be characterized, however, to
ascertain the impurity level.

Fracturing of solids to expose internal grain boundaries is a well established method in the study of materials. Care must be exercised when employing surface
characterization techniques that sample the first few atom
layers (e.g. AES), as handling-derived contaminants may
obscure the real surface composition.

FUTURE DIRECTIONS

There are many facets to the question of where the
future developments will lead us. For those who are
interested in practical tools, the pace will be set by commercial developments. The nature of the tools is such
that they tend to be expensive, even when produced in
quantity. Therefore, instrument manufacturers will be
predominantly guided by market potential, which implies
broadness of application. For that reason it should not be
expected that the more esoteric, special purpose tools
will become production items in the near or foreseeable
future.

In spite of the above, however, developments are
occurring that hold considerable hope for the user. One

factor is cost reduction. As some of the sophisticated
instruments mature, advanced engineering is reducing
costs substantially. In addition, new capabilities are
being added to some of the more conventional tools to give
them broader utility.

To be specific, the following basic instruments will
show expansion of applicability in the future. (No significance should be attached to the order of appearance).

> Scanning Electron Microscope
> Auger Electron Spectrometer
> Electron Spectrometer for Chemical Analysis

The author believes that these basic instruments will
be adapted to include more simultaneous or sequential
operations so that there may emerge a "universal" surface
analysis tool that will give compositional, topographical,
and structural information in a convenient, well engineered
package. Such developments may occur within the next
five to ten years.

CONCLUSION

The brief review contained in this paper is intended
to serve two purposes. First is to point out the variety
of physical processes that may take place while analyzing
the surface. The second is to indicate which of these processes have been developed to instrument stage. The
instrument situation should be viewed as a dynamic one,
with improvements and expansions of capability continuing
to evolve.

It should be clearly realized that as tools of analysis
become more and more sensitive to the true surface, they
become more sensitive to contaminating effects. The
origin of these effects may be sample preparation or
handling, or they may originate within the operation of
analysis.

The user of such tools should do a systems analysis of his experiment so that all operations are self-consistent in the degree of care exercised in preparation, handling, and analysis, and so that the data and conclusions produced do not depend upon unrealistic assumptions. It is extremely desirable that tool selection and experiment design be undertaken with guidance from experienced surface analysis personnel in order to avoid misleading and expensive exercises.

REFERENCES

1. Symposium Proceedings, J. Vac. Science Tech. 10 (1973).
2. Burton, J., J. Phys. Rev. 177, 1346 (1969).
3. MacRae, A.U., Surface Sci. 13, 130-3 (1969).
4. Haas, T.W., Dooley, III, G.J., and Grant, J.T., Prog. in Surface Sci. 1, Part 2 (1971).
5. Laramore, G.E., Houston, J.E., and Park, R.L., J. Vac. Sci. Tech. 10, 196 (1973).
6. Bauer, E., Techniques of Metals Research, Vol. II, Part 2, Ch. 15 (1969).
7. Taylor, N.J., J. Vac. Sci. Tech. 6, 241-5 (1969).
8. Jenkins, L.H. and Chung, M.F., Surface Sci. 28, 2 (1971).
9. Sickafus, E.N. and Steinrisser, F., J. Vac. Sci. Tech. 10, 43 (1973).
10. Melles, J.J., Davis, L.E., and Levenson, L.L., J. Vac. Sci. Tech. 10, 140 (1973).
11. Siegbahn, K., et al, Nova Acta Regiae Societatis Scientiarum Upsaliensis IV, 20, 1967.
12. Karsek, F.W., R and D, 25 (Jan. 1973).
13. MacDonald, N.C., Appl. Phys. Let. 16, 76 (1969).
14. Rutherford, Sir Ernest, Geiger, H. and Marsden, E., Phil. Mag. 21, 669 (1911).
15. Carter, G., J. Vac. Sci. Tech. 10, 95 (1973).
16. Rubin, S., Nucl. Inst. Methods 5, 177 (1959).
17. Davies, J.A., Denhartog, J., Eriksson, L. and Mayer, J.W., Can. J. Phys. 45, 4053 (1967).

18. Nicolet, M. A., Mayer, J. W., and Mitchell, I. V.,
    Science 177, 841 (1972).
19. Smith, D. P., J. Appl. Phys. 38, 340 (1967).
20. Goff, R. F. and Smith, D. P., J. Vac. Sci. Tech. 7,
    72 (1970).
21. Smith, D. P. Surface Sci. 25 (1971).
22. Ball, D. J., Buck, T. M., MacNair, D. and Wheatley,
    G. H., Surface Sci. 30, 1 (1972).
23. Liebl, Helmut, J. Appl. Phys. 38, 5277 (1967).
24. Heavens, O. S., Optical Properties of Thin Solid Films,
    Academic Press, New York (1955).
25. Frankl, D. R., J. Appl. Phys. 34, 3514 (1963).
26. Kruger, J., in: Ellipsometry in the Measurement of
    Surfaces and Thin Films, Symposium Proceedings,
    E. Passaglia, R. R. Stromberg and J. Kruger, Eds.,
    Natl. Bur. Stds., Misc. Publ. 256, Washington, D. C.,
    1964.
27. Fehlner, F. P. and Mott, N. F., Oxidation of Metals 2,
    59 (1970).
28. Paik, W. and Bockris, J. O. M., Surface Sci 28, 1
    (1971).
29. Vedam, K. and So, S. S., Surface Sci 29, 2 (1972).
30. Carter, G. and Colligon, J. S., Ion Bombardment of
    Solids, American Elsevier Publ. Co., Inc. 1968.

DISCUSSION

H. C. Gatos (MIT): The matrix of incident vs. emitted
electrons, ions and photons presented is very useful indeed in summarizing the interactions on which nearly all
modern techniques for the study of surfaces are based
(the interactions with electric and magnetic fields are
excluded). All those reviewed have been most useful to
the physicist in that the various interactions involved
provide direct guidelines to the energetics and electronic
configuration of surfaces. The terms "clean" and "real"
are perhaps better (and already widely used) than the
corresponding terms "natural" and "artificial" used by the
author.

R. W. Rice (Naval Res. Lab.): One of the important surfaces for ceramists to analyze are grain boundaries. By
fracture at an appropriate temperature we can form surfaces that consist primarily of grain boundaries, but of
course are not flat on a local scale. The surface analysis
must therefore be relatively insensitive to surface roughness.

Author: The sensitivity of a surface analysis tool to surface roughness depends principally on the depth of origin
of the observed emission. Auger electrons originate within the first few atomic layers and therefore are affected
only by "line-of-sight" roughness features. X-rays generated in either the SEM or electron microprobe may
originate as deeply as many microns. Scanning Auger
spectroscopy holds the greatest hope of answering the
questions you raise.

J. Kraitchman (PPG): Recent work has shown that information about the binding state and chemical environment of
atoms, such as is obtained by electron spectroscopy (ESCA),
can also be had from Auger spectroscopy. Since smaller
incident beam size favors the Auger, do you consider it
the more useful and universal technique?

Author: Owing to atomic level width considerations, ESCA
is inherently far more sensitive to chemical binding effects.
Auger spectroscopy (AES) has detected chemical effects
where binding energy shifts were large. From this standpoint ESCA is preferable. However, when the question of
spatial resolution is raised, developments in the AES field
suggest that probe diameters less than the presently available 5 μm will be appearing. If the question of universality
is to be answered without a detailed problem specification,
AES is a reasonable choice.

D. Dove (Univ. Florida): Could you comment on the spatial
resolution attained with various techniques.

Author: The present state of the art with respect to probe
sizes can be summarized: AES/SEM, 10 nm; scanning AES,
5 μm; conventional AES, 1 mm; ion scattering, 1 mm; ESCA,
about 1 cm. However, spatial resolution may be larger than
the probe diameter, especially for the AES/SEM case.

J. Kraitchman (PPG): In depth-profiling, complications

may arise because ion etching can remove various constituents at different rates. Even if steady-state conditions
are attained, these may not represent the true composition.
Author: Experiments performed employing both Auger
analysis and low energy ion scattering have reasonably well
established that for low energy heavy ion sputtering the
effect of compositional variation is small. (This surprised
me also!)

H. Bach (Schott): We observe distinct differences between
the start of sputtering and after a steady state is reached.
Sputtering efficiency is a function of the ratio of the mean
effective target mass to the ion mass, the relative bond
energy and the angle of incidence. The angle dependence
of energy transfer to the target atoms induces angle-
dependent concentration profile changes for different
constituents. While the more weakly bonded atoms will
be first to be knocked out of the first atomic layers, the
measured ratio will be correct at steady state if the target
is uniform.

J. Waner (GE): In ion milling or etching for in-depth analysis, even at the "gentle" rates of 0.1 nm per second, isn't
structural order at the surface apt to be highly disturbed,
e.g., crystals can be transformed to glasses and vice versa?

Author: Yes, none of the surface analysis techniques discussed (along with preparation or removal techniques) can
be considered nondestructive. The experimenter must
evaluate what threshold for damage is acceptable. Generally, when compositional analysis is undertaken, especially
when combined with surface removal techniques, structural
information is lost.

THE INFRARED SPECTRA OF SEVERAL N-CONTAINING COMPOUNDS ABSORBED ON MONTMORILLONITES

S. D. Jang, G. A. Garwood, J. E. Rourke and
R. A. Condrate Sr.
New York State College of Ceramics
Alfred University
Alfred, New York

Earlier investigations have shown that infrared
spectroscopy provides an ideal non-destructive tool
for investigation of adsorbed species present when simple compounds are adsorbed on ceramic materials[1].
In this study, we have looked at the infrared absorption
spectra of several more complex compounds that can be
adsorbed on montmorillonites. These compounds relate
to biologically important materials that can be prepared
on ceramic surfaces or in their interlamellar spaces.
We will first discuss the spectra observed for adsorbed
ammonia and gradually proceed to a discussion of the
spectra observed for an adsorbed dipeptide, glycylglycine.
The latter compound is the simplest compound which pos-
sesses a peptide linkage similar to proteins. The types
of complexes formed on the clays depend strongly upon
such factors as the time of exposure of the clay to the
compound, the nature of the heat treatment of the clay
complex and pH at which the clay complex was prepared.
Analyses of the infrared spectra in this study distinctly
indicate the exact nature of the types of complexes present
in the montmorillonites.

EXPERIMENTAL PROCEDURES

Preparation of Clay Films

Wyoming montmorillonite films were prepared from particles whose equivalent diameters were less than two microns, collected by sedimentation from a 5% clay suspension in distilled water. The cation-substituted montmorillonite was prepared by saturating natural clay with a 1-N solution of the appropriate metal chloride, and then washing several times with distilled water to eliminate free cations. H-montmorillonite was prepared immediately prior to use from a 1% suspension of natural clay by passing it through a column of Amberlite IR-120 (H-form). The concentrations of the resulting clay suspensions were adjusted so that the clay made up approximately 2. 5% by weight.

In the case of amino acids equal amounts of the appropriate clay suspension and a 0. 05M amino acid solution were mixed together. The pH was adjusted to the appropriate value by adding 0. 1 N hydrochloric acid or sodium hydroxide solution. Self-supporting thin films (approximately 1 in. diameter, 2-4 mg/cm^2) were prepared by evaporating 2 ml of the suspensions on an aluminum foil supported by a flat glass plate. Since most exchangeable cations react with aluminum foil, the foil was lined with a thin film of collodion to protect the clay. The specimens were dried over P_2O_5 in a slightly evacuated desiccator. The dried film was easily stripped from the foil by drawing the foil over a sharp edge. X-ray diffraction analysis indicated that the amino acids formed single-layer complexes in the clays.

The clays films used for ammonia and ethylene diamine (EDA) were prepared similarly to those prepared with amino acids except the complexing agent was not added until the clay film was formed. Ammonia was adsorbed from the gas phase onto the film, while EDA was adsorbed from the liquid phase.

Infrared Studies

The infrared adsorption spectra (4000-1200 cm^{-1}) were obtained by placing the self-supporting thin films in the sample beam of a Perkin-Elmer Model 621 double-beam grating spectrophotometer. An air purging unit was employed to eliminate adsorption bands due to atmospheric water and carbon dioxide. Calibration of the spectrophotometer was carried out using polystyrene bands and a wavenumber accuracy of ± 3 cm^{-1} was obtained.

The clay films were mounted in an evacuated cell with KCl windows (Fig. 1). They were held in the speci-

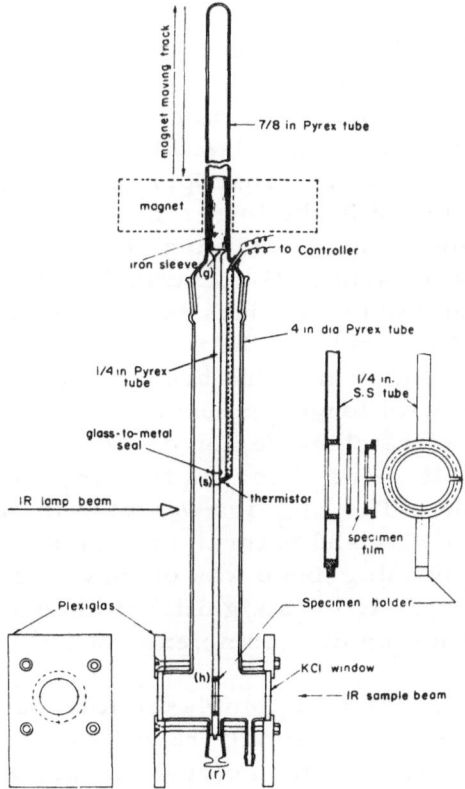

Fig. 1. The Infrared Cell.

ment holder unit by two stainless - steel rings. The sample could be rotated to any angle with respect to the incident infrared beam by turning the cock (r). The upper end
of the specimen holder unit was enveloped with an iron
sleeve. A permanent magnet was used to draw the iron
sleeve along with the rest of the unit to the top of the cell.
In this position the film could be heated by an external
infrared lamp without damaging the KCl windows, and
without breaking the vacuum in the cell and exposing the
films to air. A thermistor touching the specimen holder
ring in the heating section of the cell was used to control
the temperature of the specimen within ± 5°C.

RESULTS AND DISCUSSION

Ammonia and Ethylene Adsorbed on Montmorillonites

When ammonia is adsorbed on a montmorillonite clay,
the species present in the interlamellar spaces can be
determined by infrared spectroscopy. Table I illustrates
band maxima observed in the infrared spectra for various
ammonia- and ammonium-containing compounds. The
dominant species present in H-montmorillonite exposed
to ammonia fumes for three hours is the NH_4^+-ion. An
intense band at 1432 cm^{-1} may be assigned to a bending
mode of the NH_4^+-ion. This band broadens on the high
wavenumber side with longer exposure. The bands observed in the NH-stretching region of the spectra were
also consistent with those observed for other NH_4^+-con-
taining compounds. The NH_4^+-ion was also the dominant
species present in Ca- and natural-montmorillonite. In
these cases, the bending mode was observed at 1425 and
1420 cm^{-1}, respectively. No significant changes were
observed by heating the clay complexes at 50°C.

The infrared spectra of complexes formed by exposing
montmorillonite films containing transition metal ions to
ammonia vapor indicate that the types of species present
in their interlamellar spaces strongly depend on the time
of exposure of the sample. Fig. 2 illustrates spectra

Table I. Infrared Band Maxima of Various Ammonia and Ammonium Compounds (cm^{-1})[2].

Compound	Phase	NH Stretching Modes		Overtones & Combination Modes	NH Deformation Modes	
		Asymm.	Symm.		Asymm.	Symm.
NH_3	Gas	3444	3336	--	1628	--
NH_3	in CCl_4	3417	3315	3230	1615	--
$NH_4^+Cl^-$	Solid		3138	$\left\{\begin{array}{c}3044\\2870\end{array}\right.$		1403
$NH_4^+PF_6^-$	Solid		3330	2920		1433
$Cu(NH_3)_4Cl_2$	Solid		3270	--	1596	1245
$Zn(NH_3)_4I_2$	Solid		3290	--	1600	1242
$Ni(NH_3)_6(ClO_4)_2$	Solid	3397	3312	--	1618	1236
$Co(NH_3)_6(ClO_4)_3$	Solid	3320	3240	--	1630	1352

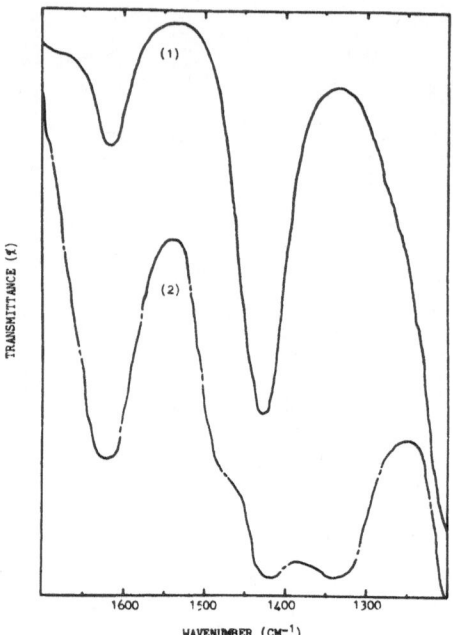

Fig. 2. The Infrared Spectra in the 1200-1700 cm^{-1} Range for Ammonia-Co-Montmorillonite Complexes Prepared at Two Different Exposure Times: (1) 5 minutes and (2) 24 hours.

observed for the NH_3-Co-montmorillonite complexes formed by exposures of five minutes and twenty-four hours. It appears that the NH_4^+-ion was the dominant species present after short exposures of the film while the Cobalt hexammine ion was a major species after longer exposures. Similar observations were noted for Cu-, Ni- and Zn- complexes. The band at 1340 cm^{-1} can be used to identify the cobalt hexammine complex. In the case of the Cu- and Zn-montmorillonite complexes, the bands at 1250 and 1270 cm^{-1} can be used to identify the respective metal tetrammine ion complexes. The nickel hexammine complex can be detected in Ni-montmorillonite by a band at 1250 cm^{-1}. The spectra in the NH-stretching region are also interpretable on the basis of the species present in the interlamellar spaces.

Similar observations were made for ethylene diamine ($H_2NCH_2CH_2NH_2$) on cation-substituted montmorillonite films. Ethylene diamine (EDA) initially exists on all montmorillonites as a free unchelated complex. However, if the clay samples containing transition metal ions were allowed to stand for a long period of time, metal chelate complexes formed in the interlamellar spaces. Cu-, Zn- and Cd-montmorillonites form square-planar complexes in which two EDA ligands chelate to a metal ion in the interlamellar space. Ni- and Co-montmorillonites form octahedral complexes in which three EDA ligands chelate to a metal ion. The bands observed for these complexes related closely in both intensities and locations to those observed for the same complexes occurring in crystals[2]. A comparison of the wavenumbers of bands observed for the octahedral complex formed in Ni-montmorillonite and a crystal is given in Table II. These chelate complexes can be formed more rapidly in the transition metal ion-substituted montmorillonites by heating the samples.

Table II. Band maxima of $\left[Ni(EDA)_3\right]^{2+}$ in Ni-montmorillonite and $Ni(EDA)_3PtCl_4 (cm^{-1})$.

Ni-EDA-Montmorillonite	$Ni(EDA)_3PtCl_4$[3]
3352	3342
3310	3292
3245	--
3170	3188
3140	--
2955	2947
2940	--
2892	2897
1590	1596
--	1581
1466	1463
1399	1399
1358	1373
1325	1332
1282	1282

Amino Acids Adsorbed on Montmorillonites

Factors such as the concentration of the amino acid
in the aqueous suspension, and the pH of the suspension
strongly influence the nature of the species present in the
interlamellar spaces of the clay films. In these cases, a
larger variety of complex ions is possible. For instance,
one may observe ions such as protonated complexes
(NH_3^+-R-COOH), monodentate complexes (M-NH_2-R-COO$^-$)
and chelated complexes $\quad$ NH_2 $\quad$ depending upon the
$$(M \underset{O-CO}{\overset{NH_2}{<}} R)$$
amino acid used and the conditions of preparation of the
clay films. The simplest amino acid, glycine (NH_2CH_2COOH),
can be used to illustrate the effects due to conditions of preparation of the films. Fig. 3 illustrates the infrared spectra
of glycine-Cu-montmorillonite films prepared at various pH's
using low concentration of glycine ($\approx$20 meq glycine/100 g.
clay) in the initial suspension. Interpretation of the spectra
on the basis of related ions in crystals[2] leads to the following
conclusions. At pH values near the isoelectronic point of

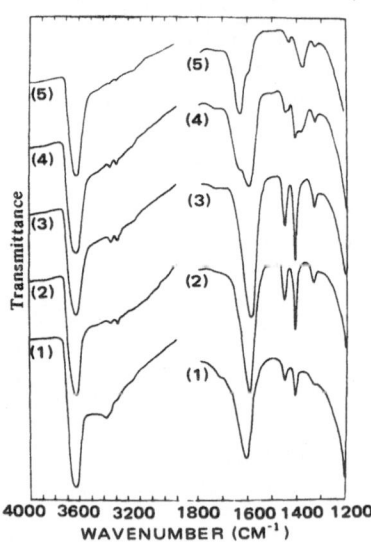

Fig. 3. The Infrared Spectra of Glycine-Cu-montmorillonite
Complexes Prepared at Several Different Equilibrium pH's:
(1) 2.9, (2) 4.9, (3) 5.5, (4) 6.8, and (5) 8.0.

glycine (6.0), the clay contains two different species, the monodentate complex (A) and the bidentate chelated complex (B). Below the isoelectric point, the monodentate complex predominates, while the bidentate complex predominates above the isoelectronic point. At very low pH's, the protonated species appears in the interlamellar spaces. Band assignments of the observed peaks and shoulders are listed in Table III. As stated earlier, the concentration of the amino acid is critical in determining the species present in the clay. Fig. 4 illustrates the infrared spectra of several Cu-clay complexes prepared with different concentrations of glycine. As the concentration of glycine in the suspension is increased, the protonated species becomes a major ion in the interlamellar spaces. The COOH mode at ca. 1745 cm^{-1} can be used to monitor this species. Similar observations in the infrared spectra have been noted upon substitution of other transition metal ions or amino acids into the clay films. Amino acids such as α-alanine, β-alanine, valine and 1-lysine have been studied. In the case of H-, Ca- and natural-montmorillonite films, the protonated amino acid is the dominant species in the infrared spectra.

Table III. Band assignments of glycine-montmorillonite complexes prepared at several different pH values.

pH 2.9	Cu-clay complexes 4.9	6.8	8.0	Assignments
3390				OH str.
	3340	3350	3355	NH$_2$ asym. str. (A, B)
3284(vw)	3286	3295	3298	NH$_2$ sym. str. (A, B)
1720(w, sh)				COOH str.
		1632(sh)	1632(s)	COO$^-$ asym. str. (B)
1605	1593	1595(s)	1600(sh)	COO$^-$ asym. str. (A), NH$_2$ bend (A, B)
1450	1450	1450		CH$_2$ scissor. (A)
		1432	1432	CH$_2$ scissor. (B)
1410	1410	1410		COO$^-$ sym. str. (A)
		1378	1378	COO$^-$ sym. str. (B)
1330	1330	1330	1328	CH$_2$ wagg. (A, B)

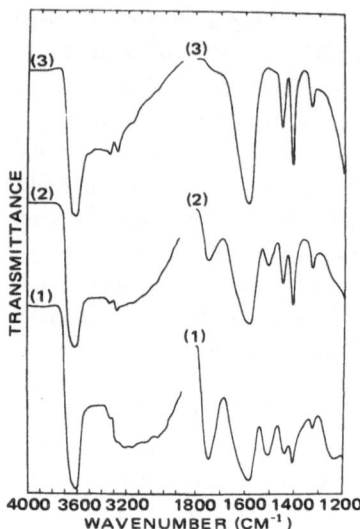

Fig. 4.   The Infrared Spectra of Cu-montmorillonite Containing Various Amounts (meg/100 g) of Glycine: (1) 75, (2) 65, and (3) 36.

Heat treatment of amino acid-transition metal ion-montmorillonite clay films can convert the complex ions present in the interlamellar spaces. Fig. 5 illustrates the spectra observed for a glycine-Cu-montmorillonite complex heated at several different temperatures.   The clay film used in this study was prepared at a pH below the isoelectronic point (5. 5), and its initial spectrum was shown in Fig. 3.   Heating this film at 150° C for 2 hours in the cell dramatically changed the structure of the complexes.   Spectrum (2) in Fig. 5 demonstrates the changes, which can be simply interpreted on the basis of the earlier observed spectra.   It appeared that monodentate complex ions convert to bidentate complex ions during the heating process.   This interpretation is reasonable because a decrease was noted in the intensity of the band assigned to hydrated water at ca. 3400 cm^{-1}.   The Cu^{2+}-ion, which was bonded to hydrated water, can bond with a carboxyl O-atom upon dehydration forming a chelated ring.   Heating of the sample at higher temperatures caused a decrease

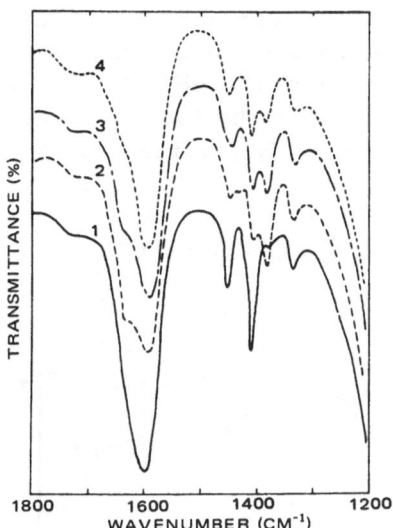

Fig. 5. The Infrared Spectra of a Glycine-Cu-Montmorillonite Complex after Heating: (1) 25°C, (2) 150°C,
(3) 170°C, and (4) 190°C.

in band intensities. A sharp band appeared at 2340 cm^{-1}
which indicates the appearance of adsorbed CO_2. Heating the sample at higher temperatures caused the decomposition of intercalated glycine, a process in which
CO_2 is given off. Similar results were observed with
heat treatment of clays containing other amino acids.

Glycylglycine Adsorbed on Montmorillonite

Proteins or polypepticles should possess more bands
in the infrared spectra than amino acids because they contain peptide linkages (- N - C -) besides being formed by
polymerization of amino acids. However, their spectra
should be similarly interpretable by systematic investigation and analysis. Spectra observed for the simplest
dipeptide, glycylglycine (NH_2CH_2C $NHCH_2COOH$), adsorbed

on Cu-montmorillonite at various pH's, can be used as an
example (Fig. 6). A consistent set of band assignments
can be obtained for each of the pH's if we assume that three
major species can occur in the interlamellar spaces of the
clay. We suggest that glycylglycine adsorbs as a glycylgly-
cinium cation,

$$NH_3^+ - CH_2 - CO - NH - CH_2 - COOH \qquad (A)$$

a tridentate complex,

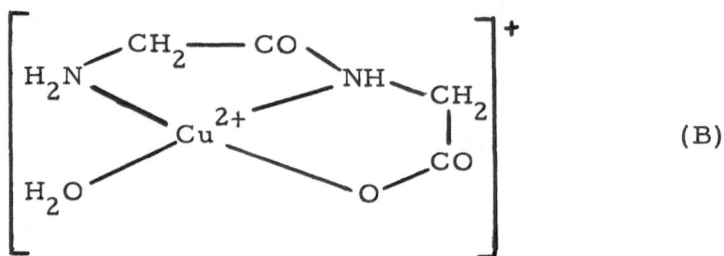

$$(B)$$

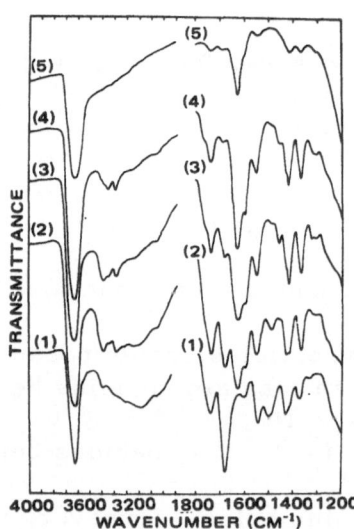

Fig. 6. The Infrared Spectra of Glycylglycine Complexes
with (1) H-montmorillonite and (2) - (5) Cu-montmorillonite:
Equilibrium pH's (2) 3. 0, (3) 4. 2, . (4) 5. 9, and (5) 7. 5

Table IV. Band Assignments of Glycylglycine-Montmorillonite Complexes Prepared at Several Different pH-values.

pH 3.0	4.2	5.9	7.5	H-clay Complex	Assignments
	Cu-clay Complexes				
3390(m)	3390(m)	3390(w, sh)		3400	OH str.
3350(sh)	3350(m, sh)	3350			NH_2 str.
3290(m)	3290(m)	3290(m)		3295(sh)	NH str. (-CO-NH-).
					OH str.
3170(b)	3170(b)	3170(b)		3220(sh)	NH_3^+ str.
				3090(b)	
2960(sh)	2960(sh)	2960(sh)		2970(w, sh)	CH str.
1740(s)	1740(m)	1740(m)	1740(vw)	1740(m)	COOH str.
1680(s)	1680(m)	1680(vw)	1680(vw)	1680(s)	CO str. (-CO-NH-).
1630(vs)	1630(vs)	1630(vs)	1630(s)		CO str. (-CO-NH-Cu).
				1615(m)	NH_3^+ deg. def.
1598(s)	1598(s)	1598(s)	1600(sh)		COO^- asym. str. (COO-Cu).
1552(s)	1552(s)	1552(m)	1550(w)	1545(m)	NH bend. (-CO-NH-).
1490(m)	1490(w, sh)			1500(m)	NH_3^+ sym. def.
1457(m)	1457(m)	1457(w)			CH_2 def.
1429(s)	1428(sh)	1428(sh)	1425(w, b)	1432(m, b)	CO str. plus OH bend.
1418(s)	1419(s)	1419(s)			CH_2 def.
1370(s)	1370(s)	1370(s)	1370(w)	1373(w)	COO^- sym. str. (COO-Cu),
					CH_2 bend.

and a bidentate complex,

$$\left[\begin{array}{c} H_2N \diagdown CH_2 \diagup CO \diagdown NH \diagdown CH_2 \\ Cu^{2+} COOH \\ H_2O \diagup \diagdown OH_2 \end{array} \right]^{2+}$$

(C)

The bands observed for the clay complexes prepared to various pH's can be understood from the spectral assignments for similar species present in aqueous solutions and crystals[2] (Table IV). At the lower pH's, the three species with structures (A), (B) and (C) could be detected in the clay film, at higher pH's, only (B) and (C). It is interesting that the copper complex which has been noted in aqueous solutions[4], and which is formed by the deprotonation of the N-atom belonging to the peptide linkage in structure (B) cannot be detected by infrared spectroscopy in the various Cu-clays.

REFERENCES

1. M. L. Hair, Infrared Spectroscopy in Surface Chemistry, Marcel Dekker, New York (1967).
2. K. NaKamoto, Infrared Spectra of Inorganic and Coordination Compounds, 2nd edition, Wiley Interscience, New York, (1970).
3. D. B. Powell and N. Sheppard, "The Assignment of Infrared Absorption Bands to Fundamental Vibrations in Some Metal-Ethylenediamine Complexes," Spectrochim. Acte 17, 68-76 (1961).
4. M. K. Kim and A. E. Martell," Copper (II) Complexes of Glycylglycine, " Biochem. 3 (8) 1169-1174 (1964).

DISCUSSION

James Reed (Alfred): What is the potential of IR spectroscopy for studying the structure of polymer molecules on
ceramic surfaces?
Author: It is useful when the ceramic material provides a
window in the infrared region where useful bands of the
polymer occur. Montmorillonites provided such a window
region for the compounds that interested us.
D. Dove (Univ. Florida): How sensitive are the results to
surface preparation?
Author: The quality of spectra obtained depends critically
upon the method of preparation of the clay film. Two important factors are film thickness and adsorbed water present in the film.
G.H. Frischat (Clausthal): At what temperatures are your
N-containing complexes stable?
Author: Amino acids adsorbed on montmorillonites and
heated in vacuum did not decompose on heating up to 170°C.
At higher temperatures decomposition began. For instance
glycine decomposition was noted in a Cu-montmorillonite
clay at 190°C.
L. L. Hench (Univ. Florida): How do you distinguish dissolution effects of the media from the adsorption effects of
species on the surface? For example, alanine and proline
require acidic media to go into solution. Such acidic media
may produce changes in the montmorillonite surface, such
as cation desorption, independent of the presence of the
amino acid in solution. Can these interfering contributions
be eliminated in the interpretation?
Author: We have looked at the species present on the clays
after the adsorption and film treatment were complete.
Processes such as the one you have mentioned will influence
the concentrations and nature of the species present in the
interlamellar spaces. One can use chemical information
such as pK-values and concentrations of metal ions present
in the clay to predict qualitatively the types of species present in the clay at various pH's and the concentrations of these
species. For instance, we would expect mainly the protonated glycine cation at low pH, and the metal-glycine

complexes at higher pH when we place a low concentration of glycine into the suspension mixture.

APPLICATIONS OF SURFACE CHARACTERIZATION TECHNIQUES TO GLASSES

G. Y. Onoda, D. B. Dove and C. G. Pantano, Jr.

Department of Materials Science & Engineering
University of Florida
Gainesville, Florida

Of particular concern to us* was the analysis of glass, following various manufacturing procedures, for chemical and structural information at the outermost atomic layers and to depths up to 1000 Å beneath the surface. Depth profiling on this scale was possible by milling away surface layers with a beam of, for example, argon ions while at the same time monitoring the surface with Auger electron spectroscopy[1,2] or other techniques depending on the information desired. Each analytical method is sensitive to a particular characteristic of the surface. Thus, ellipsometry[3] has provided a means for observing changes in refractive index in the surface region due to polishing or chemical modification. The more recently developed Auger electron spectroscopy (AES) provides identification of chemical species present at the surface, while electron spectroscopy for chemical analysis (ESCA) is particularly for analysis of chemical bonding characteristics[4]. Infrared spectroscopy as applied to glass corrosion by Hench and collaborators[5] and others[6] gives information on chemical species and bonding to greater depths, although somewhat less directly. The value of applying a variety of techniques is apparent.

---

*Work supported by the National Science Foundation and the Glass Container Industry Research Foundation.

AUGER ELECTRON SPECTROSCOPY (AES)

When a surface is bombarded with energetic electrons or other particles, some electrons are ejected from the inner shell levels of surface atoms. De-excitation may then occur by an outer shell electron dropping to the vacant inner level with emission of X-rays or with Auger ejection of another electron. The Auger electron involves the difference in energy between three electronic states and is therefore readily estimated from known X-ray energies[7]. The measurement of Auger electron energies therefore provides a direct method for identification of atomic species at or within a few atomic layers of the surface. All elements except hydrogen and helium can be detected by the presence of one or many Auger electron energy peaks.

Briefly, an Auger spectrum is obtained by bombarding the surface with an electron beam, typically having energy of a few kev. Electrons leaving the surface enter an electron spectrometer, commonly of the retarding grid or the cylindrical dispersive analyzer type. The dispersive analyzer has a considerably superior signal-to-noise ratio than the retarding grid system and offers better sensitivity and resolution at both low and high electron energies[8]. Owing to the presence of a large background of secondary and back-scattered electrons, the Auger peaks are most easily resolved by operating the analyzer in a differentiating mode.

Unfortunately, as noted above, hydrogen cannot be detected by AES and the oxygen Auger peak appears to be insensitive to the existence of hydroxyl groups so that the presence of a hydrated surface layer cannot be detected.

Spectra for glasses are obtained by keeping the primary electron beam at a glancing angle of incidence in order to prevent unstable charging of the surface. A stable charge distribution on the surface does occur and may produce a small constant shift of all peaks to lower energies.

Auger spectra for an alkali strontium silicate glass obtained by Goldstein and Carlson[9] are shown in Fig. 1. A

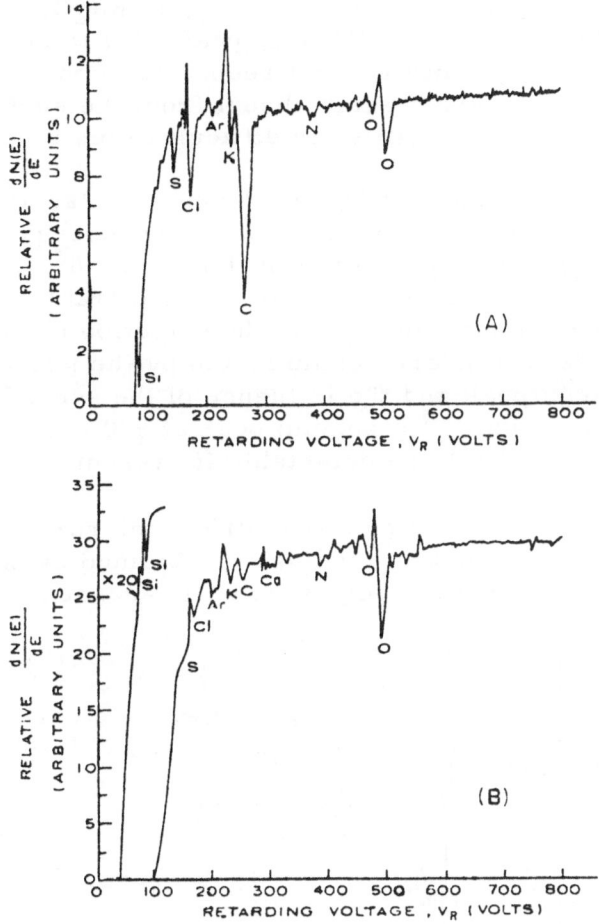

Fig. 1. Auger spectra of alkali strontium silicate glass
(A) before sputtering; (B) after 42 min. sputtering with
400-V Ar ions. From Goldstein and Carlson[9].

beam of about 10° angle of incidence was employed with an
energy between 1.5 and 3 kev. The spectrum for the glass
after initial 'cleaning' in trichlorethylene showed the presence of carbon, sulphur, chlorine, and traces of nitrogen
and argon. After ion bombardment, silicon and oxygen
peaks predominated. Although the composition of the glass

was given (in weight percent) as 61 SiO_2, 13 Na_2O, 9.9 SrO, 8.3 K_2O, 4 Al_2O_3, 2.35 CaO, peaks due to sodium, strontium and aluminum were not seen. The conclusion that these elements are largely absent from the surface, although a distinct possibility, is difficult to justify.

Strontium has peaks at 65 and 160 ev and also at 1380 and 1717 ev. Similarly, aluminum has a low energy peak at 68 ev and high energy peaks from 1329 to 1396 ev. With the retarding grid analyzer there is a considerable lack of sensitivity at high energies, while the detection of the peaks at low energies is rendered complicated by the presence of the large silicon peak and the influence of the secondary electron background. The sodium peak at 990 ev, although commonly weak, should be detectable if present.

The advantages of using a cylindrical mirror analyzer can be seen in the spectrum in Fig. 2, obtained by Weber[10] from a glass of unquoted composition. In addition to the

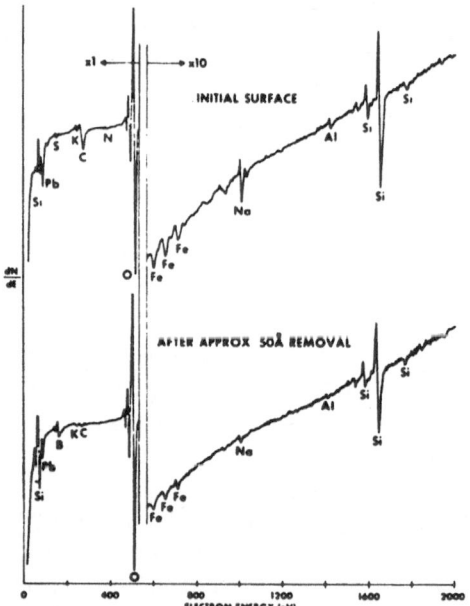

Fig. 2. Auger spectrum of glass before and after rare gas sputter etching. From Weber[10].

low energy silicon peak, the high energy peak at 1619 ev
can be clearly distinguished. This is particularly useful
when there are other competing peaks at low energies that
may interfere with the interpretation of the low energy
silicon peaks. A sodium peak at 990 ev can also be seen
in the spectrum.

     After ion bombardment to remove approximately 50
$\overset{\circ}{A}$ of material from the surface, the sodium peak becomes
much smaller and some of the surface contaminants dis-
appear. The question arises whether the apparent decrease
in sodium content is real or whether it has arisen in some
way from the effect of the ion beam or possibly electron
beam. It is known that the ion beam sputtering process
may remove certain constituents from a surface more
rapidly than others, and work is needed to investigate the
relative importance of such effects in ion milling to obtain
chemical profiles. In addition, it must be recognized that
significant electric fields produced by surface charges
could give rise to ionic migration, indicating the necessity
for evaluating data carefully, and of employing complementary techniques where possible. A particularly promising approach is to combine AES with mass spectrometry
of the secondary ions during ion beam profiling, as employed by Morabito in studying diffusion profiles in silicon[11].

     The use of AES to study coatings on container glasses
is shown in Figs. 3 and 4. Such layers, applied to provide
the surface with greater resistance to abrasion, typically
consist of an underlayer of tin oxide or titanium oxide approximately 50 Å in thickness and a top layer of organic
material. Fig. 3 shows a spectrum of a glass with such
a duplex coating, measured with a retarding grid system[12].
The predominant peak is that of carbon from the organic
coating. Some tin is also to be seen, indicating that the
underlayer is incompletely covered by the organic overcoat.
By ion bombardment the upper layer is removed and the
underlayer is exposed; the tin and oxygen peaks become
predominant while the carbon peak has sharply diminished
(Fig. 4). Further ion bombardment completely removes
the tin oxide coating and exposes the glass. Profiling studies

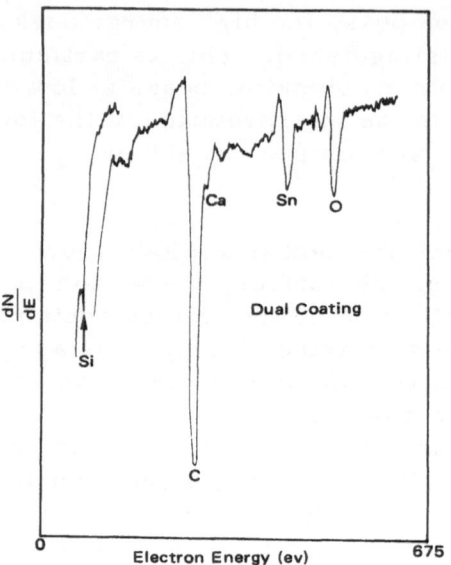

Fig. 3. Auger spectrum of commercial container glass with dual-end coating.

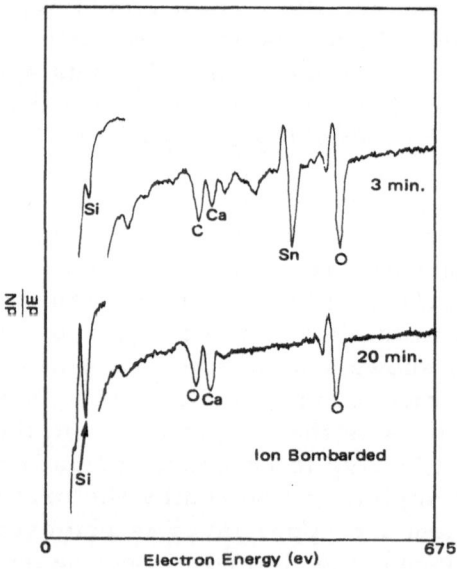

Fig. 4. Auger spectrum of glass with dual-end coating after etching away the organic film (upper) and the tin oxide film (lower).

of such coated surfaces should lead to a greater understanding of the mechanisms of surface protection.

SECONDARY ION MASS SPECTROMETRY (SIMS)

When a surface is subjected to bombardment by an
energetic ion beam, material is removed in the form of
atomic and molecular species of various degrees of ionization. Mass spectrometric monitoring of the secondary
species, particularly ions, provides surface analysis of
extremely high sensitivity. The material removed during
the period of measurement may be a small fraction of a
monolayer[13] or may be more if chemical profiling is the
objective. SIMS systems range from a straightforward
combination of ion source and mass spectrometer[14] to
very elaborate systems that provide high spatial resolution (several microns) and compositional imaging and
profiling.

An example of analysis possible with an ion source/
mass spectrometer system is shown in Fig. 5. As the
surface of a lithium aluminum silicate glass was removed
by ion bombardment, the lithium and aluminum ions leaving the surface were monitored. The results substantiated
the suspicion that the lithium content was depleted at the

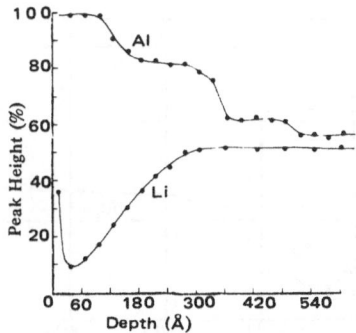

Fig. 5. Secondary ion mass analysis profile of Li and Al
on the surface of glass. (Courtesy of Commonwealth
Scientific Corporation).

at the surface region while the aluminum content was enhanced.

The use of an ion microprobe spectrometer is illustrated in the work of Heyndryckx[15] who studied the surface of glass following polishing and diffusion treatments (Fig. 6).   With cerium polishing, sodium was found to be depleted at the surface, increasing in concentration to a value of about 12%, 5000 Å beneath the surface.   Less sodium depletion, on the other hand, was found after polishing with chantereine.

In studies of surface composition following diffusion treatments Heyndryckx obtained the concentration profiles shown in Fig. 7, where diffusion of copper was monitored as a function of depth.   Sodium ion diffusion was shown to be significant at depths of several microns.

Coupled with AES or SIMS, ion milling provides a powerful tool for chemical profiling in the surface region, although difficulties of interpretation of data may be difficult if selective removal of certain species from the surface by the ion beam is a possibility.   Once a steady state surface composition has been reached during ion bombardment, the surface species may be expected to be removed in proportion to their volume concentration.   However, it is possible that the relative distribution of mole-

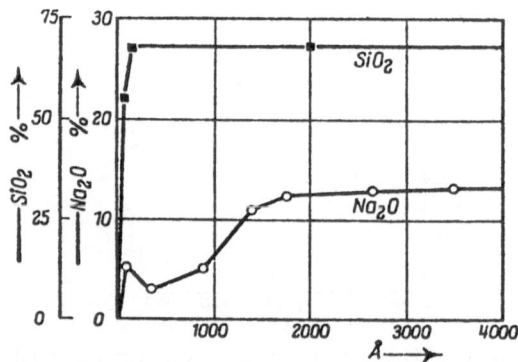

Fig. 6.   Ion microprobe profiles of SiO_2 and Na_2O in glass after polishing with cerium.   From Heyndryckx[15].

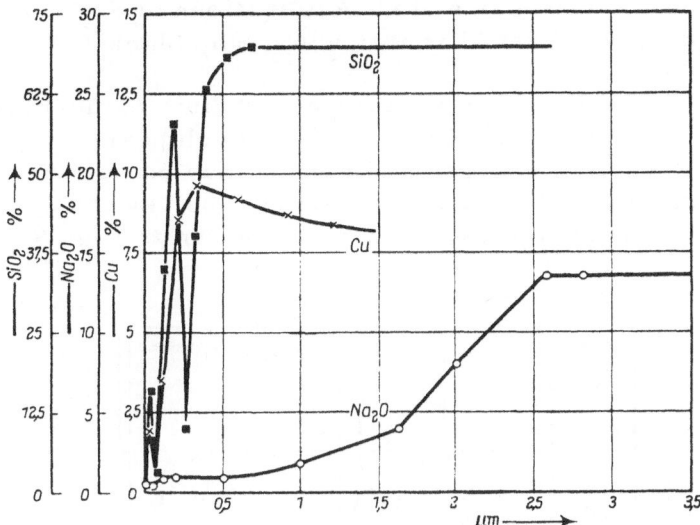

Fig. 7. Composition profiles resulting after diffusion of Cu into glass. From Heyndryckx[15].

cular and atomic species may depend on surface condition. A full SIMS analysis would include a measurement of neutrals and positive and negative ions emitted, although valuable information can frequently be obtained from simply monitoring the intensity of a particular species.

ELECTRON SPECTROSCOPY FOR CHEMICAL ANALYSIS (ESCA)

In the ESCA technique[16] the surface to be analyzed is bombarded in vacuum with a beam of X-rays of well-defined energy, and the energy of photoelectrons ejected from the surface is determined with a high-resolution electron spectrometer. The resulting spectrum gives information on both composition and chemical binding.

Although ESCA has not yet been extensively applied to the study of glasses, its potential there is evident, particularly the possibility of clarifying many of the structural

features of the vitreous state. Applications of ESCA are
discussed in more detail in this volume by Slack[17].

Fig. 8 shows a spectrum for a vanadium oxide-barium
oxide glass obtained by Hickson[18], in which peaks from
vanadium, barium, oxygen and carbon may be seen. A
slower scan over the vanadium region resolves the vanadium peaks from the oxygen peak as shown in Fig. 9.

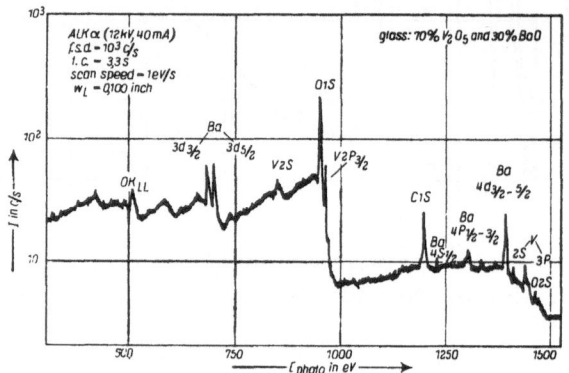

Fig. 8. Oxygen KLL Auger line and photoelectron lines
from barium, vanadium, oxygen and carbon of a vanadium
borate glass. From Hickson[18].

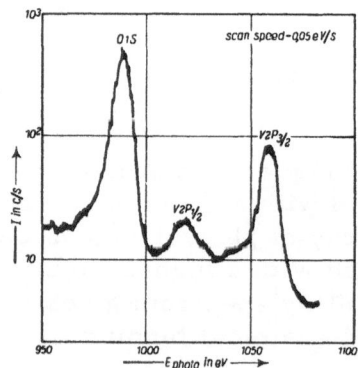

Fig. 9. Resolved vanadium VL_{II} and VL_{III} lines from the
oxygen line in the photoelectron spectrum of a vanadium
borate glass. From Hickson[18].

Studies on lead silicate glasses indicated that the lead in the glass possessed a much greater covalent bonding character than lead in crystalline PbO. Similar effects have been observed in a lead borate glass. The application of ESCA to the quantitative analysis of glasses has been discussed by Escard and Brion[19]. These authors also remark on a change in sodium concentration following prolonged X-ray bombardment.

ELLIPSOMETRY

The optical method of ellipsometry provides a means for determining the thickness and refractive index of films upon a substrate. Description of applications are to be found in several publications[20-22].

A number of ellipsometric studies on glass surfaces have indicated a difference in refractive index between surface region and bulk following various treatments. As early as 1931 it was found by Philpot[23] that the surface of various glasses had initially, in almost all cases, a layer having a higher refractive index than the bulk glass. Possible explanations are that the surface may be under stress or that there may exist a concentration of modifying ions near the surface.

Vasicek[24] has shown that new or aged glass surfaces are never completely free from surface layers with different optical properties. For example, when a Schott glass (BK7) was polished, a surface region was found which had an index of refraction of 1.47 to a depth of 10-20 Å while the bulk glass had an index of 1.516.

More recently Yokato et al[25] found that polishing leads to higher or lower indices of refraction depending upon the glass chemical susceptibility. Glasses inclined to weathering acquired a surface index lower than that of the bulk while the opposite result was found with chemically durable glasses such as Vycor, Pyrex, and silica.

Recently, Vedam and White studied the surface changes of vitreous silica after the surface was mechanically polished with diamond paste of progressively decreasing grain size[26]. The index of refraction of bulk fused silica was 1.460; at the surface, they found a value of 1.530 to an effective depth of 954 Å. It was felt that this increase was due to permanent densification of the surface by polishing. They also showed that HF etching of the surface reduces the thickness of the damaged surface layers, but never completely removes all of the surface region having refractive index different from the bulk. Polishing with cerium oxide did not leave a significantly damaged layer as detectable by ellipsometry, possibly related to the fact that the hardness of cerium oxide is very close to that of vitreous silica.

In the authors' laboratories, ellipsometry is being used to measure ultrathin organic and oxide films applied to glass containers. Optical film thickness measurements are particularly of value in calibrating ion milling chemical profiles using AES or SIMS.

INFRARED REFLECTION SPECTROSCOPY (IRRS)

The observation of absorption bands in the infrared wavelength region is an important tool for chemical analysis particularly of organic compounds. Several papers discuss the reflection of infrared radiation from glass surfaces and relate the observed spectra to the bending or stretching of silicon-oxygen bonds and other glass constituents[27].

More recently IRRS has been found by Hench and collaborators[28, 29] to have a particular utility in the study of glass corrosion, since spectral information comes typically from a thin surface region, typically 1μ using a total reflection angle of $25°$. The silicon-oxygen-silicon stretching (1100 cm^{-1}) and rocking modes (500^{-1}) may be used to determine the SiO_2 concentration, while the silicon-oxygen-alkali stretching peak at 900 cm^{-1} gives the alkali concentration. Spectra of a 33 mole % LiO_2-SiO_2 glass and of vitreous silica were obtained by Sanders and Hench[29]. Exposure

to water changed the relative concentrations due to preferential leaching of the alkali ions, producing a change in the
spectrum as shown in Fig. 10. The surface film appears
similar to vitreous silica. The IRRS technique is useful
in situ observations on glass structure and composition,
detecting the effects of variations in processing or other
procedures[29, 30].

CONCLUDING REMARKS

This review has stressed the new level of glass surface characterization that is now possible with recently
developed techniques. As with most analytical techniques,
there are definite limitations that must be recognized, including a considerable amount of interpretive judgment that
must be exercised.

The physical information obtained by the techniques is
some form of "average" over the volume of observation.
Inhomogeneities along the plane of the surface having a
dimensional scale smaller than the area of observation
cannot be distinguished, and the potential exists for mis-
interpretation in terms of a uniform distribution state
yielding the same physical response. Similar difficulties
exist for inhomogeneities normal to the surface if the
dimensional scale is smaller than the depth of observation.
Also, the surface interaction inherent in the techniques
can perturb the surface and provide misleading results,

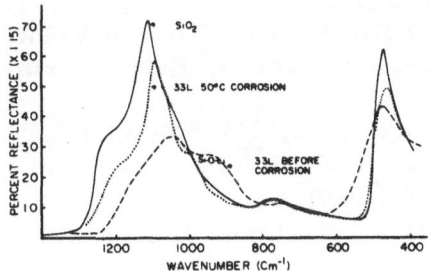

Fig. 10. Typical IR reflection spectra. From Hench and
Sanders[29].

as in the case of preferential sputtering by ion bombardment. A confident interpretation of characterization information, therefore, is not always possible by application of a single technique, but several techniques applied to the same problem can resolve uncertainties.

ACKNOWLEDGMENTS

The authors are pleased to acknowledge permission to reproduce figures from D. E. Carlson, B. Goldstein, L. L. Hench, P. Heyndryckx, K. Hickson and D. M. Sanders, and to acknowledge research support from the National Science Foundation and the Glass Container Industry Research Corporation.

REFERENCES

1. J. Khan, "Characterization of Surfaces & Interfaces," this volume.

2. P. W. Palmberg, J. Vac. Sci. and Technol., 9, 160 (1972).

3. A. Vasicek, OPTICS OF THIN FILMS, Amsterdam: North Holland (1960).

4. Siegbahn, K., et al, ESCA, ATOMIC MOLECULAR AND SOLID STATE STRUCTURE STUDIED BY MEANS OF ELECTRON SPECTROSCOPY, Almquist & Wilksells, Uppsala, (1967).

5. D. M. Sanders, W. B. Person and L. L. Hench, Appl. Spec., 26 (5) 530 (1972).

6. I. Simon, MODERN ASPECTS of the VITREOUS STATE, J. D. Mackenzie, ed., vol. 1, Butterworths, London 120 (1960).

7. Chang, C. C. Surf. Sci., 25, 53 (1971).

8. P. W. Palmberg, G. K. Bohn & J. C. Tracy, Appl. Phys. Letters, 15, 254 (1969).

9. B. Goldstein, D. E. Carlson, J. Am. Ceram. Soc., 55, (1) 51 (1972).

10. R. E. Weber, Physical Electronics, Industries, Inc., Edina, Minnesota.

11. J. M. Morabito and J. C. Tsai, Surf. Sci., 33, 422 (1972).
12. G. Y. Onoda, Jr., D. B. Dove and C. G. Pantano, Jr., 75th ACS Meeting, Cincinnati, Ohio (1973).
13. A. Benninghoven, Z. Phys., 230, 403 (1970).
14. R. Schubert and J. C. Tracy, Rev. Sci. Instr., 44, (4) 487 (1973).
15. Paula Heyndryckx, Glastechn. Ber., 44 (12) 543 (1971).
16. T. A. Carlson, Physics Today, 25, 31 (1972).
17. L. H. Slack and L. R. Durden, "Observation of Electronic Spectra in Glass and Ceramic Surfaces," this volume.
18. K. Hickson, Glastechn. Ber., 44, 537 (1971).
19. J. Escard and D. Brion, C. R. Acad. Sc. Paris, 276, B-945 (1973).
20. K. Vedam and S. S. So, Surf. Sci., 29, 379 (1972).
21. "Recent Developments in Ellipsometry," in Conf. Proc. at Nebraska Center, Univ. of Nebraska, August 1968.
22. J. Kruger, Corrosion, 22 (4) 88 (1966).
23. A. J. Philpot, Brit. Sci. Instrum. Res. Assoc., Rep. R. 74, (1931).
24. A. Vasicek, J. Opt. Soc. Amer., 37, 979 (1947).
25. H. Yokota, et al, Surf. Sci., 16, 265 (1969).
26. K. Vedam, "Characterization of Surfaces," to be published.
27. J. Wong and C. A. Angell, Appl. Spec. Rev., 4, 155 (1971).
28. D. M. Sanders and L. L. Hench, J. Am. Cer. Soc., 56, 373 (1973).
29. D. M. Sanders and L. L. Hench, Am. Cer. Soc. Bull., 52, 662 (1973).
30. D. M. Sanders and L. L. Hench, Am. Cer. Soc. Bull., 52, 666 (1973).

DISCUSSION

H. Bach (Schott): Did you measure the respective sputtering rates of the thin carbon layers or TiO_2 layer removed by ion sputtering during your surface analysis? I guess the sputtering rate of the carbon layer to be very low because of the low collision diameter of carbon. Besides this a doping of the TiO_2 layer on the carbon or the glass substrate with Ti could be possible at every angle of incidence by energy propagation in the collision cascade below the surface. This would smear out the interface.

Author: The sputtering rates have not been measured as yet. We intend to do so very shortly, utilizing both interferometer and ellipsometric measurements. From the work which we have carried out thus far, it appears that the sputtering rate of the carbon films is considerably higher than that of the TiO_2 and glass. Hence, a correction factor must be utilized in plotting any composition profiles.

D. H. Buckley (NASA): You mentioned the surface charge effects with insulators. Since charge build-up produces alteration in peak height intensities, does this inhibit quantitative analysis?

Author: Surface charging was handled by using a glancing angle of incidence between the electron beam and the sample. In this way, the secondary electron yield can be greater than unity and the surface attains an equilibrium potential.

Charge build-up on the surface shifts the Auger peak to higher or lower energy, depending upon how (positive or negative) the surface charges. This presents no problem, however, because all peaks are shifted the same amount and one can subtract out the effect.

N. J. Binkowski (Corning): Have you compared spectra from alkali-containing glass surfaces which have been cleaned with an ion beam versus that cleaned by neutral atoms? Have you identified the carbon compound seen on glass surfaces?

Author: Our laboratory does not have "neutral" bombardment capability. The results of such a comparison would prove very interesting, in order to answer the question of field effects upon the mobile ions in the glass.

It is rather difficult to use Auger data to identify chemical compounds, because peak height ratios and peak shapes are difficult to interpret. A study of these carbon compounds utilizing ESCA would lead to positive identification.

J. Wong (GE): Comment: Perhaps it is appropriate here to mention the name L. H. Harris who made the Auger Effect

possible as a sensitive analytical tool for surface characterization. <u>Question</u>: (1) What is lowest concentration detectable by Auger spectroscopy? (2) Has the Auger technique been put on a quantitative basis yet?

<u>Dr. H. C. Gatos (MIT)</u>: Detection limit by the Auger tech-
nique is of the order 10^{18}/cm^3 or surface concentration
10^{12}/cm^3, i. e. , 100 p. p. m. range.

<u>Author</u>: The quantitative aspects of Auger Spectroscopy
are relatively straightforward. The problem arises in the
complexity of the experimental factors. Only after the
experimental factors are carefully standardized and the
necessary theoretical calculations are made can one extract
quantitative information with confidence. The quantitative
information is available; it merely requires careful, dili-
gent effort in order to obtain accuracy. At present there
are no completely generalized techniques to do so; therefore, investigators using AES must develop procedures
which are best applicable to their own systems and surfaces.

OBSERVATION OF ELECTRONIC SPECTRA IN GLASS AND CERAMIC SURFACES

L. H. Slack and L. R. Durden

Division of Minerals Engineering
Virginia Polytechnic Institute and State University
Blacksburg, Virginia

ESCA, an acronym for Electron Spectroscopy for Chemical Analysis, is best described as X-ray photoelectron spectroscopy. Electronic spectra, obtained from ESCA, provide information as to the kinds of atoms present, their valence state and coordination number as well as being a direct measure of the shape of the valence bond. Sampled depth varies from 4 Å, for heavy elements such as gold, to 100 Å for complex molecules such as lead stearate. The emission of electrons is strongly dependent on the difference between the energy of exciting radiation and the binding energy of the emitted electron. This technique has been reviewed by others[1,2].

Only a few surface studies have been carried out using ESCA. Some of these studies will be described in this paper. Its primary application has been in the study of bulk materials but such studies require severe care in surface preparation. Inadvertent contamination of a cleaned sample surface will completely negate the ESCA results. ESCA, still a young and developing branch of technology, is a natural tool for surface studies and deserves attention for that purpose.

BASIC MECHANISM OF ESCA

The basic principle of ESCA is described with reference to Fig. 1.   The sample surface is bombarded with
monochromatic X-radiation.   The electrons whose energies
are characteristic of the atoms in the sample and their
chemical environment are ejected and passed through the
source slit into a lens system where they are retarded.
The electrons then pass into the double-focusing hemispherical analyzer which acts as an electron spectrometer.
Depending on the analyzer and the lens potentials, the
electrons of discrete energies are refocused at the collector slit prior to collection by the electron multiplier,
whose output pulses are amplified and fed into a rate
meter.   In the simplest form of display, the resulting
signal is sent to an X-Y recorder.   By scanning the lens
and analyzer potentials at the same time, the complete
energy spectrum of the electrons leaving the sample may
be directly recorded.   The retardation of the electron
beam by the lens system is necessary in order to keep the
size of the spectrometer reasonable, and to facilite analyses over a very wide energy range.

The electronic spectrum may be considered in two
parts, i.e., the spectrum of the core electrons, and that

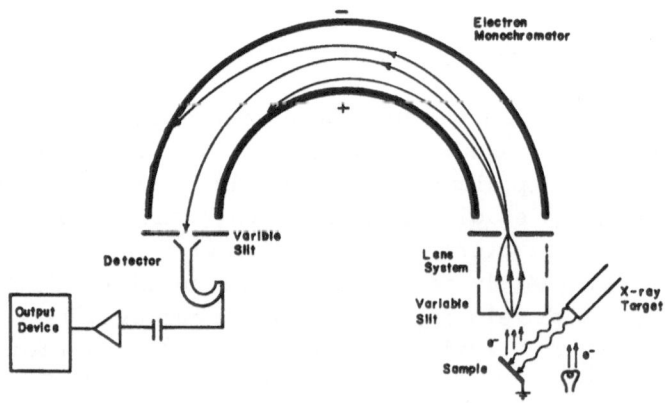

Fig. 1.   Schematic of an ESCA system.

of the outer, more loosely bound electrons. The core
electrons are localized and therefore have discrete crisp-
ly defined energies. The valence electrons are delocalized and have, by the requirements of the uncertainty
principle, energies which are diffused into a band. Every
atom has its own discrete electronic energy spectrum.
Therefore, this energy spectrum is a means of identification of the atoms present in the material being bombarded
by X-rays. Changes in valence state or coordination number cause the electron energy levels to be shifted. Therefore, ESCA can be used to determine such things as coordination numbers and type of bonding present, whether
metallic, ionic, or covalent.

Practical considerations of concern in the study of
glass and ceramic surfaces include the surface penetration,
i. e. , the amount of material being analyzed, the energy of
the exciting radiation and the resolution. The depth of analysis is limited by the mean free path of the photoelectron
which depends on the composition of the sample and the
energy of emitted electrons. As an example, Al $K\alpha$ radiation with an energy of 1486. 6 eV penetrates aluminum
oxide to a depth of about 10^4 Å, but photoelectrons with
energies in the range of 500 to 1480 eV have mean free
paths in the range of 10 to 30 Å. The depth of analysis is
thus limited by this smaller mean free path; photoelectrons
excited at greater depths cannot escape. In ESCA surface
studies where it is important to detect as few atomic layers
as possible, it is best to use low-energy exciting radiation,
and to study the lower kinetic energy electrons which are
the core electrons. Yttrium has monochromatic radiation
in the soft X-ray range and is of considerable interest for
excitation in surface ESCA studies. Both ESCA and Auger
Electron Spectroscopy provide approximately the same
depths of analysis, i. e. , 20 to 50 Å. Ion Scattering Spectroscopy, to be discussed later, identifies only the top
monolayer.

Resolution in ESCA studies is limited by the width of
the exciting X-ray line. In an X-ray target atom, when an
inner electron shell is ionized, the vacancy is filled by

electrons from higher energy levels in approximately 10^{-16}
seconds. The uncertainty in this time is very small; consequently, the uncertainty in the energy of the emitted X-
ray must be large and is of the order of an electron volt.
A further detriment to resolution is the occurence of the
$K\alpha_1$ and $K\alpha_2$ line together. As an example, the aluminum
$K\alpha_1$ and $K\alpha_2$ lines each has a natural width of about 0.7 eV.
Consequently, the total width of the Al $K\alpha$ line is 1.0 eV.
The line width varies from level to level and from element
to element[2].

APPLICATIONS

Tin Oxide Thin Films

Tin oxide thin films have practical importance because
they are electrically conducting and have a high infrared
reflectance, of interest for films on architectual glass.
The glass is heated to near the softening point, about 700° C,
and sprayed with an alcoholic or aqueous solution of
$SnCl_4 \cdot 4H_2O$. The chloride decomposes, leaving a conducting layer. At the outset, it was not known if the pyrolytic decomposition of the chloride left a layer of oxide or
if the top layer of the glass was doped with tin ions. ESCA
was used to analyze the top surface of the film for the substrate elements such as silicon, sodium, boron, aluminum,
etc. None of the glass components were present in the top
surface; only the tin and oxygen spectra were present. If
there is any interdiffusion of the substrate and the film, it
must occur near the interface. (Incidentally, ESCA should
be valuable for study of diffusion at such interfaces. The
top layer may be removed by systematic chemical or sputter etching and the composition of each revealed surface
determined. Sputter etching can be carried out in the high
vacuum chamber of the ESCA system, eliminating the danger of surface contamination.)

ESCA was then used to determine the valence state of
the tin ions. Calibration oxides were used to determine
the binding energies of the stannic and stannous ions by

observing tin's $3d_{3/2}$ and $3d_{5/2}$ peaks for powdered SnO_2 and powdered SnO (Fig. 2). It is evident that the surface of the SnO grains had been oxidized to SnO_2 in air, and the peaks were found to be doublets. Using a computer decon- volution procedure the $3d_{3/2}$ doublet was separated into two peaks centered at binding energies of 498.70 eV and 497.78 eV, the first being characteristic of the stannic ion and the second of the stannous ion. The electronic spectrum of the pyrolytically deposited tin oxide films was identical to the stannic oxide calibration, with no evidence of stannous ions present.

Annealing the tin oxide films in H_2 caused reduction of the surface stannic ions to stannous. This was accom- panied by a several order increase in electrical conduct- ivity, which suggests that conduction occurs by a hopping mechanism between the stannous and stannic ions.

Adsorbed Monolayers

ESCA measurements[4] by Wolstenholme and Kraitehman on a silicate glass substrate coated with monolayers of Sn

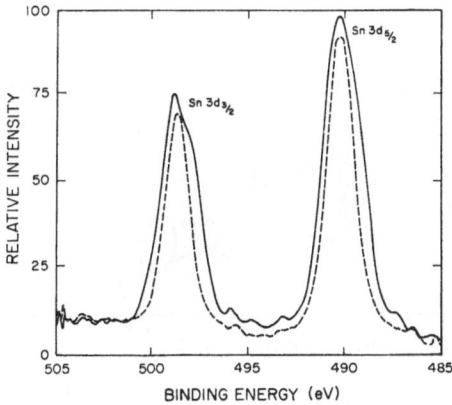

Fig. 2. Portions of tin's electronic spectra for SnO_2 (dashed line) and a mixture of SnO_2 and SnO (solid line).

and Pd demonstrated that the peaks of these elements were
as intense as the Si as shown in Fig. 3. These layers
were known to be monolayers from Langmuir adsorption
measurements. These observations prove that ESCA is a
powerful tool for monolayer compositional studies. ESCA
is sensitive to the top layer which is between 20 Å and 50
Å thick, but is very effective in detecting the top one or
two monolayers since the probability of photo-excitation
and the probability of photoelectron escape is highest for
the top monolayers.

Molybdenum/Aluminum Oxide Catalyst

Miller, Atkinson, Barber and Swift[5] have applied ESCA
to a catalytic surface, prepared by precipitating calcium
molybdate on finely powdered boehmite (γAl_2O_3) which was
then calcined to 500°C. The Mo $3d_{3/2}$ and $3d_{5/2}$ levels
were examined by ESCA and it was found (Fig. 4) that the
peaks were either doublets, or very diffuse. For comparison, the same peaks in MoO_3, cobalt molybdate, sodium
molybdate, and an evaporated and dried ammonium molybdate

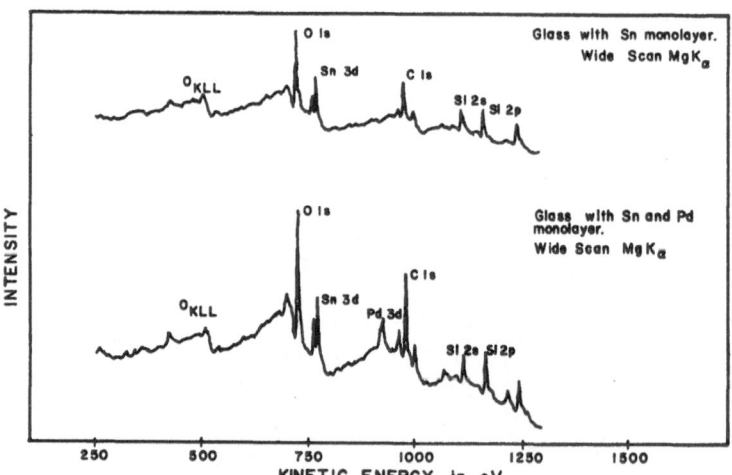

Fig. 3. Wide scan electronic spectra of silicate glass
coated with a monolayer of Sn atoms (top) and same glass
coated with a Sn and Pd monolayer[4].

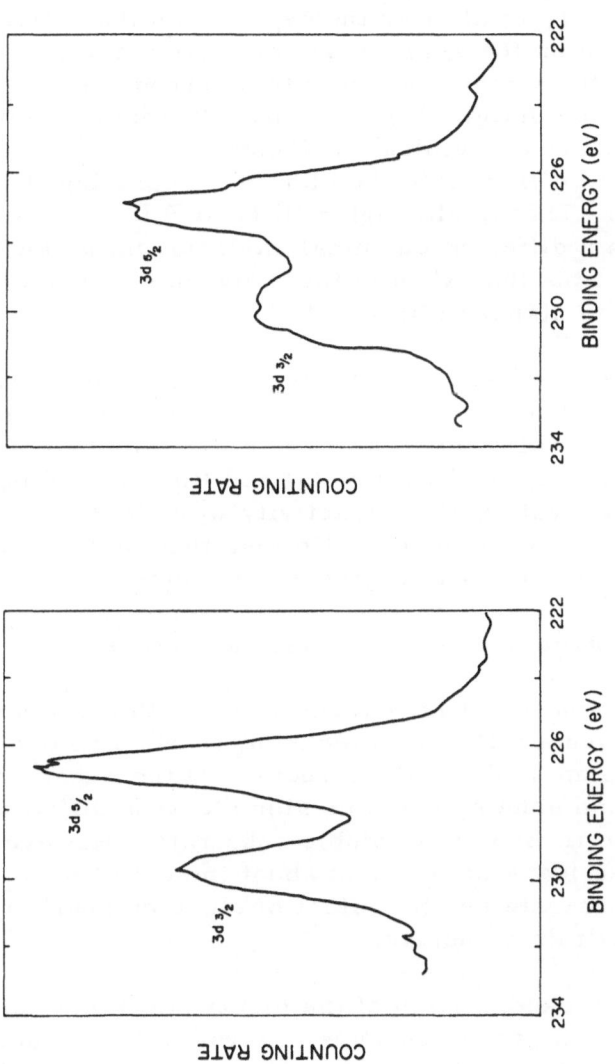

Fig. 4. Portion of molybdenum's electronic spectrum in (a) MoO_3 coated on calcined γ-Al_2O_3 powder, and (b) MoO_3.

solution were found to be well defined and identical. It is
concluded that the bonding environment for molybdenum
(certainly coordinated with oxygen ions) is the same. In
the molybdenum-coated aluminum oxide catalyst, the
molybdenum ions are in at least two different bonding
situations. Miller, et al offer the explanation that active
sites are created on the alumina by calcination and molybdenum ions in the vicinity will donate electrons and assume
a different valence state. A similar but alternate explanation for the different chemical environments is that some
molybdenum ions diffuse into the alumina, replacing the
aluminum ions. These, although still bonded to six oxygen
ions, will be in a different chemical environment as an
impurity in alumina that when in the molybdate, because
the interatomic distances differ.

In a catalyst consisting of rhodium adsorbed on char-
coal, both Rh_2O_3 and Rh metal were found to exist on the
surface[6]. The metal oxide ratio was determined by comparing the intensities of the Rh metal and Rh^{+3} ion 3d peak
intensities. Catalysts with high activity were found to be
characterized by a metal/oxide ratio less than unity while
for poor catalysts the ratio is greater than unity.

Crystal Structure vs. Electronic Structure

The determination of structural order in thin films is
difficult. The usual diffraction techniques probe so deeply that one ends up studying the structure of the substrate.
Studies have been done by transmission electron diffraction
by removing the film from a soluble substrate. However,
such procedures leave some doubt about the effect of
changing the substrate and the effect of electron beam
heating on the film's structure.

The valence-band portion of the electronic spectrum
has been found to be very sensitive to structure. In these
laboratories[7] Si, Ge, and AsTe were prepared both in the
bulk crystalline state and the amorphous thin-film state,
the latter by r.f. sputtering on cooled substrates. The
bulk samples were polished and treated with dilute nitric
acid to remove the amorphous layer formed by polishing.

Valence electron spectra for amorphous and crystalline germanium and silicon are shown in Figs. 5 and 6. Crystalline spectra contain well defined peaks at 2.8 eV for Ge, and 3.1 eV for Si. These energy levels, corresponding to Ge 4p and Si 3p levels, are the least tightly bound electrons and as such probably participate in electron transport processes. The binding energy of the outermost occupied level of Si is o. 3 eV greater than for Ge, as expected, since Si is more resistive than Ge. The amorphous spectra resemble the crystalline spectra in that there are remnants of the Ge 4s and 4p levels and a reduction to very low noise levels near the zero of binding energy. They are notably different in that amorphous spectra do not have well defined peaks for binding energies less than 15 eV. (This is the precise characteristic difference between crystalline and amorphous Si and Ge, theoretically predicted by Cramer[8] and Jungk[9] and confirmed using optical spectroscopic techniques[10].

The densities of states for amorphous and crystalline AsTe are represented by Fig. 7. The polycrystalline sam-

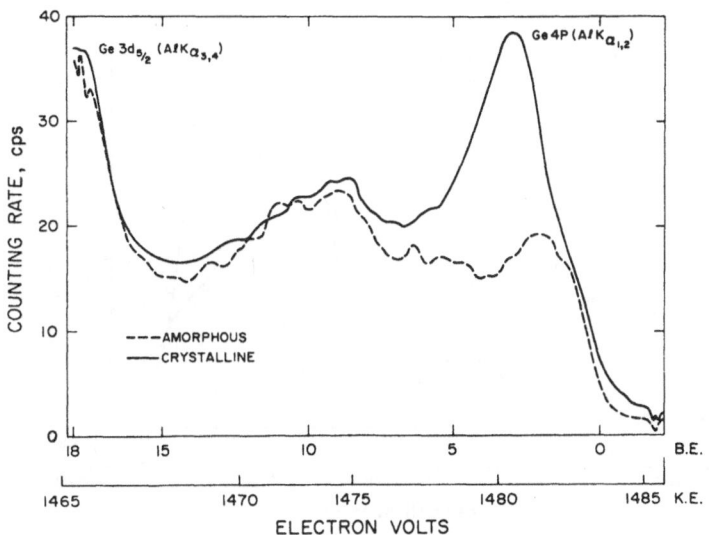

Fig. 5. Low binding energy portion of germanium's electronic spectrum.

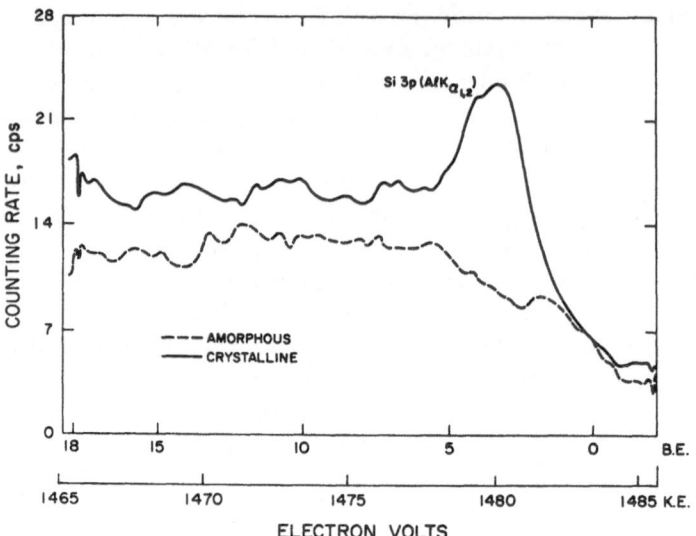

Fig. 6. Low binding energy portion of silicon's electronic spectrum.

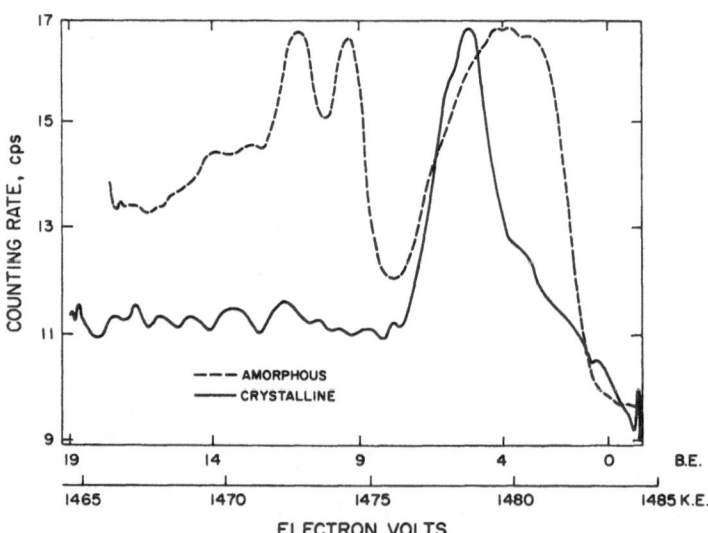

Fig. 7. Low binding energy portion of AsTe's electronic spectrum.

ple shows a shifted peak beginning at 7.3 eV and ending
near zero with its center at 5.2 eV. Its half-height width
is 2.3 eV, indicating that it is a multiple peak whose average binding energy is greater than either the 4p levels of
As (3 eV) or the 5p levels of Te (2 eV). However, the
general shape indicates that there is a strong peak centered at about 5.1 eV. The amorphous spectrum contains
three peaks, two that are part of a doublet at 10.0 eV and
8.5 eV, and a broad peak extending from 15.8 eV to zero
with its center at 2.9 eV and having a half-height width of
4.8 eV. Such broadening could indicate several peaks
whose individual widths are narrow compared to that of
the incident radiation. This is similar to the phenomena
noted for Si and Ge where the outer levels were reduced
in intensity.

Somewhat indirect but very meaningful structural
studies can be carried out on thin films using ESCA if one
can prepare bulk crystalline and amorphous samples and
determine their structure by diffraction. Then the shape
of the valence band can be determined by ESCA and associated with its structure. With this calibration procedure,
the structure of very thin films, or surfaces, may be
studied and even the degree of crystallinity ascertained.

ION SCATTERING SPECTROSCOPY

ESCA was shown above to be a tool for determining the
composition of Sn and Pd monolayers. However, lower layers also contribute their electronic spectra. Ion scattering
spectroscopy[11] analyzes only the top monolayer. The basic
principle involves first the ionization of an inert gas by
electron bombardment. These ions are accelerated to an
energy of between 1 and 3 keV and are directed at the sample surface. Some of these ions make a "head-on" collision
with the monolayer surface atoms and recoil at an angle
equal to the incident angle (Fig. 8). Only these inert gas
ions enter an electrostatic spectrometer, similar to the
one used in ESCA, and their energy distribution is determined. The ions which recoil in other directions are lost

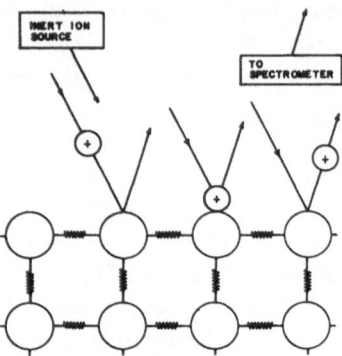

Fig. 8. Schematic of inert gas ions colliding with surface atoms in ion scattering spectroscopy.

to the spectrometer. The inert gas ions which penetrate the top surface are neutralized and many of them recoil but are lost to the spectrometer, even if they leave the surface in the proper direction and enter the spectrometer because they will not be deflected by its electrostatic fields.

In glass and ceramics, the surface becomes charged as a result of the neutralization of the inert gas ions[12]. This accumulated charge would affect the energy of the recoiled inert gas ions so it is necessary to neutralize that charge by directing a stream of electrons onto the surface.

The energy of a properly recoiling inert gas ion is related to the mass of the surface atom in the collision by:

$$\frac{E_1}{E_0} = \frac{(m_2 - m_1)}{(m_2 + m_1)}$$

provided $m_2 > m_1$, where E_1 = kinetic energy of the inert gas ion after scattering through $90°$, E_0 = initial kinetic energy of the inert gas ion, m_1 = mass of the inert gas ion, and m_2 = mass of the surface atom or ion. This equation was derived simply by considering the conservation of energy in a single elastic collision. The energy of the recoil ion is not significantly affected by the strength or type of atomic bonds in the surface. Therefore, if the surface is a monolayer of one type of atom or ion, the recoil inert gas ions will have a single energy and the output will consist of a single sharp peak on an intensity vs. D_1/E_0 plot. The principal advantages of ion scattering spectroscopy are the simplicity of the spectra, i.e., one peak per element detected, and the ability to detect and identify only the top monolayer of atoms or ions.

SUMMARY

Electron spectroscopy for chemical analysis, or X-ray photoelectron spectroscopy, is a valuable took for the study of glass and ceramic surfaces because:

1. It will detect and identify monolayers atoms, although the electronic spectra for the atoms from 20 to 50 Å beneath the surface contributes their spectra.

2. It will detect and identify light ions such as oxygen.

3. It will provide information from which the type of bonding and/or valence state can be inferred.

4. The shape of the valence bond is associated with the crystallinity of the surface.

A complementary technique, ion scattering spectroscopy, provides an elemental and quantitative identification of monolayers without detecting underlying atoms, but will infer very little concerning bonding, ionic valence state, or structure.

ACKNOWLEDGMENTS

This work was primarily supported by the Advanced Research Projects Agency of the Department of Defense and was monitored by Dr. C. Bogosian, U.S. Army Research

Office-Durham, under Grant DA-ARO-D-31-124-72-G72.
The tin oxide study was supported by ASG Industries and
monitored by Dr. A. H. Agett. Gratitude is expressed
to W. A. Wolstenholme, J. Kraitcham, and A. W. Miller,
et al. for permission to describe portions of their research.

REFERENCES

1. D. A. Shirley, ed., Electron Spectroscopy, Proceedings
    of an International Conference Held at Asilomar, Pacific
    Grove, Calif., U. S. A., September 7-10, 1971, North
    Holland Publishing Co., Amsterdam, London; American
    Elsevier Publishing Co. Inc., New York, 1972.
2. K. Siegbahn, C. Nordling, A. Fahlman, et al., Electron
    Spectroscopy for Chemical Analysis, Technical Report
    AFML-TR-189, Oct. 1968. (Reproduced by National
    Technical Information Service, Springfield, Va.; AD
    844315).
3. A. Rohatgi, Semiconducting Tin Oxide Thin Films on
    Glass, Ceramic Engineering M. S. Thesis, VPI&SU,
    Blacksburg, Va., 1973.
4. Private communication with W. A. Wolstenholme, AEI
    Scientific Apparatus, Inc., Almsford, N. Y., and J.
    Kraitchman, PPG Industries, Pittsburgh, Pa.
5. A. W. Miller, W. Atkinson, M. Barber, and P. Swift,
    "The High Energy Photoelectron Spectra of Molybdenum
    in Some Mo-Al$_2$O$_3$ Systems" J. Catalysis 22, 140-142
    (1970).
6. J. S. Brinen and A. Melera, "Electron Spectroscopy
    for Chemical Analysis (ESCA) Studies on Catlaysts
    Rhodium on Charcoal," J. Phys. Chem. 76 (18) 2525
    (1972).
7. L. H. Slack, L. R. Durden, and W. D. Leahy, "Non-
    crystalline Semiconductors, Electrical and Thermal
    Processes." Annual Technical Report, ARPA/AROD
    Grant No. DA-ARO-D-31-124-72-G72, February 28,
    1973.
8. B. Kramer, "Electronic Structure and Optical Prop-
    erties of Amorphous Germanium and Selenium."

Phys. Stat. Sol., (b), 47, 501 (1971).

9.   G. Jungk, "Determination of Optical Constants: Inter-
     band Transitions in Amorphous Ge, Si, and Se." Phys.
     Stat. Sol. (b), 46 603 (1971).

10. T. M. Donovan and W. E. Spicer, "Optical Properties
     of Amorphous Germanium Films." Phys. Rev. Let.
     21, 1572 (1968).

11. D. P. Smith, "Analysis of Surface Composition with
     Low Energy Backscattered Ions." Surface Science 25,
     171-191 (1971).

12. R. F. Goff, "Ion Scattering Spectroscopy," J. Vacuum
     Science and Technology 10, (2), 355-358 (1973).

DISCUSSION

W. M. Mularie (Martin-Marietta): I would question whether,
in an oil-diffusion pumped ESCA apparatus, one can be confident that "surface" properties are indeed being measured.
From simple gas kinetics, monolayer adsorption times at
10^{-8}-10^{-9} torr must be less than observation times. Secondly, the evidence cited for ESCA monolayer sensitivity should
be tempered by the considerable experimental evidence that
shows two-dimensional surface segregation of adsorbates
can give misleading results.

Author: An important capability of ESCA is that it will
probe through adsorbed monolayers. ESCA is not a true
"surface" instrument in that it detects several monolayers
in the vicinity of the surface. Our point was that ESCA is
a reasonable tool for qualitative and semi-quantitative
studies of layers which have an average thickness of one
monolayer. Secondly, the segregation of absorbates deposited by Langmuir adsorption is dependent upon the
characteristics of the surface. The clean glass surfaces
used by Kraitchman certainly represent reasonably clean
and homogenous surfaces. ESCA is a microscopic tool
which is not sensitive to any two-dimensional segregation
of adsorbates, only to the average thickness. Thank you
for pointing out an important concern for those using ESCA
in surface studies.

D. Dove (Univ. Florida): The local order in amorphous
materials may be different from that in the isocompositional
crystal structure, however this is not the case for Ge and
Si. Do you have any comments on the absence of the valence peak for amorphous Ge and Si?

Author: We cannot believe that the valence band for amorphous Ge and Si is truly absent. Compared to the crystalline phases it is very weak and diffuse. The true density
of states for the amorphous film could have a small peak
indicating the valence band but the lack of perfect resolution
in ESCA may cause such a peak to be lost.

R. Atkin (IBM): Have you been able to eliminate surface
charging of glasses in ESCA? Have you tried using a low
energy electron flood gun to eliminate charging?

Author: Yes, we have essentially eliminated charging by
depositing a very thin film of gold to drain off surface
charges. The intense core electron spectra of the glass
components are observable along with that of gold, but the
intense gold valence band completely negates any study of
the glass valence band. The electron flood gun is used in
Ion Scattering Spectroscopy to compensate the positive
charges left by the incident inert gas ions, but such a flood
of electrons probably cannot be used in ESCA because many
of these electrons would be scattered from the glass surface
entering the electron spectrometer and at least diminish
spectra resolution.

H. A. Schaeffer (Univ. Erlangen): Did you apply ESCA for
investigating glass surfaces which have been subjected to
different heat pretreatments or have undergone different
melting conditions? Positron annihilation experiments reveal differences in binding states (covalency) due to pretreatments, thus indicating changes in short range order.

Author: We have not undertaken such a study in a systematic
manner. ESCA certainly appears to be the proper instrument
for such a study. Changes in covalency would be accompanied
by shifts in the core electron levels. Our experience indicates that these shifts are greater for the lighter elements,
such as oxygen.

J. Wong (GE): A comment on the first question by Dr.
Schaeffer: The increase in bond covalency in going from
the crystalline to vitreous state has been deduced in some

recent ESCA studies in germanium chalcogenides (Betts, Bienenstock & Bates, J. Non-Crystal. Solids, 8-10, 364 (1972). In crystalline Ge-Te, Ge-Se and Ge-S, the binding energy of the Ge $3d^{5/2}$ level shows large positive shift 2 to 3 eV relative to crystalline Ge. In the Ge_xTe_{1-x} glasses with x = 0.15 to .57, the Ge $3d^{5/2}$ photoemitted electron energies shift positively by ~ 1 eV relative to that of the same reference material. The lower core binding energies in the vitreous materials arise from a reduction of net positive charge of the Ge as a result of an increase in co-valency of the Ge-Te bond.

Author: The shifts reported for Ge in the reference you have quoted are typical of the monoxides of Ge, found in this laboratory and reported by W. E. Swartz, Jr., "X-Ray Photoelectron Spectroscopic Studies of Phosphorous, Nitrogen, Selenium, Tellurium and Molybdenum Compounds," Ph. D. Thesis, M. I. T., 1971. However, if as Siegbahn[2] has shown, electrochemical shifts are related to electronegative differences between bonded elements, then Ge-Te, Ge-Se, Ge-S and Ge-O bonds, should show an increasing chemical shift in this order with the shift for the Ge-O bond being roughly an order of magnitude larger than the shift for the Ge-S bond.

L. Hench (Univ. Florida): Recent molecular orbital cal-culations by H. F. Schaake, Ph. D. Thesis, Univ. of Florida (1973), have shown only very small changes in the valence band of amorphous Si clusters of 44 and 49 atoms in comparison with crystal Si clusters of the same size. Therefore the ESCA data presented as "lack of valence band" of amorphous Si and Ge may require further analysis in order to make this interpretation.

Author: Every semiconductor, indeed every solid, must have a valence band. Our point was that the valence band intensities for amorphous Si and Ge are so low, that the valence bands appear to be missing. Apparently, the energy levels of the outermost electrons in the amorphous Si and Ge films are dispersed over a broad energy range. These observations have proved to be very reproducable in our laboratory and have also been reported by Ley, et al., Phys. Rev. Lett., 29, 16 (1972).

We interpret the marked energy dispersion of the binding electrons in amorphous Si and Ge to be due to a dispersion of the bond angles, bond length, or both. The identity
of the core electron spectra for the crystalline and amorphous Si and Ge films indicates that coordination number
has not changed with changes in structure.

SURFACE ANALYSIS OF GLASS BY CHARGED PARTICLE ANALYSIS

W. B. Crandall and J. W. Mandler

Illinois Institute of Technology
Research Institute
Chicago, Illinois

Properties of glass such as chemical durability, strength and erosion resistance are determined, in part, by its surface character. This paper deals with one means of characterizing the glass surface chemical constitution by using charged particle techniques. The purpose of this study of glass surfaces was to understand certain cases of failure occurring in glass containers after extended periods of time of exposure to corrosive environments and internal pressure.

A 2-MeV Van de Graaff accelerator has been used to examine the surface of the glass container by two means of analysis: Charged Particle Activation Analysis, CPAA, and Charged Particle X-Ray Spectrometry, CPXS. The two techniques will be described along with results obtained upon examining a glass surface.

CPXS, CPAA TECHNIQUES

Fig. 1 is a block diagram of the electronics used with the Van de Graaff accelerator. Gamma rays and X-rays are collected using the Ge (Li) and Si (Li) detectors positioned as shown. The Si (Li) detector is internal to the

75

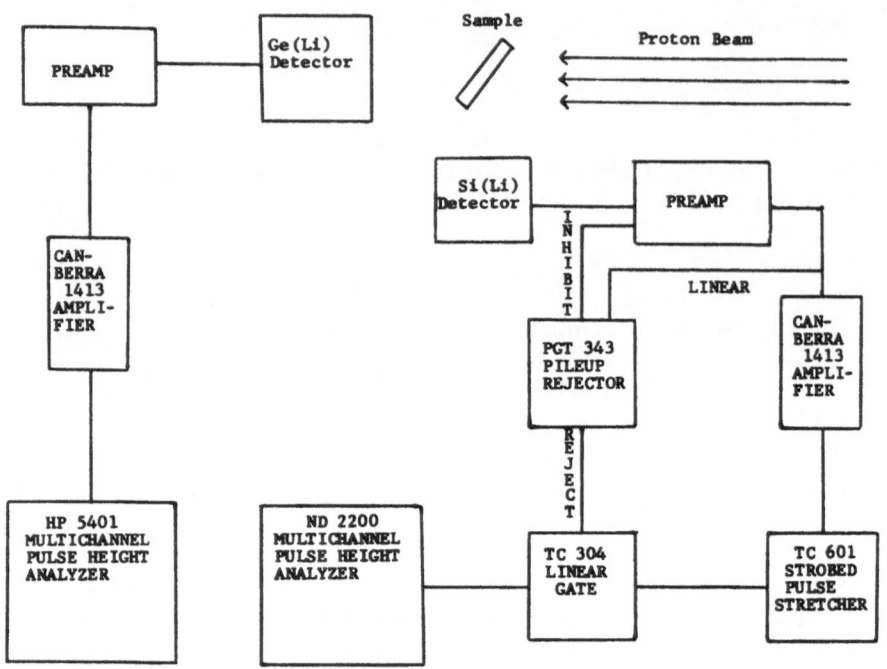

Fig. 1.　Block diagram of electronics.

vacuum system in order to reduce attenuation of the X-rays while the Ge (Li) detector is located external to the vacuum system, since high intensity gamma rays suffer little attenuation in the walls of the vacuum chamber.

CPXS, Charged Particle X-Ray Spectrometry

Charged Particle X-ray Spectrometry (CPXS), consists of bombarding a sample with charged particles (e.g., protons, alpha particles) and detecting the characteristic X-rays produced. The use of X-ray emission spectrometry for assaying small amounts of trace elements has grown rapidly in recent years. Of fundamental importance in this growth has been the introduction of high resolution Si(Li) detectors which provide fast and efficient detection of X-rays. The energies of the X-rays emitted are used to identify the elements present, and their intensities are used to determine their concentrations. Traditionally, X-rays and electrons have been used to stimulate the characteristic X-rays; however, the use of charged particles for stimulation has the advantage that the cross-section for X-ray production is large, and the background contribution from bremsstrahlung is low.

The ability to detect a given element using the CPXS technique depends on the energy of the X-rays emitted. For low-Z elements even the K X-rays are so low in energy that they are strongly absorbed by the sample material itself, the window of the detector, and any other intervening substance, even air in some cases. The practical lower limit is 1 or 2 keV. For high-Z elements (higher than about molybdenum) the L X-ray has sufficient energy to be detected and is more intense than the K X-ray. For example, the L X-ray of tin at 3.4 keV is approximately 1000 times as intense as the K X-ray at 25 keV. As a result, the X-rays of interest will usually be below 20 keV.

The high-intensity charged-particle accelerators available and the high signal-to-noise ratios obtainable allow the CPXS technique to achieve extremely high sensitivities. Duggan et al,[1] provide data on expected sensitivities relative to other analytical techniques (Fig. 2). Results of preliminary experiments and literature studies performed at IITRI[2] indicate that, in general, the elements heavier than approximately sulfur ($Z > 16$) should be detectable in concentrations of 1 to 100 ppb for a homogeneous low-Z matrix, such as Si in the present case.

The CPAA Technique

The light elements ($1 < Z < 20$) undergo a variety of reactions when bombarded with relatively high energy

(0.5 to 2 MeV) charged particles (protons, deuterons, alphas, tritons, etc.). The reactions that have proved to be of greatest interest are:

$$_{Z}X^{A} + \text{proton} \longrightarrow {}_{Z+1}Y^{A+1} + \text{gamma};$$

$$_{Z}X^{A} + \text{deuteron} \longrightarrow {}_{Z}X^{A+1} + \text{proton} + \text{gamma};$$

$$_{Z}X^{A} + \text{proton} \longrightarrow {}_{Z-1}X^{A-3} + {}_{2}He^{4} + \text{gamma},$$

where X is an atomic species, A its atomic mass, and Z its atomic number. These reactions are referred to as (p, γ), (d, p, γ), and (p, α, γ), respectively. Given the bombarding charged particle and its energy, the emitted gamma radiation is characteristic of the target element. Hence, with proper calibration, a measurement of the gamma-ray energies and intensities provide a quantitative measurement of the elements present in the sample. We term this technique Charged Particle Activation Analysis (CPAA).

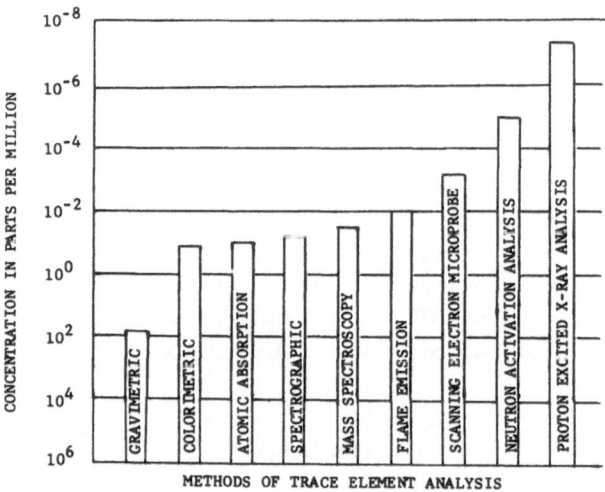

Fig. 2. Sensitivity of proton-excited X-ray analysis compared to other analytical techniques.

Since the charged particles penetrate into the sample to only tens of microns and the emitted gamma rays are highly penetrating, matrix effects are virtually nonexistent in CPAA. In addition, by the careful selection of the bombarding particle and energy, depth information can be obtained.

Preliminary investigations at IITRI yielded sensitivities obtainable using a variety of charged particle reactions (Table I). A Ge (Li) gamma-ray detector having a resolution of 3. 0 keV at 1. 33 keV and a volume of 40 cc, was used in these investigations. These are preliminary sensitivities; more elaborate experimental procedures can undoubtedly significantly lower these thresholds.

Table I. Preliminary Detection Thresholds Using CPAA

Element	Reaction	Detection Threshold (ppm)
Lithium	$Li(p, \gamma)Be^8$	0. 5
Boron	$B^{10}(p, \alpha)Be^7$	1. 0
Carbon	$C^{12}(d, p\gamma)C^{13}$	0. 2
Oxygen	$O^{16}(d, p\gamma)O^{17}$	0. 2
Fluorine	$F^{19}(p, \alpha\gamma)O^{16}$	0. 1
Sodium	$Na^{23}(p, \alpha\gamma)Ne^{20}$	1. 0

EXAMINATION OF GLASS SURFACES

As stated earlier, one of the reasons for undertaking this study was to determine if surface chemistry changes could be detected by CPXS and/or CPAA in conditions where delayed fracture of a pressurized glass container has been observed. Glass containers holding liquids with high pH were of most interest in this program.

Experimental Procedure; Glass Preparation

Typical glass containers which had had neither hot-end nor cold-end treatment were used. The specimen to

be tested was cut from the bottom of the container so as
to obtain as nearly flat a surface as possible. Both inside
and outside surfaces, with and without exposure to basic
solution, were examined to establish a method of examining the real-life examples of failure, should this testing
method prove successful. When the inside surface was to
be tested, the outside surface and bulk were ground away
until a thin wafer was obtained, 1/2 x 1/2 x 1/16 in. thick.
When the outside surface was to be tested, the inside surface and bulk were ground away in the same manner. It
was found that thin samples were necessary in order to
prevent excessive charge buildup in the Van de Graaff.

The glass was examined by standard X-ray spectroscopy and chemical methods to determine the bulk composition of the glass.

A number of the samples were exposed to 5% NaOH.
The time of exposure was 30 minutes and 200 minutes.
The pH of the test solution was 13.15, as high as any fluid
known to be contained in pressurized bottles. The time of
exposure was set for this period because 200 minutes was
the earliest known time of failure for basic fluids stored
under these conditions.

Table II. CPAA Results.

Sample	Exposure to NaOH(min.)	Depth of Analysis(μm)	Na Content (%)
1	0	0-3	10.6 ± 0.3
	0	3-6	10.6
2	0	0-3	10.6
	0	3-6	10.6
3	30	0-3	9.8
	30	3-6	10.4
4	200	0-3	7.8 ± 0.2
	200	3-6	8.3 ± 0.2

RESULTS AND DISCUSSION

The results are summarized in Table II. Tests conducted by CPXS techniques did not show any difference in chemical composition between the surfaces exposed to the NaOH solution and those not exposed. However, it was not possible, at this time, to adjust the energy conditions of the Van de Graaff to examine less than about 27 μm of depth. From the CPAA results, it is assumed that this depth of 27 μm is too great to have shown any surface effects, so it is unknown whether those elements examined above Si (Z=14) have been changed in any way by the exposure to the NaOH solution. However, some interesting comparisons may be made between the standard and CPXS data as shown in Fig. 3. Although both the standard X-ray spectrograph and the Van de Graaff X-ray spectrograph should detect the same elements, the CPXS technique should have the higher sensitivity. This increased sensitivity of the CPXS analysis may account for the slight differences in the relative intensity values obtained for the lighter elements. Also the Cl, Mn and Pb did not show up by standard methods.

The trace shown for the elemental readout for this glass is typical for both CPXS and CPAA. It is possible to see from this trace, the high signal-to-noise ratio for this type of analysis, and how this allows good analysis of the glass.

The CPAA examined the lighter atomic number atoms and in this range, a difference was found in the Na content in the sample. It may be seen from Table II that there was a significant change in Na content with NaOH treatment.

There was no variation in the amount of sodium either on the inside or the outside surface of the glass on as-received containers. However, 30 min. exposure to NaOH reduced the sodium content by 5% in the first 3 μm of surface and showed only 2% reduction in sodium content at 3-6 μm depth. It may also be seen that the 200 min. expo-

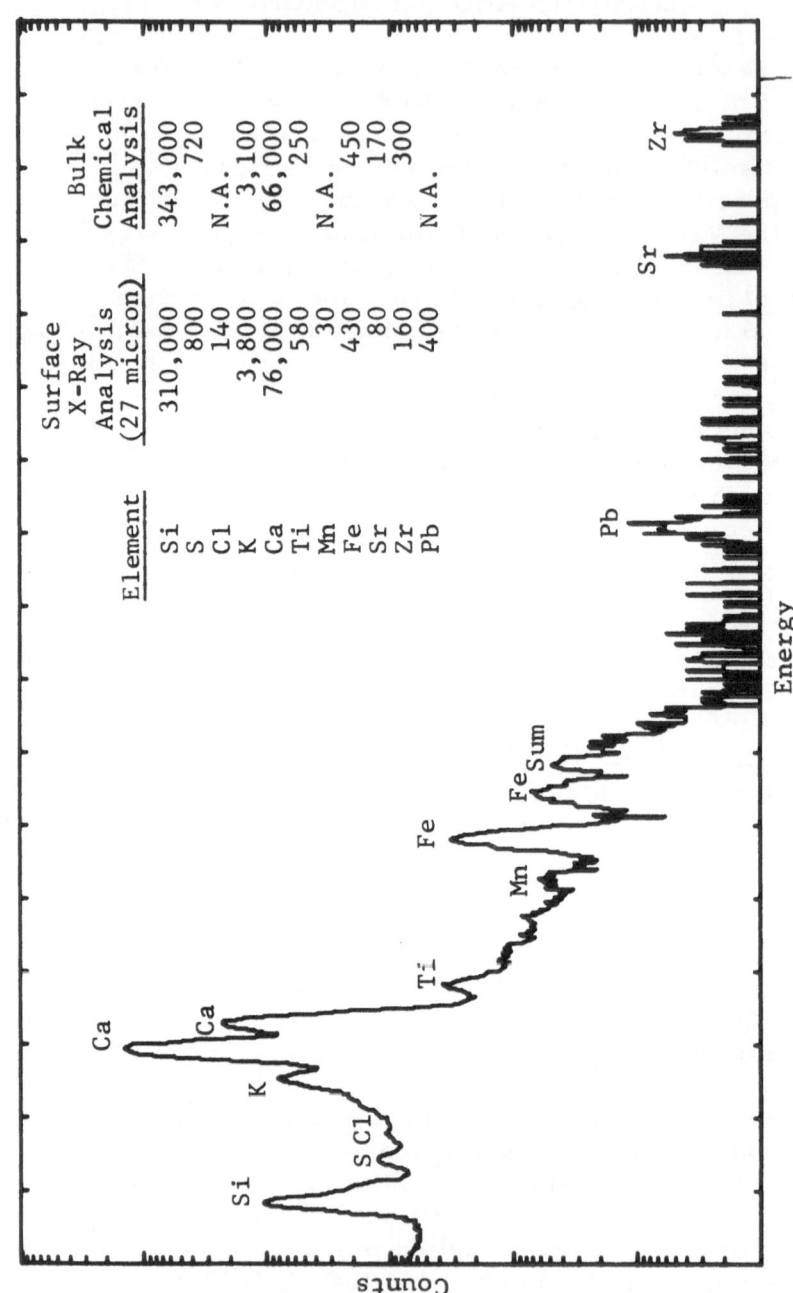

Element	Surface X-Ray Analysis (27 micron)	Bulk Chemical Analysis
Si	310,000	343,000
S	800	720
Cl	140	N.A.
K	3,800	3,100
Ca	76,000	66,000
Ti	580	250
Mn	30	
Fe	430	450
Sr	80	170
Zr	160	300
Pb	400	N.A.

Fig. 3. Surface analysis by CPXS with 1.5 MEV protons.

sure further decreased the sodium content at the outer
6 μm of surface by 20%.

The investigation is still in its preliminary stage and
explanation for the change in sodium content requires
evaluation.

REFERENCES

1. J. L. Duggan and W. L. Beck, "Studies of X-rays
   Induced by Charged Particles," Advances in X-ray
   Analysis, Vol. 15, p. 407 (1972).
2. S. T. Baker, J. W. Mandler, R. B. Moler, and J. H.
   Reed, "New Techniques for Examination of Surface
   Corrosion and Coatings," Internal Research Program,
   Final Report V1061 (1971).

DISCUSSION

J. M. Khan (Lawrence Livermore Laboratory): I wish to
underscore the nondestructive character of the CPAA and
CPXS techniques. With them it is possible to interrupt a
process, interrogate the surface, and then continue the
process of interest.

K. K. Verma (N. Y. S. College of Ceramics): If the process
is to be nondestructive, the choice of the irradiating
(probing) charged particles and their energy level must be
appropriate to the particular system under study. Recently,
even probing X-rays are being questioned for determination of structural parameters in some of glassy semiconductors such as Se, As_2S_3 and As_2Se_3.

A METHOD FOR DETERMINING THE PREFERRED ORIENTATION OF CRYSTALLITES NORMAL TO A SURFACE

Robert L. Snyder and William L. Carr

New York State College of Ceramics
Alfred University
Alfred, New York

Many polycrystalline materials exhibit anisotropy in their thermal, electrical, optical and mechanical properties owing to preferred orientations in crystallite packing. A knowledge of the directions and degree of preferred orientation is essential in understanding and predicting the physical properties of these materials. The directions of preferred orientation in a specimen are a function of crystallite shape and the process used to form the body. In extruded or rolled materials it is common to find two types of orientation, one normal to the surface of the body, the second within the surface in the rolling or extrusion direction[1,2]. Fabrication techniques based on casting, deposition or pressing, however, will introduce preferred orientations normal to the surface only, with crystallite directions within the surface at random.

The most common method of representing preferred orientation is to construct a stereographic projection of the normals to a crystallographic plane, called a pole figure[3,4]. Other techniques such as representing the angular distribution of a particular direction in a crystallographic reference frame (inverse pole figure)[5-8] or analytical methods[9,10] have been developed. The data required by these techniques for displaying preferred orientation is obtained by measuring the orientation of a large number of crystallites. This may be

85

done visually using etch-pit or Laue techniques which,
experimentally, are both tedious and difficult[11]. Most
workers take advantage of the fact that the intensities of
X-ray diffraction maxima are proportional to the number
of crystallites whose crystallographic plane normals bi-
sect the incident and diffracted beams. To measure the
data necessary to obtain a pole figure for a particular
crystallographic direction [h k ℓ] , the incident and diffracted X-ray beams are set so that each make an angle
$2\theta_{hk\ell}$ with respect to the specimen. Intensity data are
then recorded at various specimen angles with respect
to the diffractometer geometry, holding 2θ constant[12,13].
Such pole figure devices, though complex and expensive,
are in common use and often automated[14,15].

In principle all of the preferred-orientation information
for the direction normal to a surface will be contained in the
intensity differences between X-ray diffraction peaks for a
sample with a preferred orientation and one with a fully random crystallite orientation. If the sample was formed in a
way that produces only orientations normal to the surface,
an X-ray powder diffractometer tracing will contain the
needed information. If there are also preferred orientations
within the surface, then the sample should be rotated about
an axis that bisects the incident and diffracted beams, scanning 2θ, after each incremental rotation. The many experimental difficulties in obtaining the diffraction pattern of a
fully random sample may be avoided by calculating the powder diffraction pattern from crystal structure parameters.
The calculated diffraction peak intensities will correspond
to the intensities of an experimental pattern having a completely random crystallite orientation. Thus a full analysis
of crystallite orientation should be available from a comparison of an observed diffraction pattern with a calculated
"ideal" pattern.

EXPERIMENTAL

X-ray diffraction patterns were obtained with Cu Kα
radiation using an X-ray diffractometer, equipped with a

graphite monochromator.   The alumina sample used in
this study was obtained from Dr. P. H. Crayton; it had
been prepared by hot-pressing a mixture of 99. 0% Al_2O_3
and 1% MgO.   The diffraction pattern was obtained so that
the pressing direction, normal to the wafer-shaped specimen, bisected the incident and diffracted X-ray beams.
All but four low-intensity peaks in the observed pattern
corresponded to reported peaks in α-Al_2O_3 (corundum).
The four non-alumina peaks were identified as the four
strongest peaks in $MgAl_2O_4$ spinel[16] and were deleted
from the pattern.

The barium hexaferrite sample used was supplied by
Dr. J.S. Reed.   It was prepared by pressure filtration
with a magnetic field in the direction of filtration.   The
diffraction pattern of the wafer-shaped sample was obtained in the same manner as for alumina.   A number of
peaks were observed which did not correspond to the calculated pattern; these were attributed to stoichiometries
differing from $BaO \cdot 6Fe_2O_3$.   These peaks were not removed from the observed pattern but left to test the abil-
ities of our computer program to reject them.   The third
material studied was aragonite ($CaCO_3$).   The diffraction
pattern was chosen somewhat randomly from among the
orthorhombic materials for which the National Bureau of
Standards have published X-ray diffraction patterns[17].
An orthorhombic material was desired to test the procedure on a low-symmetry material.   NBS data were used
because of its high quality and the high probability of there
being as little orientation as possible in it, thus giving a
random orientation to test our procedure.

The calculated diffraction patterns were obtained using
a local modification of a program first developed by D. K.
Smith[18].   The parameters used in calculating the powder
patterns are given in Table I.   All patterns were calculated
for copper radiation using a Cauchy peak profile.   Anomalous
dispersion corrections for barium and iron were included in
the calculation of structure factors for barium hexaferrite.
The widths of the peaks at half-height were calculated in all
cases to be 0. 07° at 40. 0° 2θ.

Table I. Parameters Used in Calculating Powder Patterns.

Compound	Atom	Multi-plicity	X	Y	Z	B or β_{11}
$BaO \cdot 6Fe_2O_3$[19-20]	Ba^{+2}(d)	.125	.3333	.6667	.7500	1.0
a=b=5.889 Å	Fe^{+3}(a)	.125	0.0	0.0	0.0	1.0
c=23.182 Å	Fe^{+3}(b)	.125	0.0	0.0	.250	1.0
$\alpha=\beta=90°$	Fe^{+3}(f)	.250	.3333	.6667	.028	1.0
$\gamma=120°$	Fe^{+3}(f^1)	.250	.3333	.6667	.189	1.0
$P_{6_3}/m\ mc$	Fe^{+3}(k)	1.0	.167	.334	.108	1.0
	O^{-2}(e)	.25	0.0	0.0	.150	1.0
	O^{-2}(f)	.25	0.0	0.0	-.050	1.0
	O^{-2}(h)	.50	.3333	.6667	.250	1.0
	O^{-2}(k)	1.0	.167	.334	.050	1.0
	O^{-2}(k^1)	1.0	.50	1.0	.150	1.0
Al_2O_3[21]	Al^{+3}	.6667	0.0	0.0	.352	1.0
a=b=4.7589 Å	O^{-2}	1.0	.306	0.0	.250	1.0
c=12.9910 Å						
$\alpha=\beta=90°$						
$\gamma=120°$						
$R_{\bar{3}c}$						
$CaCO_3$[22]	Ca^{+2}	.5	.4151	.7597	.250	.0024
a=7.9792 Å	C^{+4}	.5	.7623	-.0855	.250	.0029
b=5.7499 Å	O^{-2}	.5	.9234	-.0953	.250	.0025
c=4.9677 Å	$O^{-2'}$	1.0	.6804	-.0871	.4741	.0043
$\alpha=\beta=90°$						

		β_{22}	β_{33}	β_{12}	β_{13}	β_{23}
P_{bnm}	Ca^{+2}	.0048	.0063	.0001	.0000	.0000
	C^{+4}	.0039	.0060	.0002	.0000	.0000
	O^{-2}	.0080	.0131	.0003	.0000	.0000
	$O^{-2'}$	.0078	.0061	.0005	.0012	.0002

The barium hexaferrite and alumina samples were observed with an Etec SEM. The barium hexaferrite shown in Fig. 1 is predominately in the form of hexagonal-shaped platelets with [001] normal to the six-fold symmetry axis. Though crystallites of all orientations were observed, there was a definite

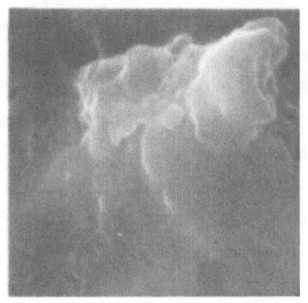

Fig. 1. $BaO \cdot 6Fe_2O_3$ specimen; SEM 13,500X.

Fig. 2. Al_2O_3 specimen; SEM 9,000X.

predominance of crystallites lying with $[001]$ normal to the direction of filtration. The alumina specimen shown in Fig. 2 has been sintered to the point of loss of identity of most crystallites, but the specimen does appear to have a layer-like structure in the direction normal to hot-pressing. Normally, alumina shows a tendency to orient with $[00\bar{1}]$ normal to the surface[1].

PROCEDURE

The problem which immediately arises, when one tries to compare observed X-ray intensities with those calculated, is that they are not on the same scale. When the observed pattern is from a randomly oriented sample, the conventional $I/I_{max.}$ values will equal those of the calculated pattern within statistical error. But when a preferred orientation occurs, the intensity of the highest peak ($I_{max.}$) will change relative to all other observed intensities, thus putting the $I/I_{max.}$ values on a new scale. In order to scale the observed-to-calculated values we would need a law of conservation of relative intensity (i.e., the sum of the fractional decreases in peak intensities equals the sum of fractional increases). Due to the fact that crystallites can orient into or away from directions whose reflections have zero intensity, such a law is impossible and we are left with having to seek a method of comparison not involving scaling.

One approach can be based on the fact that, as the fraction of crystallites oriented in a particular direction increases, not only will the reflection intensities in that zone increase, but also the intensities of those reflections whose plane normals make a small angle with the zone. This will result from the increased probability of finding crystallites almost, but not quite, perfectly oriented. Thus the slope of a plot of $(I_{obs.} - I_{calc.})$ vs. Φ, (Φ is the angle the normal to each plane (hkℓ) makes with a particular zone) should be negative at low angles for a zone of preferred orientation, first scaling the ΔI values to make those at the zone ($\Phi = 0.0$) zero. Similarly, the slope should be positive for a zone of anti-preferred orientation, i.e., an orientation which shows a decrease in the number of crystallites displaying it.

Fig. 3 shows a plot of this type for the $\begin{bmatrix}100\end{bmatrix}$ zone of barium hexaferrite and illustrates the weaknesses of this method. Statistical errors in the observed intensities and differing rates of change in $I_0 - I_c$, depending on the direction from which the zone is approached, obscure any trend. The principal weakness is the dependence on the occurence of reflections making a small angle with the zone. Thus the criteria for a procedure to analyse preferred orientation

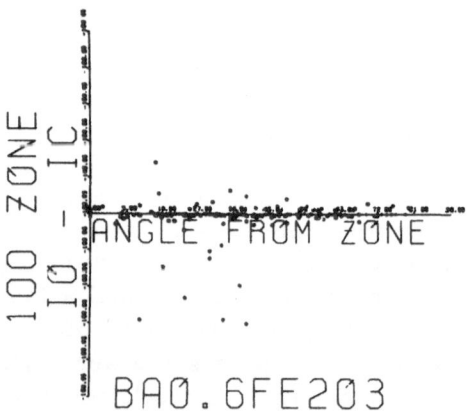

Fig. 3. $(I_{obs.} - I_{calc.})_{hk\ell}$ vs. Φ (the angle that the hkℓ plane normal makes with the zone) for the 100 zone of barium hexaferrite. Scaled so that ΔI is zero at $\Phi = 0.0$.

should be: 1.) It must be independent of scale factor, 2.) The function evaluated at each point in space must depend on a large number of intensities to minimize the large relative error in weak reflections, and 3.) The function must not be restricted to evaluation at only those points for which there are observed reflections.

In reference to the $[00\bar{1}]$ orientation in barium hexa-ferrite, Lotgering[23] has defined an orientation factor as $f = (p_o - p_r)/(1 - p_r)$ where $p_o = \sum_\ell I_{00\ell}/\sum_{hk\ell} I_{hk\ell}$ for an oriented sample. p_r is the same function referred to a randomly oriented sample. When a random sample can be prepared experimentally and care is taken to keep the diffraction pattern for it, and for the oriented sample, on the same scale, f will be zero for a completely random orientation and unity, for a complete $[001]$ orientation. Gillam and Smethurst[24], in trying to relate an orientation factor to the total magnetic moment of a sample, modified Lotgering's p factor to include the magnetic components of crystallites not completely aligned. Their "magnetic quality factor" is $q = \sum I_{hk\ell} \cos \Phi_{hk\ell}/\sum I_{hk\ell}$ where $\Phi_{hk\ell}$ is the angle between the 001 and hkℓ planes. A formula of this type can be generalized to meet the criteria outlined above.

We define a quantity $Q_{hk\ell}$ for any direction hkℓ in reciprocal space as

$$Q_{hk\ell} = \frac{\sum\limits_{h'k'\ell'} I_{h'k'\ell'} \cos \varphi_{hk\ell}}{\sum\limits_{h'k'\ell'} I_{h'k'\ell'}} \qquad (1)$$

where h'k'ℓ' refer to those reflections for which we have observed or calculated intensity data and $\varphi_{hk\ell}$ is the angle between the normals to planes hkℓ and h'k'ℓ'. A value for Q can be evaluated for any arbitrary direction hkℓ whether or not we have a reflection corresponding to that direction. Each Q value is a function of the intensities of all reflections, minimizing the effect of an error in any one intensity value and the problem of a high relative error in weak reflections. $Q_{hk\ell}$ may be thought of as the net fractional vector compon-

ent of total intensity in the direction $\left[hk\ell\right]$. Since any
scale factor between our observed and calculated intensities must be a constant which multiplies all observed
or calculated intensity values, it will factor out of the
two summations in Eq. 1 and cancel on division, leaving
$Q_{hk\ell}$ independent of scale.

The Q function can be applied to show the directions
of preferred, antipreferred and nonpreferred orientation*
in any system for which observed and calculated diffraction
patterns can be obtained. Q values from both the observed
and calculated intensities may be evaluated for a large
number of directions $\left[hk\ell\right]$. For each direction a $\Delta Q_{hk\ell}$
value may be calculated as $(Q_{observed})hk\ell - (Q_{calculated})hk\ell$.
Now the angle Φ between each $hk\ell$ direction and any zone
of interest may be calculated and a plot of $\Delta Q_{hk\ell}$ vs. Φ
may be constructed for that zone. A negative slope indicates
that the zone is a direction of preferred orientation. A
positive slope indicates a direction of antipreferred orientation. A scatter plot with low values for ΔQ indicates a
zone of nonpreferred orientation.

A computer program named PREF has been written
in FORTRAN for carrying out this procedure and is available from the first author. The program reads the observed and calculated reflections, matches them within a
preset error window and eliminates any non-matching peaks
in the observed pattern. $Q_{obs.}$ and $Q_{cal.}$ values are then
evaluated for all directions $hk\ell$ making an angle of 10°
or more with each of the seven principal zones, 100, 010,
001, 110, 101, 011 and 111. Plots of $\Delta Q_{hk\ell}$ vs. Φ are then produced for all seven principal zones on a CALCOMP plotter.

---

*The term preferred orientation is used to indicate a direc-
tion into which crystallites have oriented, antipreferred
refers to a direction out of which crystallites have oriented
and nonpreferred is used to indicate a direction which shows
no orientation effects.

RESULTS AND DISCUSSION

Figs. 4 and 5 show plots of $(Q_{obs.})_{hk\ell}$ and $(Q_{cal.})_{hk\ell}$ vs. Φ for the 001 zone of barium hexaferrite. The curves show a maximum in the region of $\Phi = 50°$. This is due to the fact that a number of very intense reflections occur in

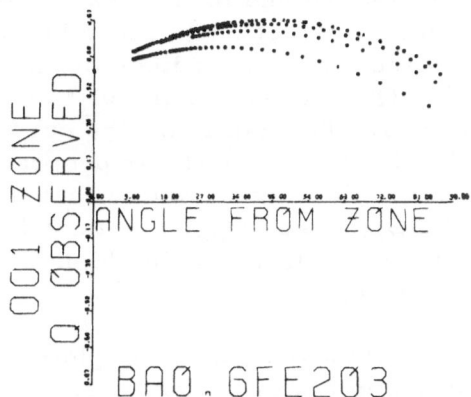

Fig. 4. $(Q_{obs.})_{hk\ell}$ vs. Φ for the 001 zone of barium hexaferrite sample.

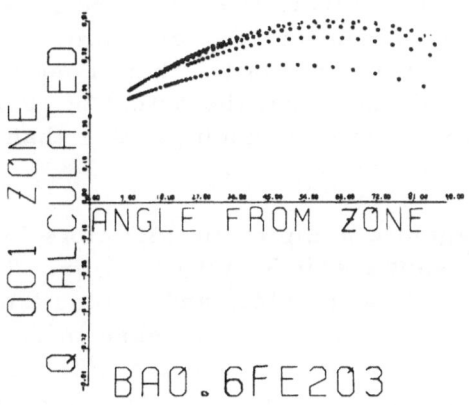

Fig. 5. $(Q_{cal.})_{hk\ell}$ vs. Φ for the 001 zone of barium hexaferrite sample.

this range. The general shape of the curves will vary from
compound to compound depending on the intensity distribution,
which is a function of the crystal structure. The magnitude
of Q for any particular class of reflections will also depend
on this intensity distribution and will thus vary with the
structure type being looked at. Each point on the lowest
curve in Fig. 5 is in the reflection class 01 ℓ, moving up
the plot the next curves are due to 11 ℓ, 02 ℓ, ending with
the top curve due to 10 ℓ directions. Since barium hexaferrite shows no reflections with indices of the type of 01 ℓ
we should expect the Q curve based on $[01]\ell$ directions to
have the lowest values. The gap in all the curves between
0° and 10° Φ is simply the result of our program not generating Q values in this interval. Notice that the only difference between the $Q_{obs.}$ and $Q_{cal.}$ plots is in the absolute
magnitude of the Q values, $Q_{obs.}$ being higher than $Q_{cal.}$
at low Φ and lower at high Φ.

Figs. 6, 7 and 8 show the ΔQ vs. Φ plots for the seven
principal zones of barium hexaferrite, alumina and aragonite,
respectively. In each of these plots the ΔQ values have been
scaled so as to make the ΔQ value at $\Phi = 0.0°$ zero. In the
cases of barium hexaferrite and alumina, the negative slope
of the $[00\bar{1}]$ zone plots indicate, as expected, that $[00\bar{1}]$ is a
zone of preferred orientation. The positive slopes of each
of the other six plots indicate that they are all zones of antipreferred orientation. In the case of alumina the magnitudes
of the ΔQ values are much smaller than those for barium
hexaferrite; this indicates that the extent of $[00\bar{1}]$ orientation in barium hexaferrite is much greater than in the case
of alumina.

The NBS aragonite sample, the ΔQ plots for which are
shown in Fig. 8, shows a different behavior. The $[00\bar{1}]$ zone
shows a symmetric scatter plot, and is thus a direction of
non-preferred orientation. The increase in the number of
points as Φ increases is an artifact due to the manner in
which the data points were generated by the PREF program.
The $[01\bar{0}]$ and $[01\bar{1}]$ zones on the other hand show clear
evidence of antipreferred orientation, while the $[\bar{1}00]$ and
$[\bar{1}0\bar{1}]$ zones show preferred orientation. The slight asym-

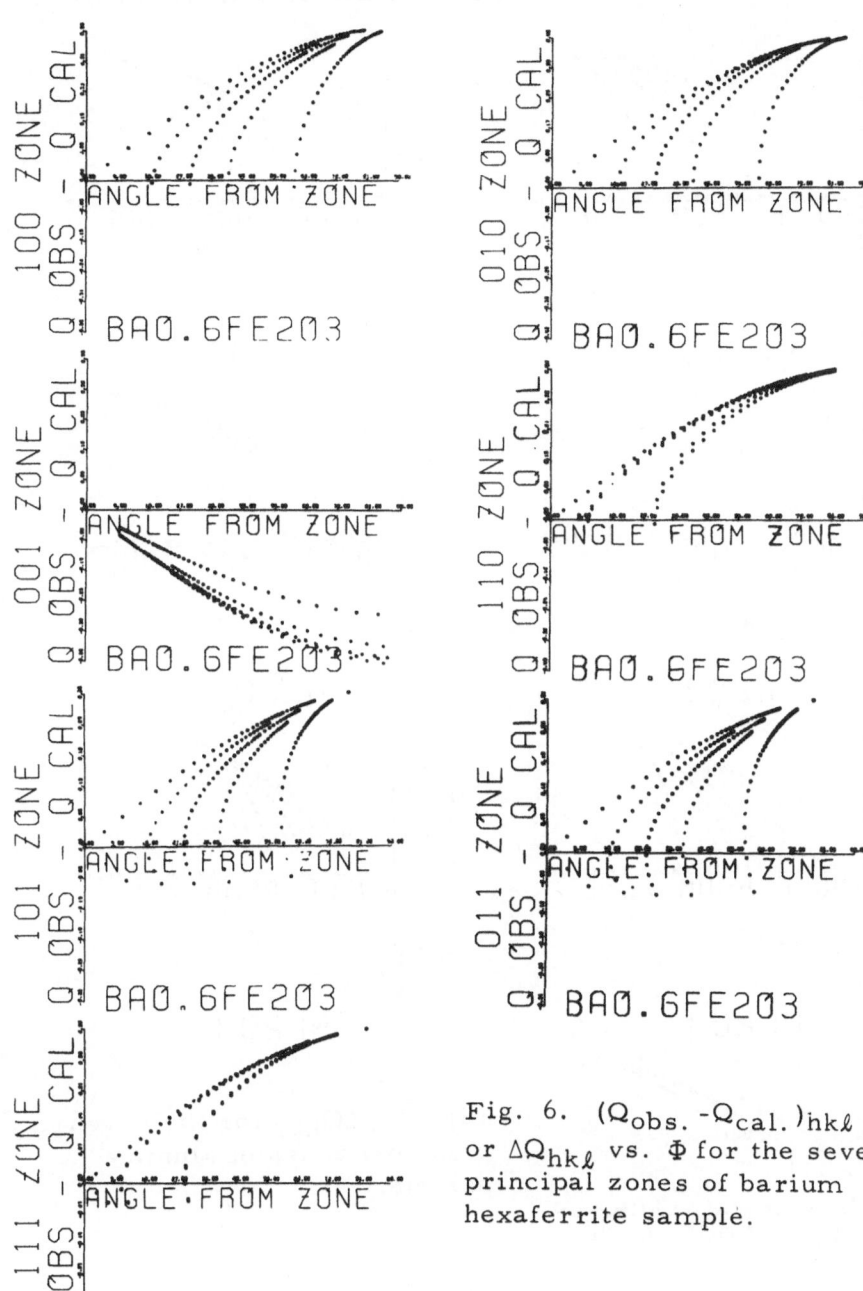

Fig. 6. $(Q_{obs.} - Q_{cal.})_{hk\ell}$ or $\Delta Q_{hk\ell}$ vs. Φ for the seven principal zones of barium hexaferrite sample.

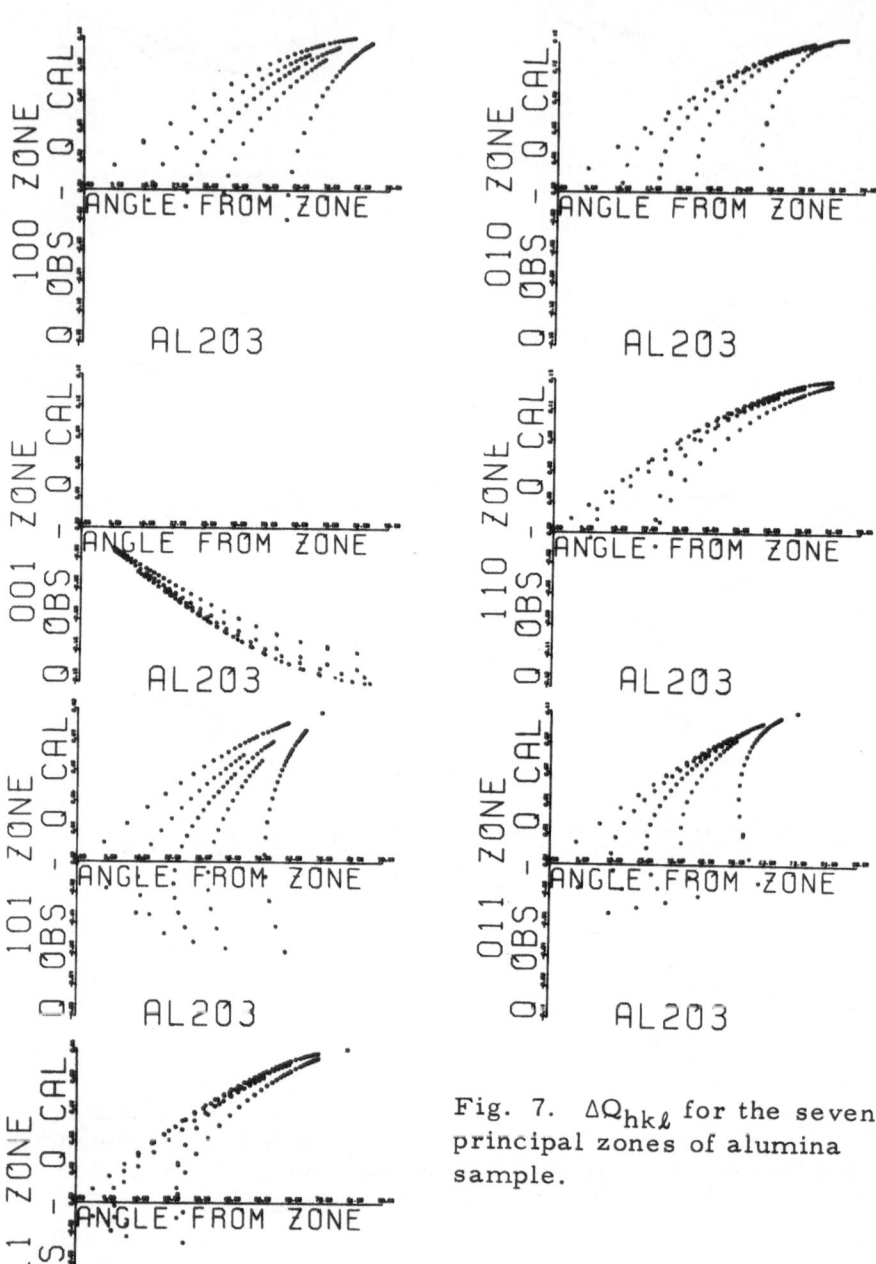

Fig. 7. $\Delta Q_{hk\ell}$ for the seven principal zones of alumina sample.

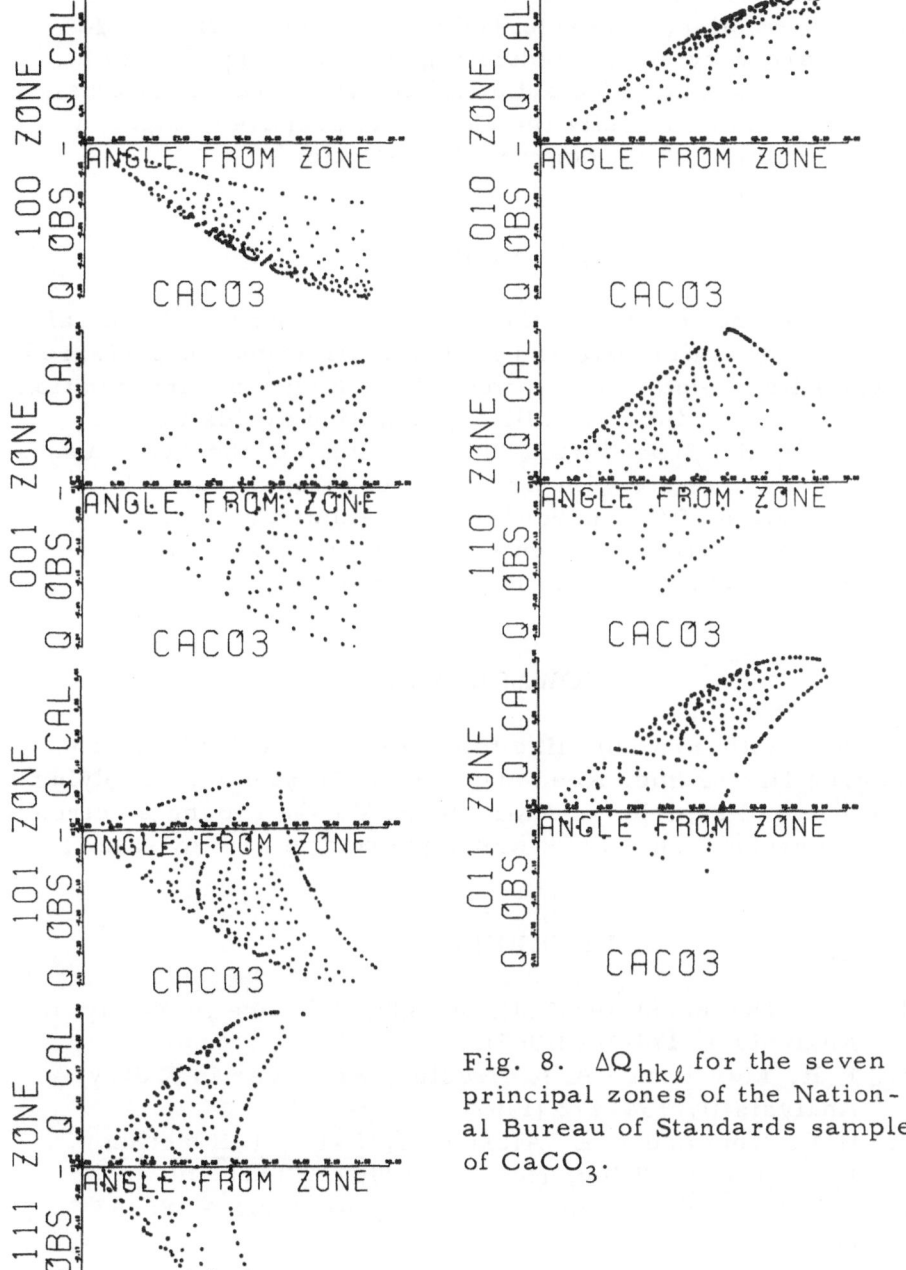

Fig. 8. $\Delta Q_{hk\ell}$ for the seven principal zones of the National Bureau of Standards sample of $CaCO_3$.

metry toward a positive slope in the $\begin{bmatrix} 110 \end{bmatrix}$ and $\begin{bmatrix} 111 \end{bmatrix}$ zones indicates a slight antipreferred orientation. Here the ΔQ values are very small, indicating that these orientation effects are slight. The authors have yet to run a material, no matter how carefully the sample is prepared, that did not show some slight degree of orientation.

CONCLUSION

Differences between the Q values for observed and calculated diffraction patterns show the directions of preferred crystallite orientation in a material. Additional information on which zones the crystallite orientation avoids is also yielded by the procedure. The magnitude of the ΔQ values appear to correlate with the extent of orientation. The quantification of this procedure to yield directional, enhancement of property coefficients is currently under study and will be the subject of a subsequent report.

ACKNOWLEDGMENTS

The authors would like to express appreciation to colleagues Drs. James Reed and Phillip Crayton for supplying the oriented samples and Mr. Ward Votava for the electron micrographs. This work was supported in part by NASA.

REFERENCES

1. J. L. Pentecost and C. H. Wright, Advance in X-Ray Analysis 4, 174-181 (1963).
2. P. R. Morris and A. J. Heckler, Advances in X-Ray Analysis 11, 454-472 (1967).
3. H. P. Klug and L. E. Alexander, X-Ray Diffraction Procedures, Wiley, New York 1954, Chapt. 10.
4. J. R. Holland, Advances in X-Ray Analysis 4, 86-93 (1963).
5. G. B. Harris, Phil. Mag., 43, 113-125 (1952).
6. R. Roe, J. Appl. Phys. 36, 2024-2031 (1965).

7. R. Roe, J. Appl. Phys. 37, 2069-2072 (1966).

8. J. A. Slane and F. Hultgren, Advances in X-Ray Analysis 14, 231-242 (1972).

9. R. H. Bragg and C. M. Packer, J. Appl. Phys. 35, 1322-1328 (1964).

10. M. N. Klenck, Advance in X-Ray Analysis 11, 447-453 (1967).

11. C. G. Dunn, J. Appl. Phys. 30, 850-857 (1959).

12. B. D. Cullity, Elements of X-Ray Diffraction, Addison-Wesley Co., Reading, Mass., 1956, Chapt. 9.

13. L. G. Schulz, J. Appl. Phys. 20, 1030 (1949).

14. S. L. Lopata and E. B. Kula, Trans. Am. Inst. Mining, Met., Petrol. Engrs. 224, 865 (1962).

15. J. R. Holland, N. Engler and W. Powers, Advances in X-Ray Analysis 4, 74 (1961).

16. H. Saalfeld and H. Hagodzinski, Z. Krist 109, 87-109 (1957).

17. H. E. Swanson, R. K. Fuyat and G. M. Ugrinic, NBS Circular No. 539 Vol. III p. 53.

18. C. M. Clark, D. K. Smith and G. G. Johnson, Dept. of Geoscience, Pennsylvania State University, September, 1973.

19. F. Bertaut, A. Deschamps and R. Pauthenet, C. R. Acad. Sci., Paris, 246, 2594-9597 (1958).

20. V. Adelsköld, Ark. Kem. Mineral. Geol., 12(A), 9 (1938).

21. R. E. Newnham and Y. M. DeHaan, Z. Krist, 117, 235-237 (1962).

22. W. L. Bragg, Proc. Roy. Soc. London, 105, 16 (1924).

23. F. K. Lotgering, J. Inorg. Nucl. Chem. 9, 113-123 (1959).

24. E. Gillam and E. Smethurst, Proc. Br. Ceram. Soc. 3, 129-137 (1964).

FRICTION AND WEAR BEHAVIOR OF GLASSES AND CERAMICS

Donald H. Buckley

Lewis Research Center
Cleveland, Ohio

There are certain fundamental characteristics of metals, carbons, polymers, glasses and ceramics which these widely different material classes have in common with respect to adhesion, friction and wear. Their surface topography is generally irregular, they contain surface and bulk defects, adhere in the clean state to themselves or other materials and are extremely sensitive to surface contaminants with respect to adhesion, friction and wear.

Certain properties of glasses and ceramics set them apart from metals and polymers with respect to adhesion, friction and wear behavior. In general, metals and polymers deform plastically, while glasses and ceramics are normally brittle and exhibit little evidence for plastic flow. Plasticity will affect the real area of contact for two solid bodies pressed together. In turn, the real area of contact affects adhesive forces, friction forces, and the propensity for adhesive wear to occur.

In a wide variety of situations, glasses and ceramics are not in contact with themselves but rather with other materials. It is important to understand which of the materials in contact is contributing to friction and wear and via what particular mechanism.

The objective here is to review those mechanical, chemical and physical characteristics of glasses and ceramics which exert an influence on their adhesion, friction and wear behavior. In addition, those factors which dictate the friction and wear characteristics for glasses and ceramics in contact with metals will be reviewed.

BACKGROUND

Solid surfaces generally contain surface irregularities called asperities. Because of these irregularities, when two solids are brought together the real area of solid contact is only a small portion of the apparent contact area. Consequently, when one surface is loaded against another, very high local stresses can develop. With metals, this means that relatively small applied forces are necessary to exceed the elastic limit in the area of real contact and result in plastic deformation. Such plastic deformation has also been observed in ceramics such as magnesium oxide[1] and aluminum oxide[2].

When solids are pressed together under a load, first elastic and then plastic deformation will occur and establish the real contact area. Where surfaces are atomically clean, adhesive bonds will develop across the interface. Adhesion force is simply that force necessary to separate the two surfaces once the load is removed. Two factors are important in determining the adhesive force, the strength of the interfacial adhesive bond or the cohesive bond in the cohesively weaker material and the real contact area.

The larger the real contact area the greater the number of potential adhesive bonds and consequently the larger the force required to separate the surfaces. In instances where a load has been applied, elastic recovery on removal of the load may result in the fracture of many of the adhesive bonds This occurs readily with high elastic moduli ceramics.

If two adhered surfaces are pulled in tension normal to the interface, at some applied force the surfaces will separate

Separation will occur at the interface when these bonds are weakest. Very frequently, particularly for atomically clean surfaces in contact, fracture will occur locally in the weaker of the two materials.

Those things which reduce the number of bonds that form across an interface affect adhesion. Thus, even fractions of a monolayer of surface contaminants have an effect[3]. Ordinary surface oxides on metal surfaces are effective in reducing adhesion to a small fraction of the value for those same metals in the clean state. Glass is one of the few materials which actually exhibits an increase in adhesive force when a water surface film is present.[4]

Friction is strongly dependent upon adhesion. Generally the stronger the adhesive force the greater the friction force, i.e., the resistance to tangential motion for two bodies in contact. The friction force, F, may be expressed as

$$F = AS \qquad\qquad (1)$$

where A is the real contact area and S the shear strength of the interfacial junctions. Eq. 1 will apply where glass is in contact with glass or ceramic is in contact with ceramic. Where glass or ceramics are in contact with soft metals or polymers, another factor must be considered, namely the large disparity in deformation characteristics. The metal or polymer will deform readily, whereas the glass or ceramic becomes partially buried in the metal or polymer. When the glass or ceramic is moved tangentially, it physically plows metal or polymer and this action offers resistance to tangential motion. A plowing term must therefore be incorporated in the friction expression for such situations and

$$F = AS + P \qquad\qquad (2)$$

where P is the plowing term.

Adhesion is also important to wear. One of the most severe types of wear is adhesive wear, which occurs when

fracture takes place in the cohesively weaker of the two
materials in contact. Adhesive wear occurs frequently for
metals in contact with ceramics or glasses.

GLASSES

Mechanical Factors

The load or force with which two glass surfaces are
pressed into contact affects the real contact area and corresponding friction force. In Fig. 1 the friction force for

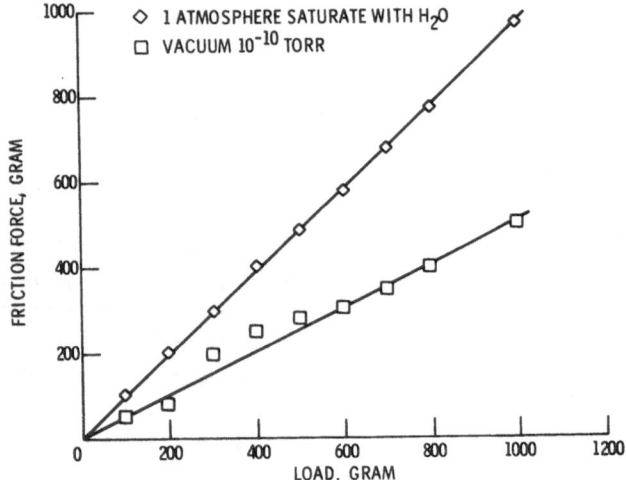

Figure 1. - Friction force as a function of load for glass sliding on
glass. Sliding velocity 30 cm/min, load 100 grams and 23° C.

glass sliding on glass is presented as a function of load in
two environments, air saturated with water vapor and a
vacuum of 10^{-10} torr. Friction is proportional to load in
both environments. This basic law of friction was first
recognized by de Vinci (1452-1519).

In Fig. 2, friction force is plotted as a function of load
for aluminum sliding on glass in vacuum. The curve can be
superimposed over the one obtained in vacuum in Fig. 1.

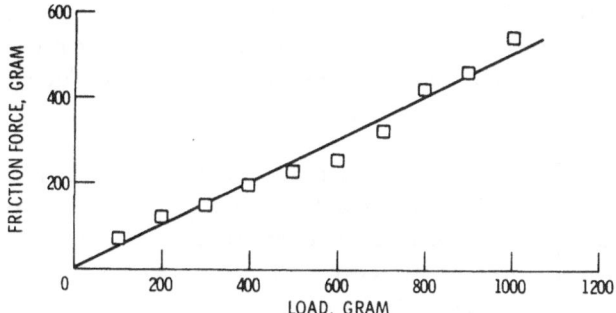

Figure 2. - Friction force as a function of load for aluminum sliding
on glass. Sliding velocity 30 cm/min, load 100 grams and 23° C.

The friction force at any particular load is essentially the
same for glass sliding on glass and aluminum sliding on
glass. Similar results have been obtained with other metals
such as iron sliding on glass. This is explained by examination of the surfaces after sliding. When sliding on metal,
the glass undergoes wear just as it does sliding on glass.
Microscopic examination of the metal surface indicates
transfer of glass into the metal surface. Thus, in a vacuum
the metal surface becomes charged with glass and ultimately
glass is sliding on glass. This is because initially the metal
adheres to the glass. With tangential motion, fracture takes
place in the weakest zone. Both the adhesive bond and the
shear strength of aluminum are greater than the force necessary to fracture glass. Thus glass transfers to the metal.
What would appear to be an abrasive wear process from an
examination of only the glass surface is in fact an adhesive
wear process. Besides the load effect, other mechanical
parameters, such as sliding velocity, affect friction behavior.

Environment

Most materials are extremely sensitive in their adhesion, friction and wear behavior to the environment[4-8].
Glasses are no exception. At 1000 grams load the friction
force of glass on glass in vacuum is one half the value
obtained in saturated air (Fig. 1).

The results in Fig. 1 are unusual, however. For most materials adhesion, friction and wear are greater in a vacuum environment; this is the case with metals[9], carbons[10] and ceramics[11].

In Fig. 3 the coefficient of friction (friction force over the normal load) is plotted as a function of ambient pressure for glass on glass. From 10^{-10} torr to 1 torr, the friction coefficient remained unaffected. On increase from 1 torr to atmospheric pressure, the coefficient of friction increased from 0.5 to 1.0. The air contained normal moisture content; none was deliberately added.

The anomalous behavior of glass with respect to friction can be explained on the basis of increased adhesion of glass in the presence of water vapor. The adhesion force for glass in the presence of water is more than three times that for glass in the presence of octane[4].

Glass-Metal Interactions

Metals in sliding contact with glass in moist air are observed to transfer to glass . Friction coefficients are then typically from 0.5 to 0.7, depending on the shear properties of the metal involved. In vacuum, glass transfers to the metal, and friction coefficients are approximately 0.5. Thus, while the friction coefficients are not markedly different in the two cases the mechanism is. The difference lies in the fracture properties of glass, which are strongly affected by water[12]. Water impedes fracture and is a man-

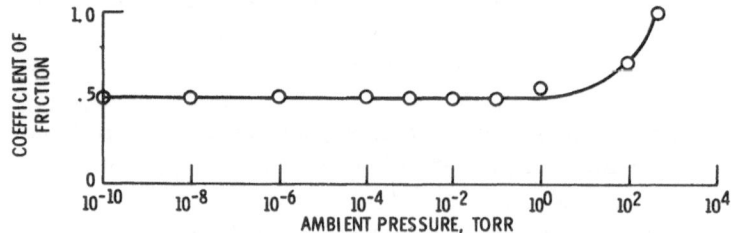

Figure 3. - Friction coefficient for glass sliding on glass as a function of ambient pressure. Sliding velocity 30 cm/min, load 100 grams and 23^0 C.

ifestation of the Joffe effect in an amorphous solid. From
the transfer characteristics observed with metals sliding
on glass, it must be concluded that the strength of glass
under these circumstances is less in the absence of water
vapor.

CERAMICS AND OTHER IONIC SOLIDS

Mechanical Behavior

Ceramics and other ionic solids like glass are load-
sensitive with respect to their friction behavior. The
friction coefficient for two orientations of single-crystal
aluminum oxide in vacuum are presented in Fig. 4 for
various loads. Aluminum oxide differs from glass in that
the coefficient is not directly proportional to load. For
both the basal (0001) and prismatic (10$\overline{1}$0) orientations, the
friction coefficient decreases with increasing load, reaches
a minimum and then begins to increase. With glass (Fig.
1), the coefficient is independent of load.

The marked difference in coefficient with orientation
of aluminum oxide indicates anisotropic behavior.

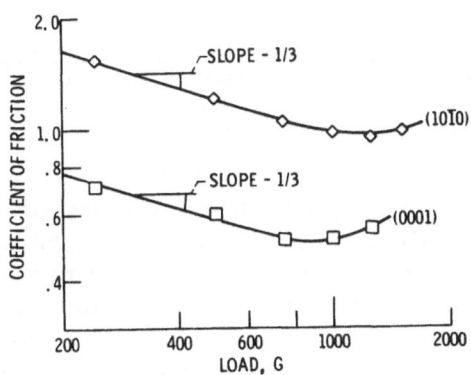

Figure 4. - Friction for two orientations of sapphire
with load

Properties of Ionic Solids Related to Friction

Certain basic properties of solids can be expected to relate to adhesion, friction and wear. For example, the greater the cohesive energy of a solid, the greater its elastic modulus and hardness, and the smaller the real contact area under a given load. This in turn should lead to lower adhesion and friction forces. The data of Table I for various ionic solids with a rock salt structure substantiate this.

TABLE I. Properties of Ionic Bonded Crystals with Rock Salt Structure

Crystal	Cohesive Energy (kcal/ mole)	Elastic modulus (10^{11} dynes/ cm^2 (100)	Relative hardness (kg/mm^2)	Coeff. of frict.[a] (100) [110] air
MgO	940.1	24.5	400 [100] 700 [110]	0.07
LiF	240.1	7.35	100	0.24
KCl	164.4	4.80	18	0.71
NaCl	153.1	4.37	17	0.70
KBr	140.8	3.70	7	0.85

[a]5 gm-load, 0.02-0.04 cm/sec, diamond 12.7 micron slider, 760 Torr.

In Table I magnesium oxide has the greatest cohesive energy, elastic modulus, hardness and, correspondingly, the lowest friction coefficient. Potassium bromide has the lowest cohesive energy, elastic modulus, and hardness and it has the highest coefficient. Note in Table I that a diamond slider was used. Thus, the deformation is taking place in the ionic solid.

The comparisons made in Table I are valid only for solids with the same crystal structure type. Crystal structure can exert a considerable influence on the friction behavior of solids[13].

Anistropy

As in metals, the adhesion, friction and wear characteristics of ionic solids are anisotropic. Table II indicates the anisotropic nature of aluminum oxide, not only with respect to planar orientation but to crystallographic direction as well. Friction is least on the preferred slip direction.

TABLE II. Coefficient of Friction for Sapphire Sliding on Sapphire in Vacuum (10^{-10})*

Plane	Direction	Coefficient of friction
Prismatic	$[11\bar{2}0]$	0.93
$(10\bar{1}0)$	$[000\bar{1}]$	1.00
Basal	$[11\bar{2}0]$	0.50
(0001)	$[10\bar{1}0]$	0.96

*Load 1000 g, sliding velocity 0.013 cm/s.

Analogies have been drawn between the crystallographic deformation behavior of hexagonal metals and the rhombohedral crystal structure of aluminum oxide[14]. The preferred slip planes and crystallographic directions for both hexagonal metals and aluminum oxide also result in the lowest friction.

In addition to friction, the wear for ionic solids is highly anisotropic, as demonstrated for single-crystal rutile in Fig. 5[15]. Wear varies with orientation by a factor of at least six. Thus, an understanding of the anisotropic

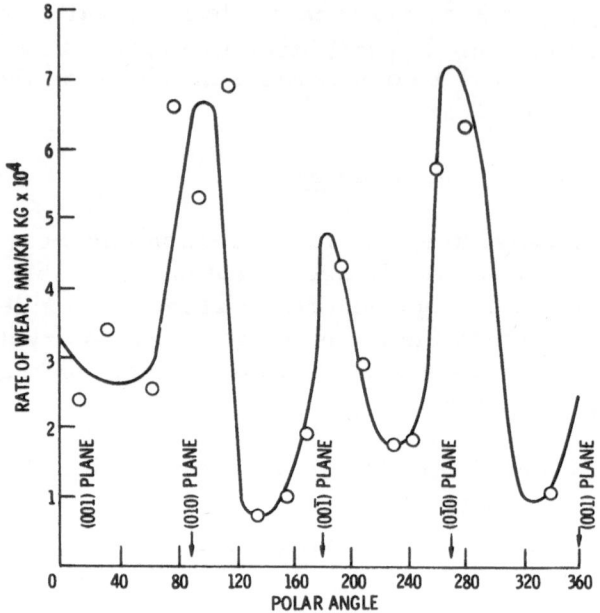

Figure 5. - Rate of wear of a rutile single-crystal sphere on a great
circle in the plane of the a- and c-axes. The c-axis is normal to
plane of sliding at 0 and 180°. Slide direction in plane of the great
circle. [13]

behavior of ionic solids can result in appreciable reduction
in friction and wear when proper orientation is selected.

Temperature Effects

With metals, increasing the temperature in air, generally results in increased surface oxidation. This increased oxide surface coverage generally reduces metal-
to-metal contact resulting in a reduction in friction coefficient.

As shown in Fig. 6, the friction coefficient for polycrystalline aluminum oxide increases with temperature.
At 400°C it reaches a maximum value of 0. 8 and then begins to decrease. The coefficient of 0. 2 at room temperature is considerably less than that obtained with metals;
the value at 400°C is comparable to that obtained with
metals.

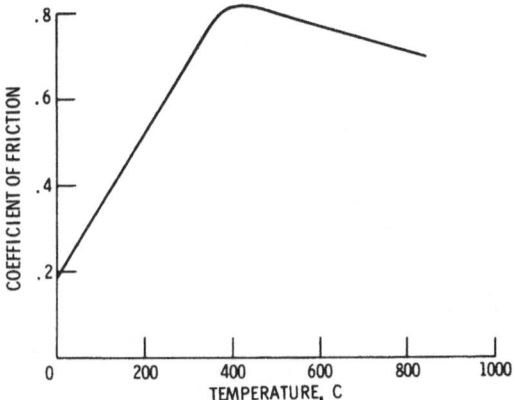

Figure 6. - Effect of temperature on friction coefficient of Al₂O₃.

The marked increase in friction coefficient seen in
Fig. 6 is due to the presence of adsorbed water. The
experiments were conducted in dry air. If a small amount
of water vapor were admitted to the system at 400°C, the
friction coefficient immediately decreased. Water then,
is a lubricant for aluminum oxide. Further evidence for
this was obtained in vacuum studies. When aluminum oxide
was simply heated in vacuum, evolution of water was detected by mass spectrometry. On subsequent cooling to
room temperature in vacuum, a friction coefficient comparable to that measured in air at 400°C was obtained.

Environment and Surface Films

The effect that environmental constituents such as
water vapor have on friction coefficient was observed
(Fig. 7) simply on increasing the load. When the load
was increased above 1000 grams, the friction coefficient
rose dramatically, owing to penetration of the adsorbed
water layer by the aluminum oxide asperities.

The presence of water and organics on the surface of
ceramics influences the mechanical behavior of these materials[15-21]. If these films influence such properties as the
deformability of the surface, then they will influence friction. The presence of surface-active agents on ceramics

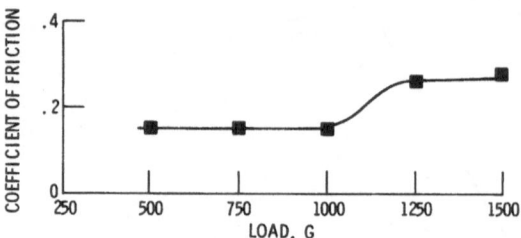

Figure 7. - Coefficient of friction as a function of load for
sapphire sliding on sapphire in air (760 torr). Sliding
velocity, 0.013 cm/s; ambient temperature, 25° C.

can arrest brittle fracture during sliding. This increase
in the ability of surfaces to deform plastically in the presence of surface-active species in the Rehbinder effect.

Sliding friction experiments have been conducted with
lithium fluoride to determine the influence of surface films
on friction and deformation. A sapphire ball was slid
across a freshly cleaved lithium fluoride (100) surface.
The lithium fluoride specimen was then cleaved normal
to the sliding track and subsequently etched. The subsurface deformation and the development of cleavage cracks
are shown in Fig. 8(a). Examination of that figure reveals
that slip has taken place along the $\{011\}$ and $\{101\}$ planes,
and cleavage cracks, originating at the surface, developed
along the $\{011\}$ slip bands. Cracks can form in lithium
fluoride at the intersection of 011 slip planes according
to the equation: $1/2\ a\ [011] + 1/2\ a\ [10\bar{1}] = 1/2\ a\ [110]$.

Equivalent experiments were conducted with lithium
fluoride. Crystals were cleaved in water and friction
experiments were conducted with water present. The
crystals were then cleaved normal to the wear track and
etched. Track subsurface deformation is shown in Fig.
8(b). While slip bands are evident from the dislocation
etch pits along the (110) plane, a subsurface crack has
formed in a (001) plane in the crystal. In dry air (Fig.
8(a)) the crack formed at the surface along (110) planes
rather than at the subsurface. With plastic deformation
of lithium fluorde, cracks can develop along a (100) plane

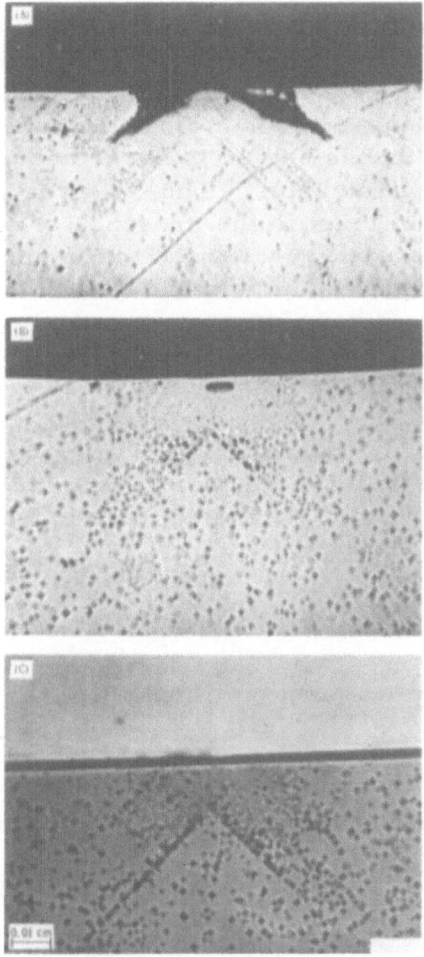

Figure 8. - Cross section of wear tracks on LiF in sliding friction experiments.
Load, 200 g; rider, 1.6-mm-diameter sapphire ball; temperature, 20° C; sliding
velocity, 0.005 mm/s. Ball made a single pass across surface covered with (A) dry
air, (B) water and (C) water with myristic acid.

with the intersection of $\{110\}$ slip bands in accordance with
the equation: $1/2\ a\ [110] + 1/2\ a\ [1\bar{1}0] = a\ [100]$. The crack
in Fig. 8(b) was the result of both compressive forces acting on the crystal surface in the form of normal load, and
tangential forces associated with sliding.

Fig. 8(c) is a sliding friction track in cross section after an experiment conducted in 5.0×10^{-6} normal myristic acid. There was no evidence of either surface or subsurface crack formation. In Fig. 8(c) the subsurface depth to which the (011) slip bands extend is appreciably greater than in the other two environments. Thus, a greater degree of plasticity appears to exist in the presence of myristic acid. The energy associated with sliding friction appears to have been absorbed completely in plastic behavior.

The influence of environment is further shown in some sliding friction experiments conducted on the (111) cleavage face of calcium fluoride. Fig. 9 shows that with decreasing concentrations of dimethylsulfoxide or increasing concentrations of water, the width of the wear track increased. This increase in width may be attributed to an increase in surface plasticity.

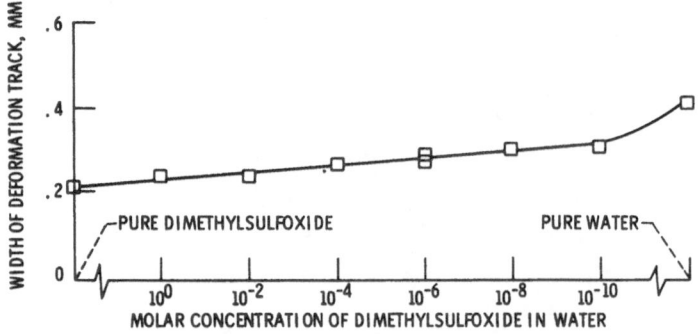

Figure 9. - Dislocation track width for sapphire ball sliding on (111)-cleavage surface of CaF$_2$ in various concentrations of dimethylsulfoxide in water. Sliding velocity, 0.005 cm/s; load 300 g (2.9 N); ambient temperature, 20°C.

The foregoing discussion indicates that the presence of surface films on ionic solids not only influences surface behavior, but subsurface behavior as well. The ability of surface films to influence deformation behavior will not only influence friction (because it determines true contact area) but also wear. With repeated traversals over the surface, the presence of surface or subsurface cracks can give rise to the formation of wear particles. This has been demonstrated with the ionic solids lithium and calcium fluorides.

Comparison of the friction behavior of various solids in three different environments is made in Table III. For all materials except glass, an increase in friction coefficient is observed when the environment is changed from air to vacuum. The increase for the ionic solids is by a factor of about two to four. With diamond, the material exhibiting the lowest coefficient in air, there is a nine-fold increase in friction and with copper it is in excess of one hundred. Other metals behave like copper. Thus, while glasses and ionic solids are sensitive to environment, their sensitivity is not as great as observed for diamond and metals.

TABLE III. Coefficient of Friction for Various Solids

	COEFFICIENT OF FRICTION		
Material Combinations	AIR (Moisture)	VACUUM $(10^{-9} - 10^{-10}$ torr)	LUBRICATED (Mineral Oil)
Glass	1.0	0.5	0.28
Sapphire/Sapphire	0.2	0.8	0.20
MgO/MgO	0.2	0.8	0.21
Quartz/Quartz	0.35	0.7	0.20
NaCl/NaCl	0.70	1.3	0.22
LiF/LiF	-	1.2	0.22
Diamond/Diamond	0.1	0.9	0.05
Copper/Copper	1.0	>100	0.08

When lubricated with a mineral oil, the friction coefficients are reduced from those values obtained in air with glasses and ionic solids, but the differences were not as great as that with copper, which is typical of metals in general.

It is interesting that sapphire and magnesium oxide were not influenced by the presence of lubricating oil; the moisture in the air was evidently just as good a lubricant as the oil.

Metal-Ceramic Systems

Another effect which can influence friction is plowing. The greater the deformability of one of two surfaces in contact, the deeper the second can embed in it and thereby impede tangential motion.

The marked difference in elastic and plastic deformation of ceramics and metals can result in plowing being the principal contributor to measured friction forces. This is demonstrated by Fig. 10. A spherical rider of sapphire was slid on a single-crystal copper flat. Then a single crystal copper rider was slid over a sapphire flat. The coefficient of friction for the sapphire sliding on copper was 1.5. With copper sliding on sapphire, it was 0.2. In both instances, adhesion of copper to sapphire occurred. The differences in friction coefficient are due to the effects of plowing.

Surface chemistry also plays a role. Various metals were slid over a sapphire flat with the sapphire basal plane

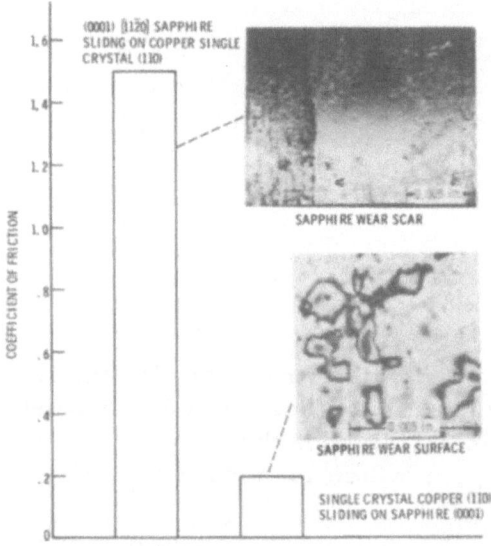

Figure 10. - Coefficient of friction for copper in sliding contact with sapphire in vacuum (10^{-10} torr). Load, 100 g; sliding velocity, 0.013 cm/s.

parallel to the sliding interface. With metals which form
stable oxides such as copper, nickel, rhenium, cobalt and
beryllium, adhesion of the metal occurred to the surface
oxygen ions of the sapphire.

With these metals fracture took place along the sapphire
basal cleavage plane. This resulted in plucking out large
particles. This indicates that the strength along the basal
plane was less than the bond strength and the metal coherence.
The friction coefficient for all of the metals with sapphire was
essentially the same, 0.2, as dictated by the cleavage strength
of the sapphire (Fig. 11).

Metals examined in sliding contact with polycrystalline
aluminum oxide showed friction coefficients greater than
obtained when metals slid on sapphire, exceptions being
rhenium and lanthanum. The reason was that shear took
place in the surface layers of the metal rather than fracture
occurring in the aluminum oxide, as was observed with the
single crystal sapphire experiments. Metal transferred to
the polycrystalline aluminum oxide disk surface. The shear
properties of the metal were therefore determining.

Differences in the friction coefficients for hexagonal and
cubic metals in Fig. 11 occurred because of the differences
in their slip and the shear behavior. In general, hexagonal
metals have fewer operable slip systems, shear more readily and do not work harden rapidly; as a consequence, they
exhibit lower friction coefficients than cubic metals. Titanium shows complex slip, making it behave more like a cubic
than a hexagonal metal, which accounts for its striking friction behavior.

If a metal does not form a stable oxide, the observed
friction coefficient is lower. With both gold and silver (Fig.
12), the friction coefficient on sapphire in vacuum was 0.1
or half that obtained with the oxide-forming metals. The
sapphire surface after sliding revealed no evidence of fracture. The lack of strong interfacial bonding between these
metals and sapphire resulted in shear of the bonds. From

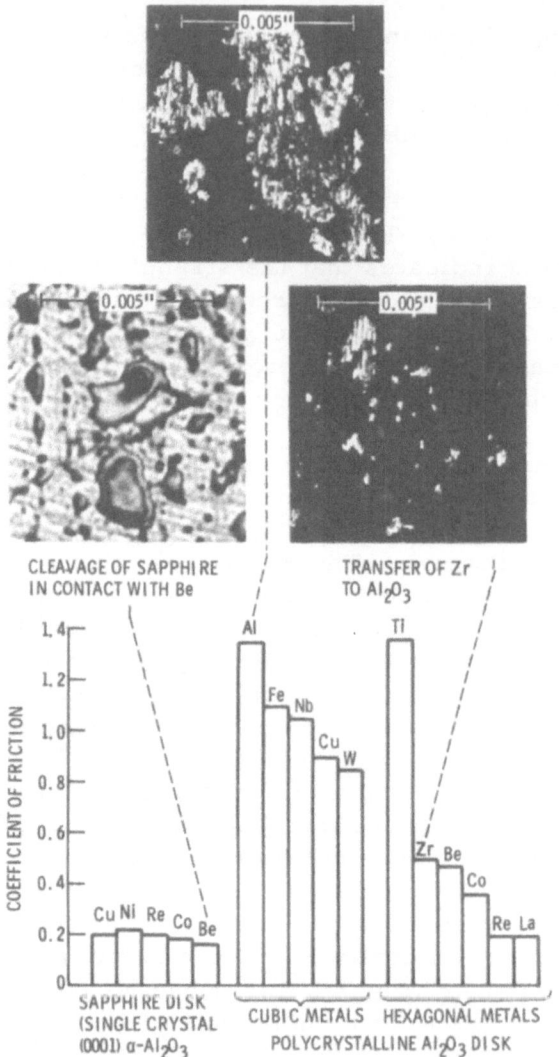

Figure 11. - Coefficient of friction for various metals sliding or Al_2O_3 in vacuum (10^{-10} torr). Load, 1000 g; sliding velocity, 0.013 cm/s; duration of experiment, 1 h.

a practical point of view, this is the most desirable area for shear, since both friction and wear are least.

The crystallographic nature of the metal exerts a marked influence on friction beyond simply the crystal

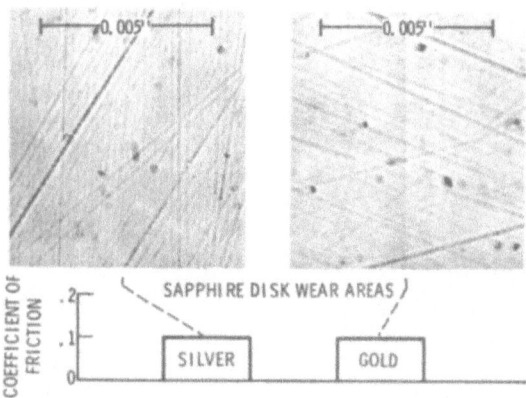

Figure 12. - Coefficient of friction for gold and silver riders sliding on sapphire in vacuum (10^{-10} torr). Sliding velocity, 0.013 cm/s; ambient temperature, 25^0 C; duration, 1 h.

structure discussed in reference to Fig. 11. Even with a single metal, changes in surface orientation with sliding and the accompanying changes in associated slip systems affect friction.

Fig. 13 indicates the marked dependency of the friction coefficient with sapphire on the orientation of tungsten grains in a 5-cm disk. In vacuum, as much as seven-fold differences in friction exist with changes in the orientation of the metal. Just as with glasses in contact with glasses or ionic solids with ionic solids, metal-ionic solid interactions are sensitive to environment. This is demonstrated by the differences in friction behavior seen in air and vacuum. The influence of orientation in the metal is less pronounced in air, where oxides and adsorbates are present, than it is in vacuum with a clean metal surface.

A change in mechanical parameters such as load or speed generally does not have as marked effect in rolling friction as in sliding. Fig. 14 shows the depth of plastic deformation in magnesium oxide as a function of the number of repeated rolling passes over the surface with a steel ball at three different velocities under constant load. The depth of plastic deformation was inferred from dislocation etch

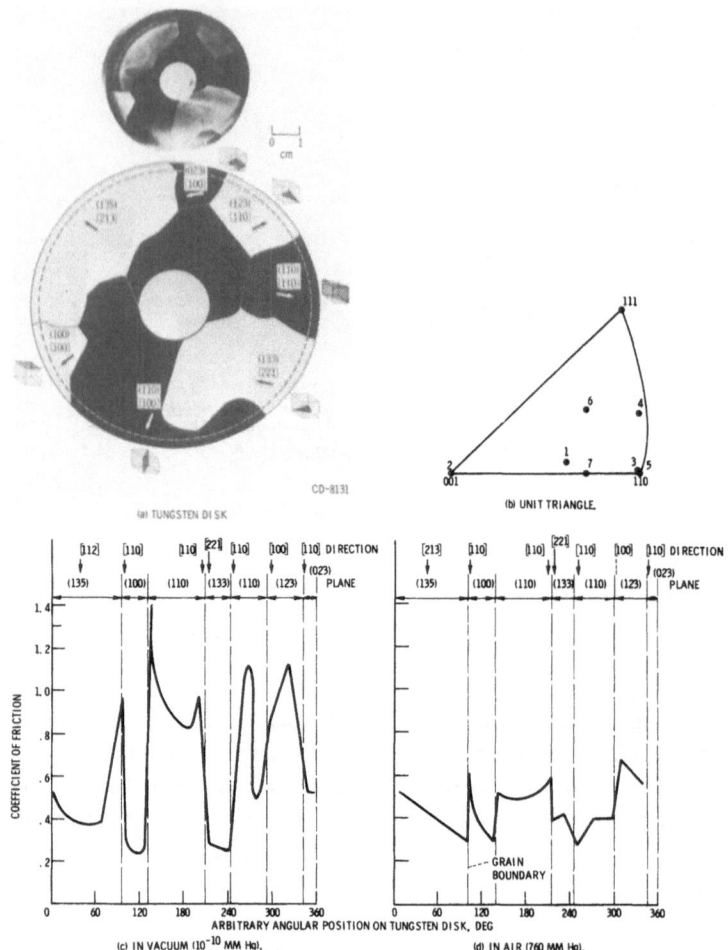

Figure 13. - Coefficient of friction of sapphire (10$\bar{1}$0) plane sliding [0001] direction on polycrystalline tungsten. Load, 500 g; sliding velocity, 0.013 cm/sec.

pits after rolling. Slip depth is the depth below the surface to which etch pits along slip bands could still be seen.

With increasing number rolling cycles, there is an increase in the depth to which plastic deformation occurs in the magnesium oxide. The depth to which deformation

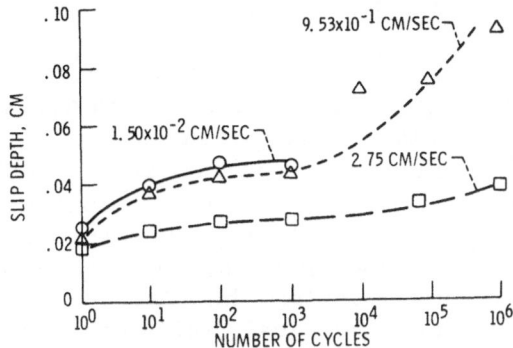

Figure 14. - Variation of slip depth with rolling-contact
cycles at three rolling velocities for a steel ball on
magnesium oxide.

occurs is strain-rate sensitive. Deformation takes place
to greater depth at slower rolling velocities. Similar behavior has been observed with calcium fluoride under
sliding conditions[22].

Rolling repeatedly over a magnesium oxide surface
with a steel ball produces strain hardening in the magnesium
oxide (Fig. 15) just as it does in metals.

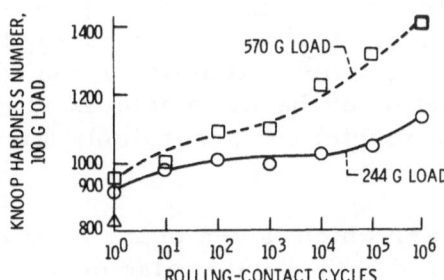

Figure 15. - Variation of track hardness
with rolling-contact cycles under dry
conditions in 100 rolling direction on
a magnesium oxide single crystal surface.

CONCLUSION

In many aspects glasses and ionic solids behave in a manner sililar to metals with respect to adhesion, friction and wear. Adhesion plays an important role in their behavior, as it does with metals. With clean metals brought into contact with themselves, the forces to fracture the adhered junctions equal the tensile strength of the bulk metal [23]. Likewise with ionic solids the adhesive forces developed across an interface are equal to that of the bulk solid. This has been effectively demonstrated with such ionic compounds as sodium chloride[24].

The change in friction with a change in environment is significantly greater for metals than it is for glasses and ionic solids. With metals, removal of surface adsorbates and oxides results in friction increase to complete seizure (friction coefficient $>$100). Glasses and ionic solids do not rise above 1.3.

Glass is anomalous as the only material which upon removal of surface adsorbates (principally water) exhibits a decrease in friction coefficient, possibly a manifestation of the Joffe effect. Water is known to affect the mechanical behavior of glass[25-26]. From sliding friction studies in air, a calculation of shear strength for glass results in values of 140 Kg/mm^2.[24] This value represents the surficial shear strength. The bulk shear strength for glass is approximately 25 Kg/mm^2. Removal of the adsorbed water results in a reduction of the value calculated from friction behavior (shear strength) and accordingly friction characteristics.

Adhesive wear, which is one of the most severe types of wear encountered with metals, also occurs with ceramics and is most pronounced where metals are in contact with glasses or ionic solids. In air, metal is generally observed to transfer to the glass[27]. In vacuum, where the surficial strength of the glass appears to be reduced, glass transfers to metal with the end result that glass is sliding essentially on glass.

The adhesive wear behavior of ionic solids in contact with metals is strongly dependent upon the particular ionic solid involved and on its form. With sapphire, for example, adhesion to metals resulted in fracture along basal planes in the sapphire and wear to the sapphire. With polycrystalline aluminum oxide, shear took place in the metal and the metal underwent wear.

REFERENCES

1. K. F. Dufrane and W. A. Glaeser, NASA CR-72295 (1967).
2. R. P. Steijn, J. Appl. Phys., 32 [10] 1951 (1961).
3. D. H. Buckley, NASA TN D-5689 (1970).
4. F. P. Bowden and D. Tabor, The Friction and Lubrication of Solids, Oxford Press (1950).
5. D. H. Buckley, Friction, Wear and Lubrication in Vacuum, NASA SP-277 (1971).
6. F. P. Bowden and A. E. Hanwell, The Friction of Clean Crystal Surfaces, Proc. Roy. Soc., A295 [1442] 233 (1966).
7. I. V. Kragelskii, Friction and Wear, Butterworths (1965).
8. E. Rabinowicz, Friction and Wear of Materials, John Wiley & Sons, Inc., (1965).
9. F. P. Bowden and T. P. Hughes, Proc. Roy. Soc., A172 263 (1939).
10. D. H. Buckley and R. L. Johnson, ASLE Trans., 7 91 (1964).
11. D. H. Buckley, ASLE Trans., 10 134 (1967).
12. S. M. Wiederhorn, Environment-Sensitive Mechanical Behavior of Materials, Gordon and Breach, 293 (1966).
13. D. H. Buckley, ASTM STP 431, 248 (1967).
14. M. L. Kronberg, Acta Met., 5 [9] 507 (1957).
15. E. J. Duwell, ASLE Trans., 12 34 (1969).
16. J. H. Westbrook, Environment-Sensitive Mechanical Behavior of Materials, Gordon and Breach, 247 (1966).
17. P. A. Rehbinder, Proc. Sixth Physics Conf., Moscow (1928).
18. A. R. C. Westwood, Strengthening Mechanisms - Metals and Ceramics, Syracuse Univ. Press, 407 (1966).

19. W.G. Johnston, J. Appl. Phys., 33 2716 (1962).

20. P.L. Pratt, R. Chang, and C.H. Newey, Appl. Phys. Letters, 3 83 (1963).

21. A.R.C. Westwood and D.L. Goldheim, J. Appl. Phys., 39 3401 (1968).

22. D.H. Buckley, NASA TN D-5580 (1969).

23. D.V. Keller, Jr., Jour. Vac. Sci. Tech., 9 133 (1972).

24. R.F. King and D. Tabor, Proc. Roy. Soc., A223 225-238 (1954).

25. S.M. Wiederhorn, J. Am. Ceram. Soc., 52 99 (1969).

26. P.A. Rehbinder and V.I. Likhman, Second Intl. Cong. on Surface Activity, III, 563 (1957).

27. W.A. Wooster and G.L. MacDonald, Nature, 160 260 (1947).

DISCUSSION

D. Dove (Univ. of Florida): Are strain rate effects in friction and wear consistent with surface deformation theories?

Author: They are, both in metals and ceramics. We have found, for CaF_2 and MgO, that the total surface deformation is a function of strain rate.

R.W. Rice (Naval Research Laboratory): (1) Have you studied any carbides or nitrides, and (2) Do you foresee practical wear applications of ceramic materials?

Author: (1) We have not examined carbides or nitrides. (2) Research is presently being conducted with ceramics such as aluminum oxide as components for high temperature ball bearings. Air bearings have already been constructed where one of the surfaces is sapphire.

P.J. Gielisse (Univ. of Rhode Island): The geometric effects you mentioned I would call secondary geometric effects. I have found in ceramic-ceramic sliding contact that the topography of the surfaces, not just the roughness, influences the type and degree of wear to a great extent.

Author: Initial surface topography can play an important role in wear, particularly abrasive wear, as the roughness of a file determines the nature and size of the wear particle of a work piece. Usually with normal engineering surfaces, the surface topographical effects are manifested only in the initial stages of wear while surfaces are being "run-in". After they have been in rubbing contact awhile, they develop a natural conformity, with initial topography being less critical. Generally one finds the initial wear rate high, levelling out after conformity has been achieved.

A. Choudry (Univ. of Rhode Island): Would it not be possible to conduct experiments in helium and thereby see possible elastohydrodynamic effects.

Author: Elastohydrodynamic effects are generally observed with the increase in liquid lubricant viscosity. Thus, at aircraft bearing pressures, a typical automotive motor oil will take on the viscosity of tar in the contact zone and support the load transmitting elastic forces to the solid on which the film rests. I seriously doubt seeing these effects with a gas such as helium.

D.E. Smith (Owens-Illinois): How were the test parameters of speed and loading selected? Are slower speeds a more extreme test?

Author: In basic lubrication studies speeds and loads are selected to assist with some particular research objective. Slower speeds are a more extreme test with respect to friction, but frequently the opposite is true with respect to wear.

L. Hench (Univ. of Florida): Large structural changes in glass surfaces accompany composition, time and pH effects in aqueous environments. Do these effect friction and wear of glasses?

Author: We have not studied these variables but certainly they should be examined.

G. I. Madden (DuPont): (1) Can your friction and wear data be used to predict the behavior of composites such as metal matrix/ceramic systems, and (2) In considering the wear of coatings such as plated metals on metallic substrates, what effect will the substrate properties, such as elastic modulus, have on the wear behavior of the coatings?

Author: (1) Partially yes, but there are other factors which must be considered with composites such as ratio of the components, size of the particles of fibers, etc. (2) In general, thin soft metal films when applied to hard metal substrates act as lubricants. The elastic and plastic behavior of the substrate determine contact area, and the shear properties of the soft metal film determines friction. This is true for soft metal films on ceramics as well.

THE WEAR BEHAVIOR OF CAST SURFACE COMPOSITES

P. S. Kotval

Union Carbide Corporation
Tarrytown Technical Center
Tarrytown, New York

Novel wear-resistant cast surface composite materials have been developed recently at our laboratories by combining conventional metal casting technology with the technology of refractory metal carbides in the form of fibers and textiles. Metals and alloys which hitherto were considered wear-prone, e. g., aluminum alloys, can be used as matrix materials for cast surface composites which exhibit wear properties comparable to those associated with traditional sintered carbide materials. These new wear-resistant materials can be produced simply by casting metal in a mold or die lined with a refractory metal carbide textile. During casting, contact of the molten matrix material with good wetting, permits interfacial bond formation and penetration of the textile by the molten alloy. After solidification, the surface has a unique duplex microstructure consisting of 75 - 80 volume percent of hard carbide phase MC (M = Ta, Ti or W) dispersed within the matrix alloy to a depth from the surface determined by the type of textile used.

To illustrate the key features of cast surface composite materials, microstructural and wear test data will be presented for some having an aluminum alloy matrix. However, the principle is applicable to various other metallic and nonmetallic matrices.

127

MC TEXTILES: PREPARATION AND STRUCTURE

In the early 1960's B. H. Hamling[1] developed a "pre-cursor" process, by which refractory metal carbide fibers and textiles could be produced. In its essentials, the process involves: (a) the impregnation of a preformed, organic polymeric material with an aqueous solution of a metal compound; (b) pyrolysis, under controlled conditions, of the loaded polymer to decompose the polymer and leave behind a skeleton of the metallic compound; (c) further heat treatment of the material under controlled atmos-pheric conditions to form metal carbide. In this process the metal carbide produced essentially adopts the form and shape of the original polymeric material but with shrinkage from the original dimensions. Most commonly, a woven textile made from cellulosic fiber, such as rayon, is employed. The type of weave can be preselected de-pending upon the depth of the wear-resistant layer required. In Figure 1, strips of TaC textile with a 5-shaft, 4/1 warp satin weave structure are shown; the 4/1 designation sig-nifies that each fill thread goes under four warp threads and over one and then is repeated. (Examples of MC tex-tiles with other weave structures are shown later.) Irres-pective of the type of weave, the MC textile is quite flexible, is strong enough to be handled, and can be cut

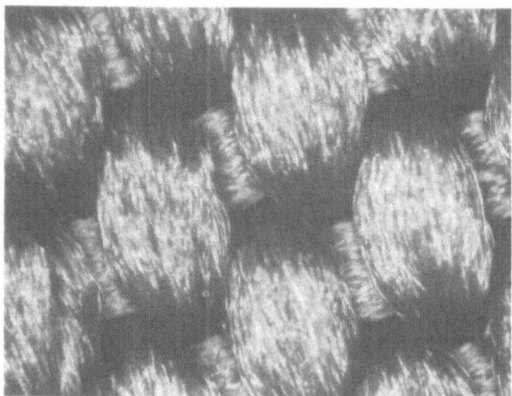

Fig. 1.   Structure of a satin weave MC textile; 21X.

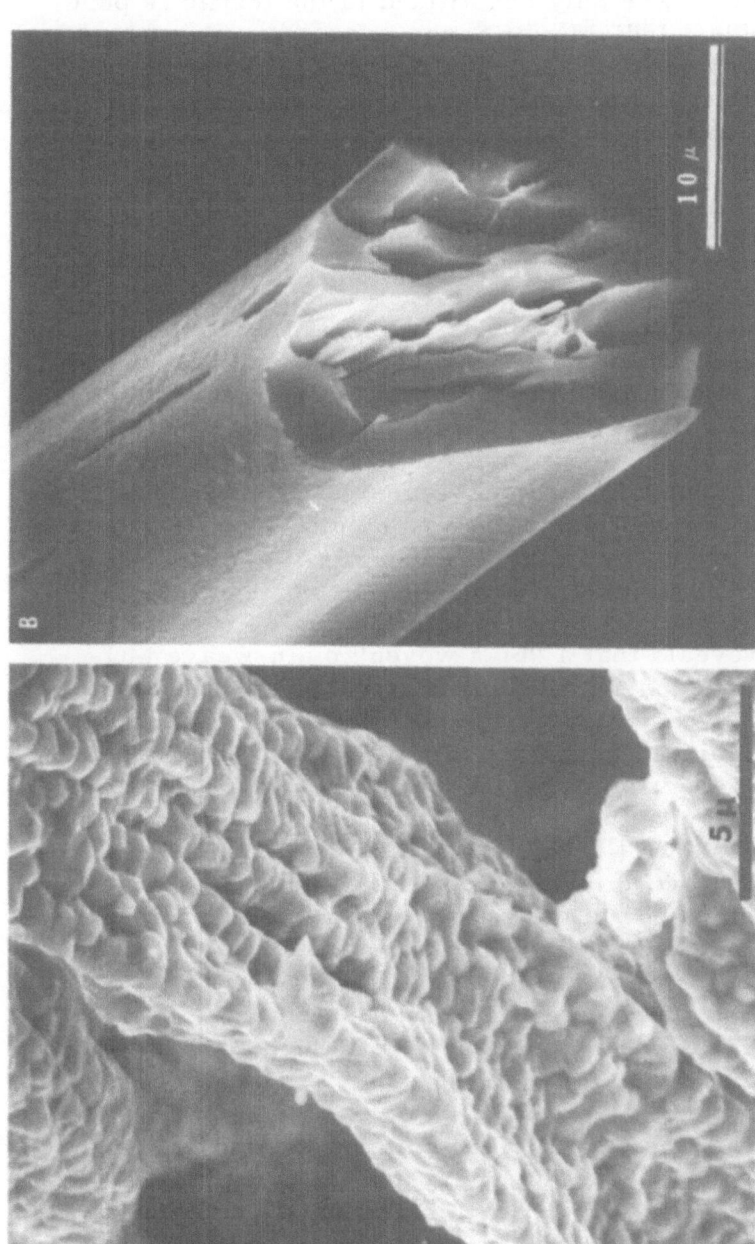

Fig. 2. Scanning electron micrographs showing MC fibers varying from (A) "corn-cob" geometry to (B) a fully dense fiber.

with scissors. The only constituent of the textile is pure
metal carbide. Individual fibers can vary in microstructure from a "corn-cob" morphology to a fully densified
solid fiber (Fig. 2) depending upon the conditions of the
preparative treatment.

MICROSTRUCTURE OF THE COMPOSITES

Figure 3 shows the microstructure of the duplex surface produced by casting 6061 aluminum alloy in molds
lined with MC textile strips of three different weaves.
Figure 3-a shows the distribution of MC carbide phase
within the aluminum alloy matrix when a satin weave
textile (such as the strip shown in Figure 1) is used. In
Figure 3-b a surface structure is exhibited where the
"islands" of the MC phase are more uniformly distributed.
This structure results from the use of a fine denier starting textile with a velvet weave. In Figure 3-c, the distribution of MC carbide phase within the duplex surface
structure reveals that a "bias woven" plain-weave MC
textile was used. All these microstructures demonstrate
that various weave structures of MC textile can be used
to cast the surface composite materials. In each case, a
high volume fraction of MC carbide phase dispersed in
aluminum alloy can be produced at the surface. The woven
MC textiles shown (in Figs. 3-a, b, c) are examples of
conventional "two-dimensional" weave structures and, as
such, the thickness of the wear-resistant layer in the
surface composite is limited to approximately 0.015 in.
However, by using a three-dimentional woven rayon textile as the "precursor" it is possible to obtain up to 0.050
in. thick duplex wear-resistant layers such as that shown
in Fig. 3-d.

In each of the cast surface composites shown, bond
formation between the MC fibers and the alloy matrix
occurs during casting. Thin-foil transmission electron
microscopy of the duplex structure has revealed no
morphologically distinct phase formation, i.e., no dis-

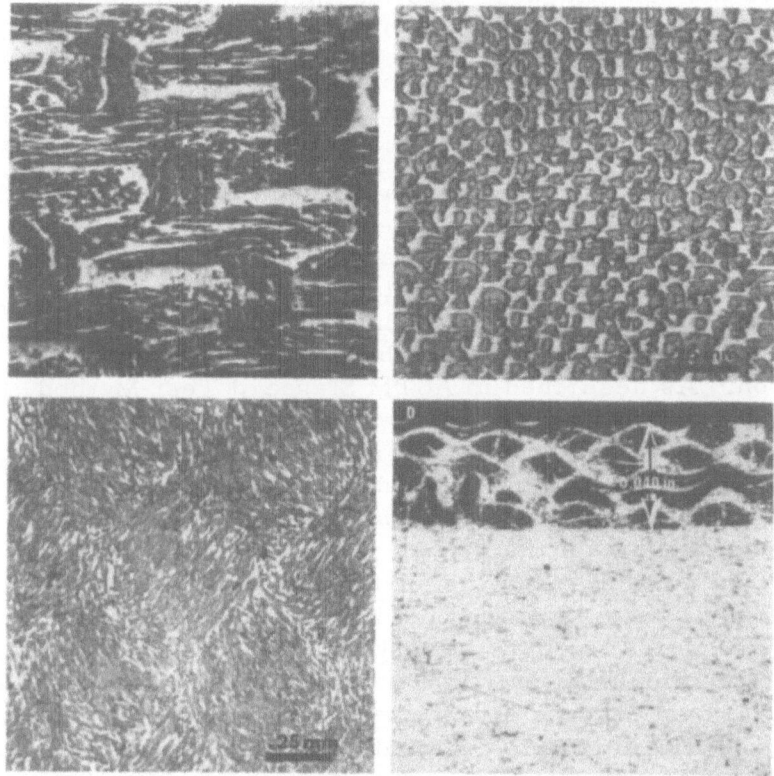

Fig. 3. Microstructures of the duplex surface of cast surface composites showing the distribution of monocarbide phase in a 6061 aluminum alloy matrix: (A) satin-weave MC textile, (B) velvet-weave MC textile, (C) biased plain-weave MC textile, (D) 3-D woven MC textile, shown in cross-section.

crete particles of a reaction phase at the MC fiber/matrix interface. However, selected-area electron diffraction of the fiber/matrix interface regions has shown evidence that an intermetallic phase of the type Al_3M does form when TaC and TiC textiles were used. The "wetting" reaction which occurs when molten aluminum alloy interpenetrates the weave structure of the MC textile during casting is evidently accompanied by a limited interaction

between the constituent phases leading to a well bonded
fiber/matrix interface.

WEAR PROPERTIES

The microstructures shown in Fig. 3 a thru c exhibit
surfaces with approximately 65-75 percent carbide phase
dispersed within a 6061 alloy matrix. Such a surface
exhibits interesting wear-resistance. The wear data
presented here was obtained using a LFW-1 model Wear-
Testing Machine manufactured by the Dow Corning Corporation. As described in ASTM Standard Test Method
D 2714-68, all tests were carried out on stationary rec-
tangular test blocks pressed by a predetermined load
against a rotating ring. The wear properties measured
were: (i) volume of the wear scar on the test surface,
(ii) weight change of the mating ring, and (iii) friction
force measured at intervals during the test. Figure
4 shows low-magnification photomicrographs of typical
wear scars generated at the test surfaces of four materials:
(a) 6061 aluminum alloy, (b) hypereutectic Al-18 w/o Si
alloy, (c) cast surface composite of MC textile in a 6061
alloy and (d) Aluminum - graphite composite material
(developed by the Toyo Kogyo Co. as a Rotor Apex Seal
for the Wankel Engine). In each of these wear tests
(Figs. 4a-d) the following conditions were used:

Mating Ring:	4620 Steel; R_C = 58-62; 8-12 micro inch finish
Lubricant:	Mobil 5606-A fluid
Load:	30 lbs.
Wear Speed:	180 r.p.m. (Ring diameter - 1.3775 in.)
Total revs:	5400.

The photomicrographs of the wear scars provide a
qualitative, and visually obvious, comparison of the wear
resistance of the materials tested. Bare 6061 aluminum
alloy shows extremely wide wear scars (Fig. 4a) which
correspond quantitatively to wear scar volumes in the
range 5600 to 6000 x 10^{-6} cm^3. By utilizing the natural

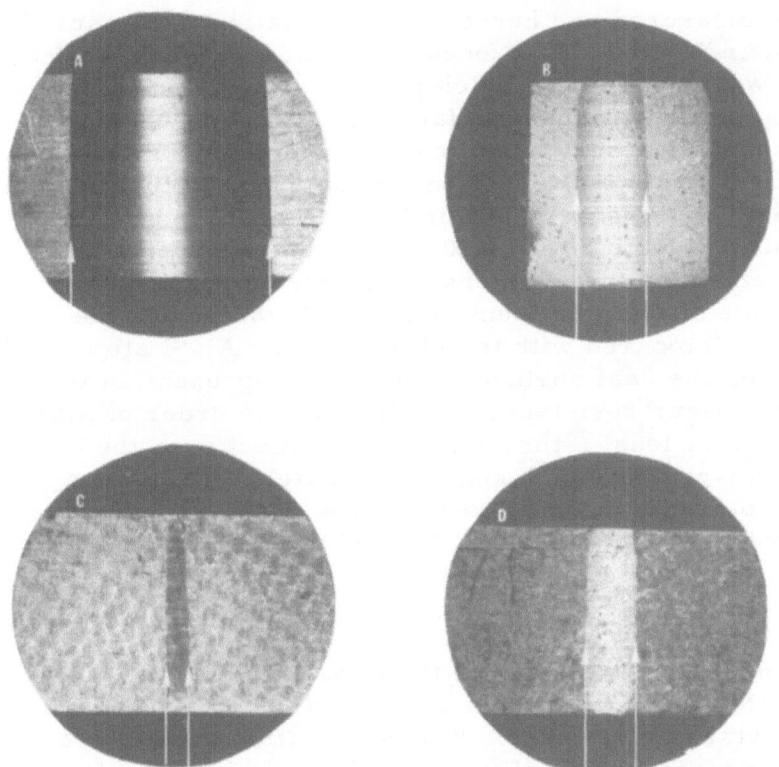

Fig. 4. Typical wear scars on test surfaces of: (A) 6061 aluminum alloy; (B) hypereutectic Al-18 w/o Si alloy; (C) cast surface composite of MC Textile in a 6061 alloy; (D) aluminum-graphite composite.

dispersion of a hard phase, such as α-Si in the Al-Si system, aluminum alloys of the hyper-cutectic Al-18 w/o Si type (Fig. 4b) reveal greatly improved wear behavior but the wear scar volumes measured still are in excess of 350×10^{-6} cm^3 for tests against a steel ring. In Fig. 4c, an example is provided of a wear scar on a surface with 75 percent MC phase in a 6061 matrix. It should be noted that in the case of such duplex (MC + Al) surfaces, the exact distribution of the phases in the region where the rotating ring makes contact influences the wear scar

volume measured and hence, some variation in measured
wear scar volume values occurs. However, given an MC
textile weave structure which permits the surface area
fraction of the MC phase to fall in the range of 65-75 percent, the duplex surfaces of cast surface composites show
variations of wear scar volume in the range 6-40 x 10^{-6} cm^3.
The width of the scar shown in Fig. 4c corresponds to a
wear scar volume of 6. 5 x 10^{-6} cm^3 and represents an
improvement of three orders of magnitude in the wear
resistance of the aluminum alloy which comprises the
matrix. Compared with the hypereutectic Al-Si alloy
(Fig. 4b), the cast surface composites represent an increase in wear resistance of well over one order of magnitude and at least a threefold improvement over the
aluminum-graphite bulk composite material (used as a
Rotor Apex seal in the Wankel engine of Mazda automo-
biles) which yields values of 120-130 x 10^{-6} cm^3 for
measured wear scar volumes (Fig. 4d shows a typical
wear scar).

A detailed discussion of the wear mechanisms of the
duplex (MC + Al) surface of cast surface composites will
be undertaken elsewhere. It should suffice here to present measured values of the wear caused on the surface
of the mating steel ring by the materials tested. Of the
materials discussed above, only the tests with bare 6061
aluminum alloy showed a predominantly adhesive wear
behavior, i. e., the steel ring showed a weight gain in
tests against bare 6061.

Tests with the hypereutectic Al-Si alloy showed ring
weight loss values of approximately 0. 5 mg. The aluminum-
graphite composite showed ring loss in the range 0. 6 to
1. 0 mg. Both composite materials showed a combination
of adhesive and abrasive wear behavior. In twenty tests
with different MC textile geometries, cast surface composites showed ring weight loss in the range of 0. 16 to
0. 7 mg. It should be emphasized that during tests on
composite surfaces, wear particles of the hard phase
cause abrasion of the mating steel ring and both hard and
soft wear particles adhere to the steel ring. There is a

considerable difference in density between the carbide
particles and the aluminum wear particles. Thus, the net
ring weight change represents the overall effect of abrasion and adhesion of particles with different densities and
therefore the weight change is somewhat ambiguous.
However, the ring weight change values do provide an
indication of the extent of wear.

REFERENCES

1.   Hamling, B.H., U.S. Patent 3,403,008, issued Sept.
     24, 1968, assigned to Union Carbide Corporation.

DISCUSSION

A.R.C. Westwood (Martin Marietta): Can you comment
on the ability of aluminum base-carbide "textile" hardened
bearings for rotary engines to hold up at engine operating
temperatures?
Author: The possible application is in the area of apex
seals for an engine based on the Wankel design. Such a
seal would probably be a sacrificial member during service and not a bearing in the usual sense. In field tests,
our composites showed no evidence of deleterious wear
caused by inability of the matrix to withstand operating
temperatures.
A. Choudry (Univ. of Rhode Island): Your testing RPM
of 180 is rather low for practical operating conditions.
Would not the wear rate change significantly with RPM?
Author: I presented data only on tests carried out with
the LFW-1 test machine. Under much higher RPM in
field tests we found that the relative ranking of our materials vis-a-vis state-of-the-art materials remained essentially the same as that indicated by the low-RPM test.

AN EXPERIMENTAL INVESTIGATION OF THE DYNAMIC AND THERMAL CHARACTERISTICS OF THE CERAMIC STOCK REMOVAL PROCESS

P. J. Gielisse, T. J. Kim and A. Choudry

University of Rhode Island
Kingston, Rhode Island

Analysis of the ceramic stock removal system is a complex process, basically involving the transfer of energy from the grinding wheel to the ceramic workpiece. This involves the various factors generally accompanying the phenomenon of material failure under ballistic impact. However, the shock front normally encountered in ballistic impact is not generated here, since the impact velocities are subsonic. On one hand, this simplifies the matter by avoiding the complications associated with shock front propagation while on the other hand theoretical schemes to analyze the fracture under subsonic impact are not as fully developed as those for the shock front propagated fracture. Furthermore, the existing theoretical models for either the static or the shock failure cannot be directly applied to the stock removal process for at least the following reasons:

1. Many major phenomena act simultaneously, such as high velocity impact, crack propagation, wear and friction processes.

2. Experimental parameters: specific conditions and their variations are often extreme. As an example, interfacial pressures may reach levels of 25 kilobars,

within one tenth of a microsecond; temperatures of around 1500° Kelvin are experienced.

3. The process is a continuous dynamic process; that is, there is a very definite time period over which the material continues to be subjected to impact parameters, after the initial impact is made.

In the grinding process, one normally encounters four major areas of concern:

1. The grinding force: This is the most critical process parameter, and its force level determines on one hand, the force required for the process, and on the other hand, the extent to which the material breaks up.

    Force level should be evaluated in reference to the prime system parameter, i.e., speed, and in order to be useful should be related to the ceramic material properties.

2. Wear: specifically on the abrasive, which for most ceramic systems is diamond. The mode extent and rate of wear determines the efficiency of the process, but also strongly affects the force level due to geometric parameters.

3. The actual grinding temperature at the abrasive material interface, as well as temperature distribution: The importance of the grinding temperature to the entire process of stock removal cannot be underemphasized, since it is related to the materials properties (Hugoniot) in a very fundamental way.

4. The specific environment in which grinding is done strongly influences force levels, the amount of material removed, the wear and the temperature. The most important question to be asked in this area is why certain environments have their particular and often specific influence.

It is the present intent to present some of the results of
our experimental work and its analysis in the previously
mentioned four areas. We further intend to have this act
as an introduction to the following chapter, which will
support these results with a more fundamental analysis.

FORCE LEVELS IN GRINDING CERAMICS

All the force results shown here have been obtained as
a result of the impact of a single point (diamond) on ceramic.
The measurements were carried out with the aid of a very
sensitive piezo-electric dynamometer, designed for this
purpose, which records the vertical and horizontal components of the impact force. The experimental set up has
been described earlier[1].

Grinding forces can be assumed to be related to two
functionals: a zero-wear parameter, and a wear-related
parameter. The separation between these two entities
facilitates the evaluation process, stresses the important
duality, and allows one to quantify the individual parameters.
The following force data have all been compiled under the
so-called zero wear condition, by replacing the specific
diamond point before the extent of wear could influence
significantly the measured force level.

The typical force data for four commercial aluminas
and silicon nitride are shown as a function of wheel speed
in Fig. 1. The trends of the force curves remain the same
for all materials and environmental conditions, except that
the vertical force level is always higher than the horizontal
force.

The same data is shown on a normalized linear basis in
Fig. 2. In general, we observe a sharp drop in force level
as speed increases, continuing toward a leveling-off point.
Theory predicts an increase in force after the minimum is
reached. The steepness of the slope and position of the
minimum is primarily determined by crack length distribution in the sample. Each material has its own characteristic

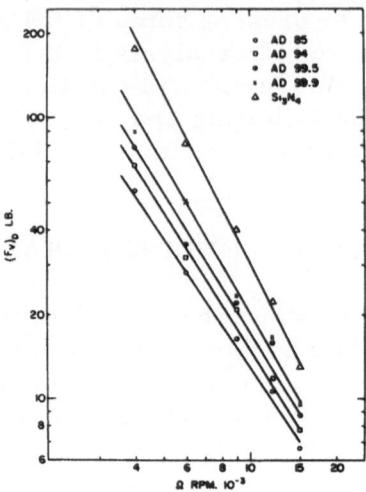

Fig. 1.   Wheel speed Ω vs. log (vertical force, F_v)

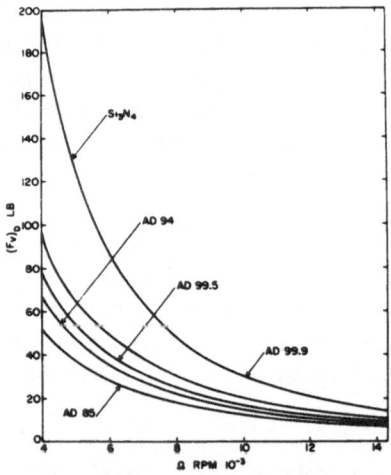

Fig. 2.   Wheel speed Ω vs. vertical force, F_v.

force level, and in general the force increases as hardness, strength, and Young's Modulus of the material increases. Silicon nitride has its own unique level, the highest among the materials investigated, but retains the same attitude which is characteristic of ceramics.

An excellent verification of the point made earlier with reference to crack size has recently been obtained from a study in which we have evaluated the force levels of various aluminas differing in grain size only. It was found that the forces were highest for the smallest grain size, with a steep decrease of force levels at sizes beyond approximately 20 microns. In aluminas displaying bimodal grain size distribution, the largest average grain size determine the specific force level.

From the data of Fig. 1, it is clear that the mathematical expression for a grinding force in ceramics is simply:

$$F = Fa / (\Omega^m) \qquad\qquad (1)$$

where Fa, m and Ω are the vertical intercept, wheel speed exponent, and wheel speed respectively. The parameters Fa and m can be related to a materials parameter, Λ (= σ^2/2E, modulus of fracture), and their functional relations are shown in Fig. 3 and Fig. 4. Substitution of the resulting functional relations into equation (1) yields a set of force equations in terms of the two significant parameters Ω and Λ, for all alumina ceramics tested, as follows:

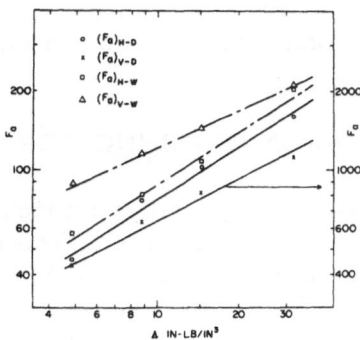

Fig. 3.   Fracture modulus Λ vs. force intercept + Fa.

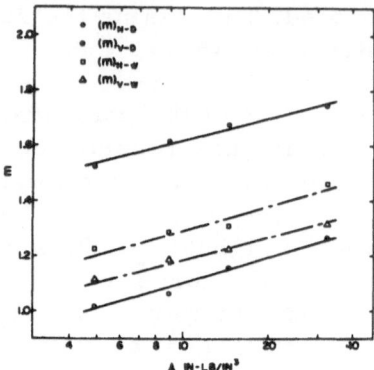

Fig. 4.   Fracture modulus Λ vs. wheel speed exponent m.

$$F = \frac{I_\varkappa \, \Lambda^n}{\Omega^{(I\nu + 0.434\nu \ln\nu)}} \qquad (2)$$

where $I_\varkappa$ and $I\nu$ are defined as (cf Figs. 3 and 4)

$$Fa = I\varkappa \, \Lambda^n$$

$$m = I\nu + \nu \log \Lambda$$

indices n and ν refer to dry and wet conditions respectively. It has thus been shown that there is a direct relationship between a force variable and the mechanical parameter, Λ, and its value governs the mode of stock removal, and that proposed force relationship offers an independent deter- mination of strength, and a modulus of an unknown ceramic.

DIAMOND WEAR IN GRINDING CERAMICS

Since the specific force level is critically influenced by the amount of wear on the diamond, the force level it- self can be used to evaluate quantitatively the extent of wear. The results of such measurements are shown in Figs. 5 and 6 for one alumina, and two wheel speeds under wet and dry grinding conditions, as a function of length, of

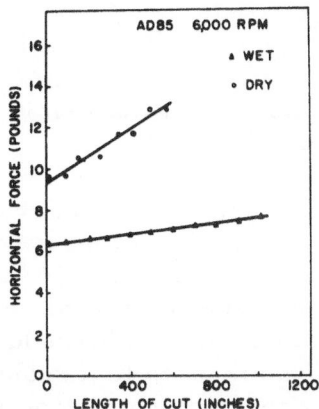

Fig. 5. Length of cut vs. horizontal force, for AD 85 at 6,000 r.p.m.

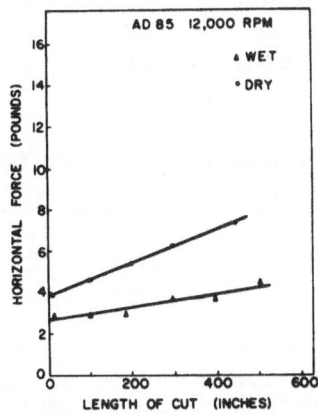

Fig. 6. Length of cut vs. horizontal force, for AD 85 at 12,000 r.p.m.

cut; it may be noted that the relationship is entirely linear. In this fashion, the slope of the curves given in Figs. 5 and 6 becomes a quantitative wear parameter,

$$w = \frac{d(F)_H}{ds} \qquad (3)$$

The quantitative data for all materials investigated so far have been compiled in Table I.

The wear parameter shows diamond wear to be primarily dependent on grinding force, properties of the workpiece and environmental conditions. For aluminas, an increase in wheel speed twice the initial value, reduced wear roughly by a factor of two. The reduction in wear under similar conditions amounts to almost 75% for silicon nitride. Relative wear in grinding silicon nitride dry is more than 25 times that of a low strength alumina such as AD 85. Wet grinding reduces diamond wear when compared to dry grinding under all test conditions. The largest influence is at low wheel speeds and for the higher strength material. At high wheel speeds there is less benefit to a wet environment, particularly for the low strength materials. It has furthermore been shown that wear is linearly related to the modulus of the material.

TABLE I. Relative Diamond Wear Rate (w) for Grinding Various Materials[*]

$$w = \frac{d(F)_H}{ds} \times 10^3 \; (lb/in)$$

Material	6,000 rpm (71.96 ft/sec)		12,000 rpm (143.92 ft/sec)	
	Dry	Wet	Dry	Wet
AD85	8.25	1.28	5.0	4.2
AD94	26.2	3.11	14.5	4.0
AD99.5	45.8	3.5	24.0	3.7
AD99.9	55.0	---	28.0	---
Si_3N_4	230.0	---	170.0	---

[*]All observations for 0.002" depth of cut, upgrinding, table feed of 1.5 inches/sec and 120 degree conical diamond.

TEMPERATURES IN GRINDING

The ceramic stock removal process is always accompanied by light emission at the cutting area. This may be due to direct thermal radiation (black body) or to a non-thermal component or a combination of both. Qualitatively, the light emission intensity normally increases in going to a higher strength ceramics. As such, light emissions of a material such as a low strength alumina is minimal, whereas light emission in grinding silicon nitride resembles an almost volcanic explosion when observed as a single impact event. Photographing of individual cutting events with infrared and visible light filters in the optical path have shown good images for both conditions, indicating the presence of a thermal as well as visible component to the light emission.

With the aid of a grating spectro-photometer, we have analyzed the spectra of individual cutting events for five different ceramics. In this way, time exposure spectra of the actual cutting event were obtained for all ceramics. Spectra of incandescent tungsten sources at various known temperatures were used for calibration. Using a laser source with a photodensitometer, we have obtained the spectra shown in Fig. 7.

It has thus been possible to obtain quantitative data indicating temperatures between 1100°K and 1600°K in the ceramic grinding process. The grinding temperature increases as the value of the modulus of fracture of the specific ceramic increases. For the low strength ceramics, the spectra follow essentially black body behavior. However, with higher strength materials, deviations from the black body curves do occur starting in the green, and apparently shifting to higher frequencies with an increase in strength. For these materials, a non-thermal radiation component is discernible.

The grinding temperatures indicated here are essentially those of the flame emission which can have a variety of causes. The first possibility is the conversion of

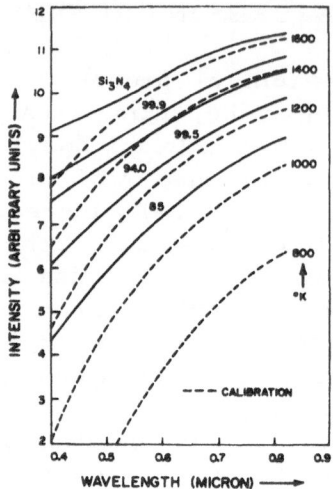

Fig. 7. Comparison of spectra during cutting with incandescent tungsten spectra.

mechanical to thermal energy at impact. Other possibilities are conversion of the elastic strain energy into the kinetic energy of fragments, and the emission of light due to the flight of high velocity fragments in air. Furthermore, the chemical reactions of the material during the removal process, or possibly, reactions with the environment may influence the specific emission characteristics. With the first three possibilities, one would expect the radiation to be essentially thermal, not leading to any deviation from the thermal curves. In the case of chemical reactions, a discrete structure in the spectra should appear. Reaction with the environment could therefore be the cause for the previously mentioned additional non-thermal component.

INFLUENCE OF THE ENVIRONMENT

At the time of writing the authors have started an evaluation of the influence of environment on the grinding process, specifically the influence of normal alcohols and alkanes. Initial indications are that a functional relation-

ship between force levels and specific parameters of the
environment, such as boiling point, melting point, and
surface energy was indicated with forces generally going
up as the molecular weight of the environmental material
increased.   The best correlation indicated by experimental
results, is between the grinding force and kinematic viscosity, which is a direct linear relationship for all environments, including water.   This indicates that the environment may have a direct influence on the mechanical and
frictional characteristics of the grinding process, rather
than on the specific material or surface characteristics of
the ceramic.   Recently, our results have indicated both
positive and negative deviations from the previously mentioned linear relationship with reference to the kinematic
viscosity.   The largest deviation appears to occur with the
alcohols around octylalcohol.   It appears that the force
levels are influenced by the wetting ability of the liquid.
It may be further noted that whereas the force level does
deviate from the linear relationship with respect to kinematic viscosity, the force ratio, that is, horizontal force
to vertical force does not significantly differ from one
material to another within any one material to another
within any one type of environment.   More work will have
to be conducted in this area before definitive statements
can be made.

CONCLUSIONS

The results obtained by a preliminary experimental
investigation of the stock removal process have shown
definite trends for correlation among various system and
materials parameters.   Among these the most noteworthy
are:

1.   Wheel speed has an inverse exponential relation-
     ship with cutting force.

2.   The cutting force is directly related to the fracture
     modulus of the material.

3. The effect of environment is more pronounced at low wheelspeeds and for lower strength materials.

4. The emission from the impact contains both thermal (black body) as well as chemical components indicating the presence of non-mechanical processes. These findings point to the complexity of the ceramic stock removal process and indicate the necessity of further fundamental investigations.

BIBLIOGRAPHY

1. P. J. Gielisse; T. J. Kim and Choudry
    Force and Wear in Analysis in Ceramic Processing.
    Naval Air Systems Command
    U. S. N. #N00019-72-C-0202

DYNAMIC ELASTIC MODEL OF CERAMIC STOCK REMOVAL

A. Choudry and P. J. Gielisse

University of Rhode Island
Kingston, Rhode Island

Considerable work has been done on stock removal
processes in metals; and on the failure properties of
ceramics under ballistics impact. The former is dominated by plastic deformation prior to fracture, whereas
the latter is dominated by the propagation of a shock
front which causes brittle fracture. For the process of
ceramic stock removal as encountered in practice,
neither of these mechanisms play dominant roles, i. e.,
neither the ceramic sample is expected to manifest plastic flow, nor do the practical machining techniques generate shock fronts. The theoretical model is thus different from commonly accepted theories of "cutting" and
"brittle fracture in ceramics". We shall describe a
model where the cutting tool is a single diamond point as
shown in Fig. 1. The conical diamond rotates with an
effective radius R and is set to cut a V-shaped groove of
depth r. The total length of the groove ℓ is traversed by
the cutting point in a large number of passes as the sample is fed with a velocity v.

We shall first consider an elementary phenomenological model to examine the interrelationships of various
macroscopic parameters (wheel speed, feed speed, depth
of cut, etc.) involved, and then develop the model to

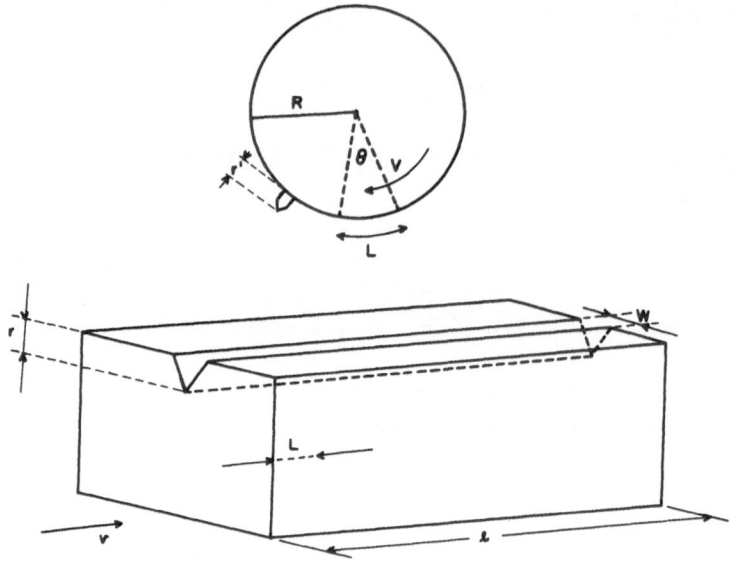

Fig. 1. Cutting wheel and workpiece geometry.

include parameters of more fundamental nature (grain size, porosity, Hugoniot properties, etc.) which would yield a dynamic elastic model.

PHENOMENOLOGICAL MODEL

This simple model is based almost entirely on the geometric and kinematic characteristics of the system, and hence is limited in its scope. However, this model will be modified to include eventually dynamic and thermodynamic aspects of the process to correspond to practical situations. The kinematical and geometrical parameters that describe a typical set up as shown in Fig. 1 are: R = radius of the diamond point trajectory (R = r' + R' cf. Fig. 1); r = depth of the groove; ω = angular velocity of the cutting wheel; v = sample feed velocity; W = width of the groove. We further introduce the following quantities;

$V = \omega R$; $\eta = v/V$; and $\alpha_o = r/R$.

Fig. 2 shows the trajectory of two successive cuts based on the approximation that during the trajectory CB, the sample moves a negligible distance.

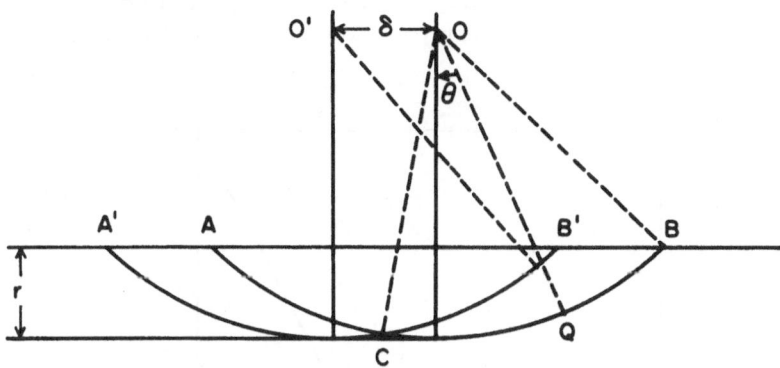

Fig. 2.   Trajectory of cutting point in two successive cuts.

The time $\tau = (\theta_o + \Phi_o) / \omega$ (1)

where the angles θ_o and Φ_o as shown in Fig. 2 are given by following expressions

$\cos \theta_o = (1-\alpha_o)$ (2)

$\tan \Phi_o = \eta\pi / \sqrt{1-\eta^2\pi^2}$ (3)

Thus δ, the feed distance per pass, is given by:

$\delta = 2R \sin \Phi_o$ (4)

During the second cut CB, the diamond removes the volume BCB' and the depth of cut PQ at any angle θ varies.   The relative depth of the cut $\alpha(\theta)$ at any angle, is given by:

$\alpha(\theta) = 1 + 2\sin \Phi_o \sin \theta - \sqrt{1-4 \sin^2\Phi_o \cos^2\theta}$ (5)

where $-\Phi_o \le \theta \le \theta_o$.

Now we make the dynamic assumption that F_D, the "cutting force", is linearly proportional to the depth of the cut, which is equivalent to assuming that the energy used in cutting is proportional to the volume removed; thus,

$$F_D (\theta) = C\alpha(\theta) \qquad\qquad (6)$$

where C is a constant. Eq. (6) gives the instantaneous force which could possibly be measured by a transducer of ~1MHz bandwidth under normal operating conditions.

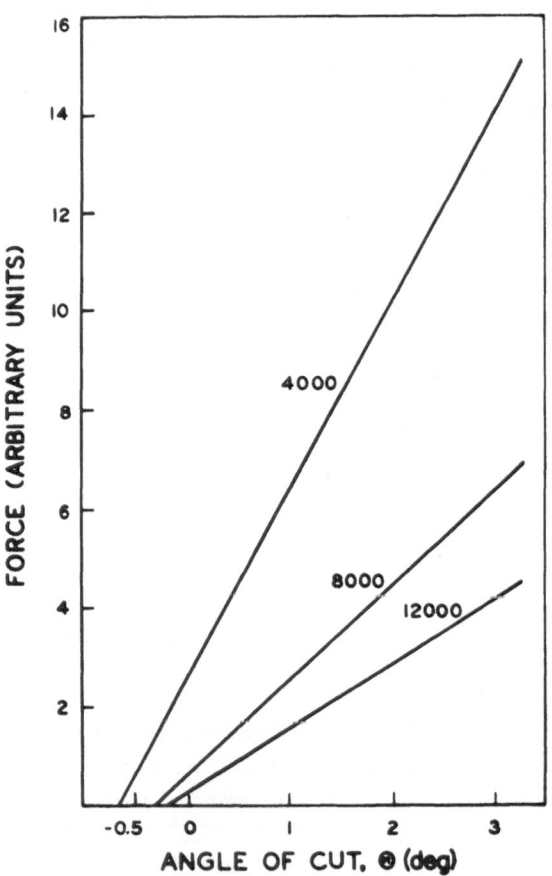

Fig. 3. Cutting force vs angle.

Fig. 3 shows the variation of F_D with θ during a cut of .002" at various rpm's. The departure of these curves from a straight line is of the order of $(\theta - \sin\theta)$. the force F_D as measured in a lower band width dynamometer would be a time integral of the force shown in Fig. 3, which is the same as the energy spent in one cut. Thus,

$$W = \int_{-\Phi_o}^{\theta_o} F_D(\theta)\, R d\theta$$

or

$$\frac{W}{CR} = (\theta_o + \Phi_o) + 2\sin\Phi_o\,(\cos\Phi_o - \cos\theta_o) -$$

$$\int_{-\Phi_o}^{\theta_o} 1 - 4\sin^2\Phi_o\,\cos^2\theta\, d\theta \tag{7}$$

The integral in Eq. (7) can best be evaluated numerically.

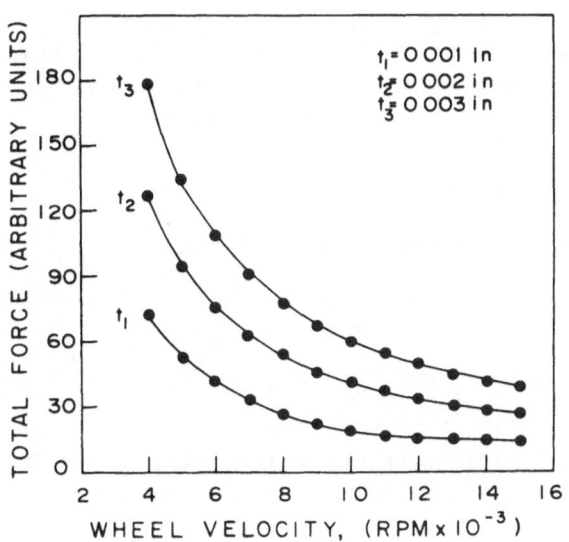

Fig. 4. Total force vs RPM.

Also we note that,

$$W = CR \int F_D(\theta) \frac{d\theta}{dt} = CR\omega \int_0^\tau F_D(\theta) \, dt \qquad (8)$$

where the time integration is done during the "dwell time" τ. Fig. 4 shows the variation of the "total force" $\int F_D(\theta)dt$, for different rpm's and cut depths. The most striking feature to be discerned in Fig. 4 is the sharp drop in the total force as the wheel angular velocity is increased. This is also the most salient feature of the experimental data[1], and permits confidence in the model and the dynamic assumption made. The model at this stage has no features regarding the details of brittle fracture in the polycrystalline ceramic as caused by the cutting point impact. To include such details, we shall first develop a one-dimensional dynamic elastic model similar to that of Steverding[2].

ONE DIMENSIONAL MODEL

Fig. 5 shows a semi-infinite rod (density, d Young's Modulus E) of brittle polycrystalline material along the x-axis; impacted at t = 0 and x = 0 by a driver D moving with constant velocity V. It can be shown that the space time stress distribution, $\sigma(x, t)$ is given by[3]

$$\sigma(x, t) = \frac{VE}{c} S(ct-x) \qquad (9)$$

where s(x) is the step function such that

$$S(x) = \begin{array}{l} 1 \; x \geq o \\ 0 \; x \leq o \end{array} \qquad (10)$$

and

$$c = E/c \qquad (11)$$

If internal friction is neglected, then it can be shown[4] that the force F_D, experienced by the driver is given by

$$F_D = \delta V^2 \sigma^2 / E \qquad (12)$$

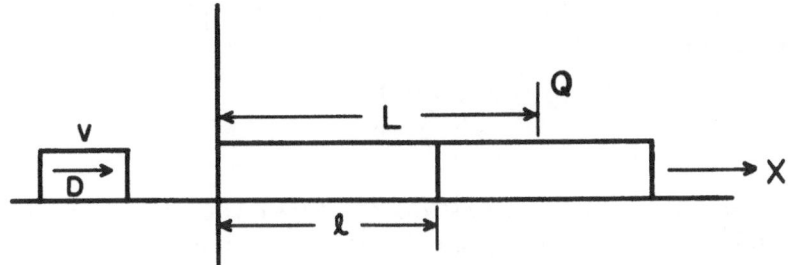

Fig. 5.   One dimensional model.

The force F_Q as measured by a transducer Q situated on the bar (of Fig. 5) will be the same as F_D except for a time delay factor, i.e.,

$$F_Q = F_D \, S(x - \frac{L}{c})$$

(13)

In writing Eq. (13) we have assumed that no inelastic modifications in the stress wave are introduced during its passage to Q. However, such would not be the case if there exists a crack which begins to propagate under the influence of F_D. In the extreme case, the crack could cause instantaneous failure and the stress would be relaxed. This point can be illustrated as follows: The bar is replaced by a spring to correspond to a Voigt body (no dashpot for brittle materials). A crack at length $\mathcal{l}$ is represented by a small connector which breaks down when the stress exceeds the Griffith[5] stress limit such that:

$$\sigma_o^2 = \frac{\gamma E}{\lambda_o}$$

(14)

where γ = surface energy/cm^2, and would include plastic deformation energy if it occurs prior to failure and λ_o = crack length. If $F_D < \sigma_o$ the connector would transmit the stress wave undistorted to Q however, if $F_D > \sigma_o$ and if we assume that λ_o fails instantly upon the application of F_D, the stress wave will not be propagated to Q. The section

$\mathcal{l}$ of the spring will become a "fractured chip". The energy of this chip is:

$$W = \frac{1}{2} \, \mathcal{l} \left(dv^2 + \frac{\sigma^2}{E} \right) \tag{15}$$

where $\mathcal{l}$ = ct and d = mass/cm^3. After the chip is fractured and free, the strain will be relaxed, however, the strain energy cannot be converted into the kinetic energy of the chip, and it would instead cause a tensile stress, and if no inelastic damping is present, the chip will execute oscillations under cyclic compressive and tensile stresses of equal magnitude[2].

The fracture strength under compressive stress is higher than that under the tensile stress. Thus, in Griffith theory, we assume that for the propagation of a crack of length λ_o there are two critical stresses σ_t and σ_c for tensile and compressive loading respectively, thus:

$$\frac{\gamma E}{\lambda} = \left\{ \frac{\sigma_t^2}{\sigma_c^2} \right\} \tag{16}$$

and $\sigma_c > \sigma_t$.

Conversely, for a given magnitude of stress, more cracks can cause fracture under tension than under compression. This is somewhat similar to the modified Griffith theory as discussed by Brace[6].

Now, if the fracture chip has cracks within it, it is possible that some of them might propagate during the tensile stress cycle and cause further fragmentation of the chip. This logic could also be applied to the relaxation of the residual compressive stress in the stock leading to "spalling" under tensile stress after the cutting point has passed. From this point of view, spalling is fragmentation under tensile stress which is generated by the post-impact relaxation of the compressive stress. If the material

exhibits inelastic behavior, e.g., viscous damping due to
plastic deformation, the post impact tensile stress will be
less than the compressive stress and hence less spalling,
as is commonly the case with ductile materials. If the
stress pulse is indeed as sharp as shown (∞ slope) then we
do not expect any spalling in the one-dimensional model
since the material beyond ℓ never gets compressively
strained. A modification of this will naturally ensue when
we consider a general three-dimensional model.

If we assume that there are n cracks/unit length distributed uniformly, then the time-average force experienced
by D, $\overline{F}_D$, can be given by[4]:

$$\overline{F}_D = V^3 d/n(c + V) = F_D V/n(c + V) \qquad (17)$$

which is based on the assumption that once a fracture
occurs at a distance from D, the cutting point then travels
the distance ℓ in the fracture field essentially without any
resistance and during this time $F_D = 0$. From Eq. (11)
and (14), we get

$$\lambda_o = \frac{\gamma c^2}{E V^2} \qquad (18)$$

which implies that at an impact of velocity V all the cracks
of length greater than λ_o will propagate. In general, polycrystalline materials have both a grain size distribution and
a pore size distribution and if it is assumed that all cracks
in the sample are due to either of these two as suggested by
Jaeger[7], then we can write down a crack length distribution
$f(\lambda)$ per unit length. Thus, dn, the number of cracks of
length between λ and $\lambda + d\lambda$ in unit length, is given by:

$$dn = n_o f(\lambda) d\lambda \qquad (19)$$

where n_o is the total number of cracks in unit length. Thus
n used in Eq. (17) is given by:

$$n = n_o \lambda \int_o^\infty f(\lambda)d\lambda \tag{20}$$

To further emphasize the difference in the compressive and tensile strengths, we note that there should be two crack length thresholds. These can be introduced from energy balance considerations of the type discussed by Drucker[8] by assigning different effective surface energies to compressive and tensile crack propagations, thus we modify Eq. (20) to the following:

$$\lambda(c, t) = \frac{\gamma(c, t) c^2}{EV^2} \tag{21}$$

where $\gamma_t < \gamma_c$.

In view of this, fracture chips formed furing a compressive impact should have an average length of $\bar{\lambda}$ where:

$$\bar{\lambda} = \left[n_o \lambda \int_{\lambda_c}^\infty f(\lambda)d\lambda \right]^{-1} \tag{22}$$

During the tensile stress cycle, the chip should fragment into a number (n_f) of smaller pieces given by:

$$n_f = n_o \bar{\lambda} \int_{\lambda_t}^{\lambda_c} f(\lambda)d\lambda \tag{23}$$

The average force $\bar{F}_D$ experienced by the cutting point is:

$$\bar{F}_D = \frac{V}{n_o (\epsilon + V)} \left[\lambda \int_{\lambda_c}^\infty f(\lambda)d\lambda \right]^{-1} \tag{24}$$

For a simple parabolic distribution of crack lengths of the form

$$f(\lambda) = \lambda(\lambda_m - \lambda) \tag{25}$$

where λ_m is some reasonable maximum crack length,

we get[4]

$$\overline{F}_D = \frac{6pE^3 v^9}{n_o (c + v) \gamma^3 c^6} \left[2 - 3\beta^2 + \beta^6\right]^{-1} \qquad (26)$$

where

$$\beta = \frac{V}{V_m} \quad \text{and} \quad V_m = c \frac{\gamma_c}{E\lambda_m}$$

For small values of V, Eq. (26) has similar form as Fig. 4 which is derived from the phenomenological model and thus this comparison serves as a test of the inner consistency of the entire model, and gives us confidence to develop it into a three dimensional mode.

THREE DIMENSIONAL MODEL

This generalization entails that we consider fracture under a three dimensional stress wave. The problem is difficult since the stress wave is not generated by pure impulsive impact as is widely encountered in the detonation of small explosive charge (of Kolsky and Rader[9]) but by a constant-velocity shear applied at one corner of the sample. The calculation of spatio-temporal distribution of stress in a closed analytical form under such conditions is extremely complex if not impossible and thus one is forced into making simplifying assumptions. First of all we note that the velocity of the cutting tool $\sim 10^3$ cm/sec whereas the sonic velocities are $\sim 10^6$ cm/sec, and thus we are justified in adopting a quasi-static approach to the problem, i.e., we shall first consider the static stress generated by the cutting point shear loading and then develop mechanisms for its propagation which eventually causes fracture. To calculate the static stress we shall first examine the stress distribution caused by a localized force on a semi-infinite elastic body. A sphere of diamond of radius ρ dents an elastic medium by a distance e as shown in Figure 7. For small indentations ($e \ll \rho$) we have

$$e \sim r^2/2\rho \tag{27}$$

If we consider the above as an exact relation, then the equation

$$e = r^2/2\rho \tag{28}$$

describes a paraboloid of revolution instead of the sphere as shown in Fig. 6. The pressure under the paraboloidal punch has been calculated by various authors[10, 11, 12] and is best given in cylindrical coordinates as

$$p(r, \theta) = \frac{2E}{\pi(1-\nu^2)} \left[\sin^2 \theta_o - \sin^2 \theta \right]^{1/2} \tag{29}$$

where

$$\sin \theta_o = r/\rho$$

and the total force is given by

$$F = \frac{4E}{3(1-\nu^2)} e^{3/2} r^{1/2} \tag{30}$$

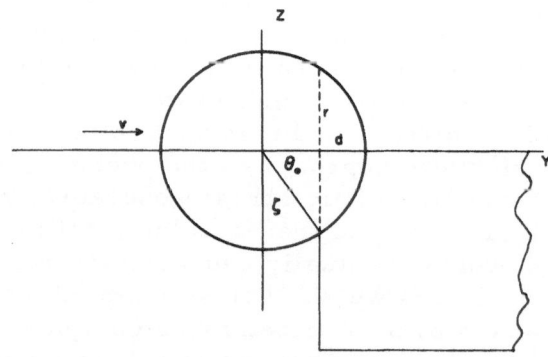

Fig. 6. Spherical tool at quadri-infinite medium.

The average pressure $\bar{p}$ and the energy W spent in making the indentation are given by

$$\bar{p} \sim \frac{4E\ e^{3/2}\ r^{1/2}}{3(1-\nu^2)\rho^2} \tag{31}$$

and with

$$e = Vt; \quad W = \int_o^t F(t)Vdt$$

$$W = \frac{8E\ r^{1/2}d^{5/2}}{15(1-\nu^2)} \tag{32}$$

To get some estimates of the orders of magnitude, Table I shows the values of $\bar{p}$ and W for the following typical values of the parameters.

$$E = 2.25 \times 10^{12}\ dynes/cm^2$$

$$\nu = 0.22$$

$$r = 4 \times 10^{-3}\ cm$$

$$\rho = 7 \times 10^{-3}\ cm$$

which corresponds to a 120° conical diamond tool set to cut a groove of 0.004 cm.

Table I

t(μs)	d(μm)	$\bar{p}$(kbar)	W(ergs)
0.1	1.0	2.5	7.5
1.0	10.0	80.7	2300
10.0	100.0	2542	5.6×10^6

This shows that within 10 μs of the contact the pressure at the contact area becomes of the order of Megabar and thus either the tool or the sample must fail. Also we note that in 10 μs the stress wave has traveled a distance of approximately 1 cm in the sample which can be considered infinite as compared to the indentation or groove depth.

The stress distribution at the contact area in the sample as described by Eq. (29) is purely compressive, and determines the time and extent of fracture.

Now we make an assumption that the stress away from the contact area falls off as $1/r$ instead of $1/r^2$ as one would expect from a purely static stress distribution. This assumption can also be taken to imply that the stress wave propagates as a two-dimensional wave (Rayleigh wave) which penetrates a very short distance perpendicular to its direction of propagation. With this assumption Eq. (29) can be rewritten in cartesian coordinates, as shown in Fig. 7 to describe the stress distribution in the strained part of the sample.

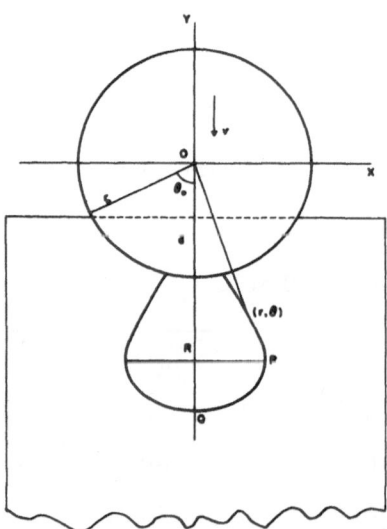

Fig. 7. Stress distribution in the sample.

$$p(x, y) = \frac{2Er_o}{\pi(1-\nu^2)} \left[\frac{\sin^2\theta_o}{x^2 + y^2} - \frac{x^2}{(x^2 + y^2)^2} \right]^{1/2} \tag{33}$$

This can be transformed to the following parametric form:

$$(x^2 + y^2) \; B(x^2 + y^2) - \sin^2\theta_o + x^2 = 0 \tag{34}$$

where

$$B = \frac{p(x, y) \; \pi(1-\nu^2)}{2Er_o} \tag{35}$$

The isobaric contours are given by B = constant in Eq. (34). A noteworthy feature of these contours is that they display a maximum width away from the contact area. If it is assumed that the fracture region is bounded by isobaric contours then the maximum width of the groove would be larger than the tool width.

It can be shown (Fig. 7) that

$$RP = \sin^2\theta_o/2B$$

$$OQ = \sin\theta_o/B$$

$$OR = \frac{\sin\theta_o}{B \; 2} \left[1 - 1/2 \sin^2\theta_o\right]^{1/2} \tag{36}$$

Now if we further assume that the compression is isentropic then one can calculate the rise in temperature through the equation of state[13,14]. Using values of pressure derived earlier a rise in temperature of 400°K - 1100°K is expected[15] for alumina of varying fracture strength. Of course the strength of polycrystalline materials depends very strongly on the porosity[16]. In our model both the grain size and the porosity will affect the stock-removal force

since both the pores and the intergranular boundaries can
serve as Griffith cracks. Thus the crack length distribution used in the one-dimensional model include both the
grain size and the pore size distribution. As the average
crack size increases, the pressure developed to propagate
them decreases and thus the temperature of the stressed
medium decreases, which could be verified experimentally.
Unfortunately temperature measurements in such processes
are not feasible to any degree of accuracy at present[17] and
the method described in Ref. 1 is a significant step towards
verifying the above conjectures.

REFERENCES

1. P.J. Gielisse, T.J. Kim and A. Choudry, Proc. Eighth
   Cer. Symp. Alfred Univ., N.Y., Aug 26-29, 1973.
2. B. Steverding, J. Am. Cer. Soc., 52 (1969) 133.
3. Timoshenko and Goodier, 'Theory of Elasticity, McGraw
   Hill, N.Y. 1956.
4. P.J. Gielisse, T.J. Kim and A. Choudry, Technical
   Report No. NOOO 19 71 C 0244 Naval Air Systems
   Command Dept. of Navy Wash. D.C. 20360.
5. A.A. Griffith Proc. Intern Congr on Appl. Mech.,
   Delft 1924.
6. W.F. Brace, J. Geophys. Res., 65 (1960), 1773.
7. J.C. Jaeger, Fracture Proc. first Tewksbury Symp.
   Univ. of Melbourne, August 1963, ed. C.J. Osborn.
8. D.C. Drucker, Fracture, An advanced Treatise, Vol.
   1, ed. H. Lieboritz Acad. Press. 1968.
9. H. Kolsky and D. Rader, loc. cit. 8
10. Boussinesg J., Compt Rend, 154, 1883.
11. Galin, L.A., Contact Problems, Ed. I.N. Sneddon,
    North Carolina State College, Raleigh, N.C. 1961.
12. H. Hertz, Collected Work, Vol VI 179 (1895).
13. Ahrens, T.J. Gust, and W.H. Royce; E.B., J. Appl.
    Phys. 39 (1968) 4610.
14. Delannoy, M., Lacam. H., Phys. Rev. 6 (1972) 3593.
15. Rice, M.H., McQueen, R.G. and Walsh, J.M., Solid
    State Physics, F. Seitz and D. Turnbull, Eds. (Academic
    Press, Inc., N.Y. 1958) Vol. 6, pp. 1-63.

16. Carniglia, S.C., J. Am. Cer. Soc. <u>55</u> (1972) 610.
17. Triemel, J. and Pahhorst, H.J., Industrial Diamond Review, March 1973, pp. 95-101.

DISCUSSION

<u>R.W. Rice (Naval Res. Lab.)</u>: I am glad to see you consider the pore size distribution. Our work indicates that the finer grain sizes control machining speeds in a manner opposite to tensile strength dependence on grain size as you showed.

<u>Author</u>: Actually from our model one derives a functional relationship between crack length, distribution, tensile strength, tool speed, etc., and it is shown that the cutting force goes through a minimum as the tool speed is increased from zero. Thus it is possible to operate in a region where a finer grain size distribution would require a smaller cutting force than a coarser one. Without looking into the numerical details of your data we can only assert in general that your observations of machining speed dependence on grain size distribution, even though contrary to what one would expect from tensile strength behavior, are compatible with our theoretical model.

<u>R.W. Rice</u>: Did you consider the change in the spectrum of flaws activated as a function of wheel speed following Steverding's work on impact?

<u>Author</u>: The theoretical model presented here is general enough to accommodate any analytical functional form of grain size distribution. It is also possible to further develop the model to include the impact activation of flaws. As an example, though we have not done it, but it is quite within the scope of the model to incorporate features like the growth of Hertzian cracks under impact as studied by Evans recently.

<u>R.W. Rice</u>: You have apparently assumed that the tensile strength of diamond is greater than the compressive strength of the diamond. Do you mean this as a real or an apparent (i.e., considering mounting etc.) effect? The former would be contrary to the expected behavior.

<u>Author</u>: We have really not considered it as an assumption but as a requirement. Even in the very simple one dimen-

sional impact studied by Steverding. It is clear in the absence of internal friction, plastic flow and other dissipative mechanisms the driver (tool) will be subjected to a cycle of equal compressive and relaxation tensile stresses. The integrity of the tool as a whole cannot be maintained unless the entire system, the diamond point, mounting, etc., can withstand the tensile relaxation. The failure (machining) of the material of course occurs during the compressive stress cycle. Here we must emphasize that in order to machine, the dynamic compressive stress produced by the driver does not have to reach the static compressive strength of the material being machined. Thus what one requires is that the tool assembly remain integral under the relaxation tensile stress corresponding to the dynamic compressive strength of the material, which depends on many factors, e.g., wheel speed, depth of cut, etc., as discussed in the text.

R. W. Rice: Did you correlate machining forces with compressives or tensile strain energy in the material? We have observed some evidence indicating better correlation with compressive strain.

Author: This is a very important point and deserves a thorough investigation. Since the work that we have presented here is mostly theoretical in nature we have avoided to include experimental details except in one case to show an experimental comparison of the machining force vs. tool speed to assure ourselves of the validity of the theoretical model. I am very glad to learn that you have found a better correlation with compressive strain rather than the tensile strain, since this is just what we expect from our model, i.e., the material failure occurs during compressive stress and spalling during tensile relaxation. If your experimental results are firm we can take them as a confirmation of our theoretical model.

CERAMICS IN ABRASIVE PROCESSES

E. Dow Whitney and Robert E. Shepler

Department of Materials Science and Engineering
University of Florida
Gainesville, Florida

The grinding wheel is a productive and versatile tool.
It may be used in fine machining to produce a high degree
of geometrical accuracy and surface finish or to provide
an effective means for heavy stock removal in snagging
operations[1]. Although grinding is the most common means
of metal removal, still much remains to be learned before
a fundamental understanding can be claimed. Research
has classically dealt with the relations between independent
wheel and machine parameters and the dependent variables,
surface finish, metal removal rate and wheel wear. The
desire for improved productivity, particularly in the fabrication of refractory hard alloys, has resulted in the need
for a better understanding of friction and wear in metal-
ceramic systems. It is now realized that the action of an
abrasive in a metal removal operation is a combination of
a mechanical, or strain-energy effect, and a chemical effect. Both contribute to the overall performance. The
abrasive grain should possess a combination of properties
such that it will not undergo excessively rapid breakdown
during grinding. Those physical, chemical and mechanical
properties of ceramics which influence adhesion, and in
turn friction and wear have already been reviewed[2], thus
providing an excellent basis for the present discussion.

Work supported by the National Science Foundation

167

In general, two types of wear are encountered in
abrasive action, i.e., adhesive wear and abrasive wear.
Adhesive wear, essentially chemical, occurs when bonding takes place across the ceramic-metal interface.
Tangential motion after bonding will result in fracture
through the material with the lowest strength. Mechanical wear occurs when two surfaces, one considerably
harder than the other, are brought into contact. The
harder material can cut into and remove material from
the softer surface.

Coes[3] distinguished between "abrasive quality" and
"durability" as separate intrinsic properties of abrasive
materials. Abrasive quality is related to melting point
and chemical reactivity, whereas durability is related to
hardness and impact strength. One may consider improv-
ing abrasives either by counteracting adhesive wear
through alteration of "abrasive quality," or by increasing
resistance to abrasive wear through improvement of "durability".

The number of very hard abrasive materials is sur-
prisingly limited. All known hard materials have melting
points of at least 1600° C. This requirement can be satisfied only by about 1.3% of all known inorganic compounds,
and these contain no more than two different elements. Of
the 3,160 binary compounds theoretically possible as determined by Berezhnoi[4], only four are of major interest in
the abrasives industry; i.e., aluminum oxide, silicon carbide, boron carbide and cubic boron nitride.

In considering durability, one is primarily concerned
with impact strength. The success of sintered alumina
abrasives, where the ultimate crystal size is of the order
of 5-20 microns, in heavy-duty snagging operations is
indicative of the role played by microstructure in impact
strength. A unique microstructure is afforded by the
alumina-zirconia system. Alumina-rich fused mixtures
of Al_2O_3-ZrO_2 when cooled rapidly result in the formation
of alumina crystals in a bonding matrix composed of ZrO_2

dendrites in a corundum mass. The matrix may exhibit a very fine texture, the dendrites approaching 1 micron in size. This results in a very tough abrasive grain whose durability exceeds that of sintered alumina.

Modern abrasives technology may be considered as beginning in 1901 with the development of silicon carbide and fused aluminum oxide[5]. Prior to this time natural abrasives such as flint, emery, garnet and diamond. served the purpose. Classically, the prime requirement of an abrasive was that it simply be harder than the material it was to abrade.

The familiar Mohs hardness scale is based upon the scratch hardness of materials. In Fig. 1 is shown the relation between Mohs hardness and Knoop indentation hardness values. A correlation is to be expected since both scratching and static indentation are determined primarily by the plastic properties[6]. There are two major difficulties with the Mohs scale. First, the scale is not linear. Corundum, with a Mohs hardness (H) of 9 and a Knoop indentation hardness (K) of approximately 2000 is not three times as hard as calcite, with H = 3 and K = 135. The second problem with Mohs's scale is that it is compressed at high hardness values so that the

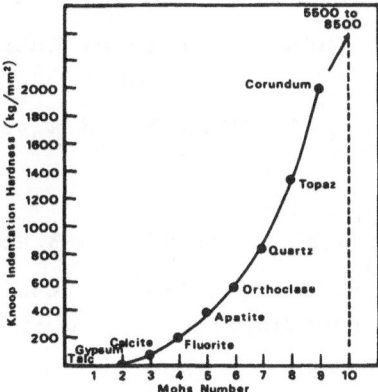

Fig. 1. Relation between Mohs hardness and indentation hardness.

very hard abrasives fall between corundum (H = 9) and diamond (H = 10).

Since 1901 only three additional materials of high hardness have been introduced into the abrasives field. These are boron carbide, introduced in 1934[7], manufactured diamond in 1955[8], and cubic boron nitride in 1957[9]. Today synthetic materials dominate the market. The reason is the reproducibility of manufactured abrasives, a quality demanded for systems employing automated high productivity machines.

Table I lists physical properties of the five hardest synthetic abrasives in use today. Fig. 2 is a combined plot of log K against H for natural abrasives up to corundum, after which log K of synthetic abrasives is plotted against Wooddel abrasion hardness[10] (W). It may be

Table I. Physical Properties of Synthetic Abrasives

	Aluminum Oxide	Silicon Carbide	Boron Carbide	Cubic Boron Nitride	Diamond
Formula	Al_2O_3	SiC	$B_{12}C_3$	BN	C
Structure Type	α-Corundum	Zinc-blende	Boron Carbide	Zinc-blende	Diamond
Crystal Structure	Hexagonal	Cubic	Rhombo-hedral	Cubic	Cubic
	a_o=4.758 c_o=12.99	a_o=4.349	a_o=5.598 c_o=12.12	a_o=3.615	a_o=3.567
Density (g/cm^3)	3.98	3.22	2.52	3.48	3.52
M.P.	2040	∿2700 (decomp.)	2430	∿3200 @ 105 kbar	∿3700 @ 130 kbar
K_{100}	2100	2480	2800	∿4500	8500
W	9.0	14.0	19.7	∿26	42.5

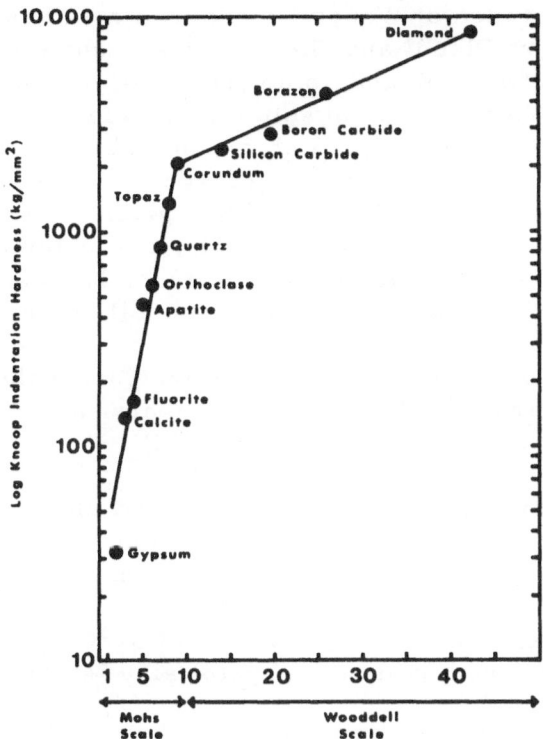

Fig. 2. Log Knoop hardness vs. Mohs-Wooddell abrasion hardness.

seen that up to corundum the relation $\log K = nH$ is obeyed[6]. The n corresponds to a ratio of hardness between each Mohs standard of about 1.6. From corundum to diamond $\log K = n'W$ where n' is approximately 1.4. Each increment on the W scale corresponds roughly to a 40 percent increase in abrasion hardness. Tabor[6] argued that the linear relation between $\log K$ and H suggests that Mohs experimented with a variety of materials until he obtained "equality of intervals." The linear relation between $\log K$ and W is based upon experiment. One is tempted to speculate that the simple relation $\log K = n'W$ may somehow reflect an intrinsic property of materials harder than corundum.

The importance of Woodell measurements has been demonstrated by Plendl and Gielisse[12] who showed a linear relationship between W and volumetric lattice energy (E_c/V). Hardness thus becomes an absolute magnitude with the dimensions (kcal/cm^3). Fig. 3 shows the relation between W and volumetric lattice energy for synthetic abrasives harder than corundum. DeVries[11] has given a corrected value for E_c/V for cubic boron nitride, and using the hardness vs. E_c/V plot of Plendl and Gielisse, has established a value of ᴧ26 for cubic BN on the Wooddell scale.

The work of Plendl and Gielisse provides a theoretical basis for preparation of new abrasive materials of both improved "abrasive quality" and "durability".

The improved performance of sintered alumina abrasives over conventional fused alumina in heavy duty stock removal dramatically illustrates the influence of microstructure. Sintered alumina is a very tough, fine-grained material with an average crystal size of 5-20 microns whereas in the fused oxide the crystal size is 400-1200

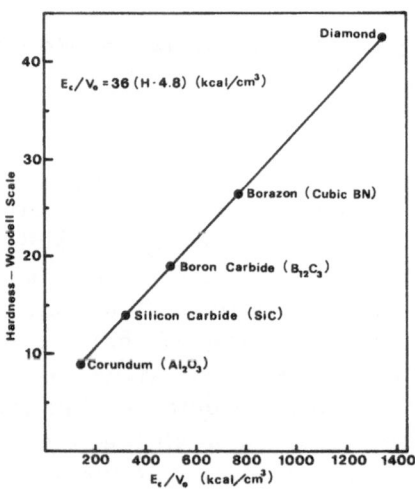

Fig. 3. Wooddell abrasion hardness (W) vs. volumetric lattice energy (E_c/V) for synthetic abrasives harder than corundum.

microns depending on the rate of cooling. A relatively new abrasive which has shown good performance in snagging operations is alumina-zirconia. The phase diagram for the system appears to be a simple binary eutectic with a eutectic at 1870°C and 42 wt. percent ZrO_2[13]. The compositions of interest for abrasives seem to be those toward the alumina side. Once the structure is characterized and basic properties known, correlation with grinding performance should optimize this abrasive.

EXPERIMENTAL

A copper hearth arc furnace was used to melt and quench Al_2O_3-ZrO_2 compositions in an argon atmosphere. Some materials were also melted in molybdenum crucibles in a tantalum resistance furnace at up to 2100°C. Microhardness measurements were made on ground and polished specimens with a Kentron Microhardness Tester using a Vickers indenter with loads of 100 and 1000 grams, and with a Knoop indenter and 100-gram loads.

Counting measurements were made on the precipated primary alumina crystals (α phase) in the matrix (β phase) for samples on the alumina side of the eutectic. A 25-point grid at scanning electron microscope magnifications from 2000X to 5000X was used for the point fraction and line intercept counts. Area tangent counts were made with a five-square grid using the raster lines as the tangent forming lines.

RESULTS AND DISCUSSION

In the discussion below the following quantities will be used:

V^α = volume fraction of the α phase (primary alumina).

$S^{\alpha\beta}$ = total surface-to-volume ratio of the interface between the α and β phases.

$M^{\alpha\beta}$ = total surface curvature of the $\alpha\beta$ interface-to-volume ratio. Total surface curvature is related to the principal normal curvatures integrated over the $\alpha\beta$ interface.

The volume fraction of primary alumina is shown in Fig. 4 as a function of composition. The fact that this plot does not result in a straight line connecting $V^{\alpha} = 1$ at 0 wt. % ZrO_2 and $V^{\alpha} = 0$ at 42 wt. % ZrO_2 means the composition of the matrix can vary over some range and still have the same basic structure. Fig. 5 shows the total surface area of the alpha-beta interface while Fig. 6 gives

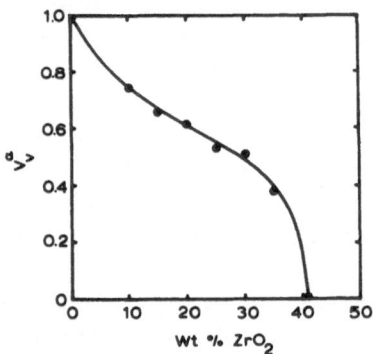

Fig. 4. Volume fraction of primary alumina in supporting matrix.

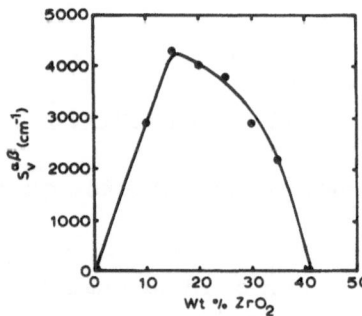

Fig. 5. Total area of primary alumina and matrix interface.

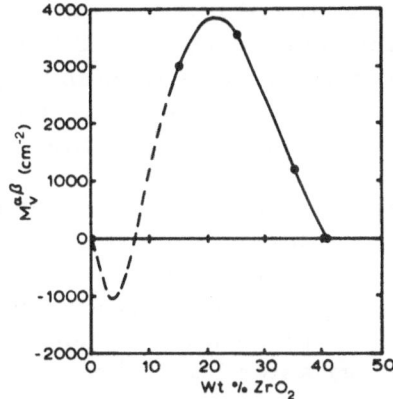

Fig. 6. Total surface curvature of primary alumina and matrix interface.

the total surface curvature as functions of composition[7]. The microstructure as described by Figs. 4-6, is not a function of composition alone but of many variables. Care was taken to assure that the data used for these three plots came from samples which differed only in composition. Each curve is but one of a large envelope of such curves. Fig. 7 is a SEM micrograph of a typical quenched Al_2O_3-ZrO_2 structure. The α phase is the dark region and the β phase is the lighter region.

Fig. 7. Scanning electron micrograph of a typical quenched arc-cast Al_2O_3-ZrO_2.

Microhardness results with 1000-gram loads are shown in Fig. 8. Each point represents about 20 tests. There appears to be almost a linear relationship between hardness and composition from pure alumina to zirconia. Knoop indenter measurements showed the same linear relation.

It is evident that microhardness is not the primary variable controlling suitability as an abrasive. If this were the case, pure alumina would be better than any alumina-zirconia composition. However, materials with high hardness generally make better abrasives than materials with low hardness, other variables being the same.

The areas labeled A, B, C and D in Fig. 9 denote the extremes in structural possibilities. Systems A can be

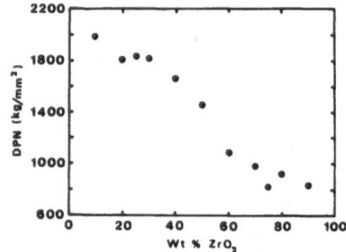

Fig. 8.   Diamond Pyramid Numbers for arc-cast Al_2O_3-ZrO_2.

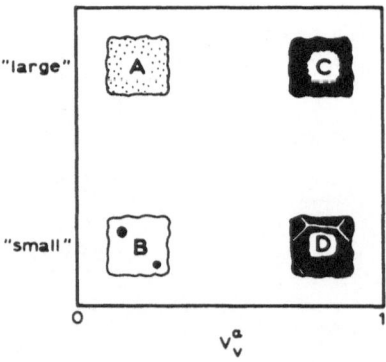

Fig. 9.   Structure extremes from V^α and $S^{\alpha\beta}$ measurements on a possible system.

represented by small spheres of the alpha phase in a matrix or of thin alpha platelets in a matrix. The alpha phase may or may not be multiply connected. The connectivity of a system is not measured in quantitative microscopy but is a topological property which, although important, is beyond the present discussion. Area B can be represented by a few large alpha spheres or thin rods in a matrix. Area C corresponds to a dense packing of small alpha particles with very complex surfaces in a matrix, or as a large number of small beta particles in an alpha matrix. Area D can be represented by large particles of alpha densely packed in a beta matrix or a few beta particles in an alpha matrix. Many other possibilities can be imagined for each of these areas. Thus, any point in the $S^{\alpha\beta}$ versus V^{α} plot will not describe a unique structure but will quantitatively describe some structure which may be modeled using other information obtained on the system.

REFERENCES

1. H. F. G. Ueltz, "New Developments in Abrasive Grains," Proceedings of Conference on Ultra Hard Tool Materials, M. C. Shaw and J. N. Brecker (Editors), Carnegie-Mellon University, Pittsburgh, Penn., May 1970.

2. D. H. Buckley, "Adhesion, Friction and Wear of Ceramics," This Volume.

3. L. Coes, Jr., Abrasives, Springer-Verlag, New York, New York (1971), p. 65.

4. A. S. Berezhnoi, "Advances in the Study of Physicochemical Systems - The Basis of Refractory Technology," Ogneupory, 3-4, 229-32 (1970).

5. L. Coes, Jr., Ref. 3, Chapt. 1.

6. D. Tabor, "Mohs' Hardness Scale-A Physical Interpretation," Proc. Phys. Soc. 67, 249-257 (1954).

7. R. R. Ridgeway, "Boron Carbide, A New Crystalline Abrasive and Wear Resisting Product," Trans. Electrochem. Soc. 66, 117-33 (1934).

8. F. P. Bundy, H. T. Hall, H. M. Strong and R. H. Wentorf, Jr., "Man Made Diamonds," Nature 176, 51 (1955).

9.   R. H. Wentorf, Jr., "Cubic Boron Nitride," J. Chem. Phys. 26, 956 (1957).

10.   C. E. Wooddell, "Method of Comparing the Hardness of Electric Furnace Products and Natural Abrasives," Trans. Electrochem. Soc. 68, 111-30 (1935).

11.   R. C. DeVries, Cubic Boron Nitride: Handbook of Properties, Report No. 72CRD178, General Electric Corporate Res. and Dev., June 1972.

12.   J. N. Plendl and P. G. Gielisse, "Hardness of Non-metallic Solids on an Atomic Basis," Phys. Rev. 125, (3) 828-832 (1962).

13.   A. M. Alper, Phase Diagrams: Materials Science and Technology, Academic Press, New York, New York (1970), p. 129.

14.   R. T. DeHoff and F. N. Rhines, Quantitative Microscopy, McGraw-Hill Book Co., New York, New York (1968).

ROLLING CONTACT FATIGUE OF HOT-PRESSED SILICON NITRIDE VERSUS SURFACE PREPARATION TECHNIQUES

H. R. Baumgartner and W. M. Wheildon

Norton Company
Worcester, Massachusetts

Hot-pressed silicon nitride is receiving interest as a potential antifriction bearing material[1]. As a result of advanced material preparation and processing techniques, bend strengths in excess of 100,000 psi are routinely achieved. In addition to its high strength, dimensional stability and corrosion resistance, silicon nitride has a density of 3.2 g/cm^3, sixty percent lower than that of steel. This light weight is especially important for high speed bearings, where the maximum stresses occur in the outer race, primarily as a result of the centrifugal force loading from the rolling elements. Inasmuch as strength tends to be sensitive to surface integrity, finishing procedures are an important aspect for bearing applications. This paper will report experiments on their effect.

EXPERIMENTAL PROCEDURE

Finishing

The material was hot-pressed from silicon nitride powder as described elsewhere[2]. The finishes were produced on 3/8 inch diameter rods by the procedures listed in Table I. All grinding was done wet with a water-

179

based fluid. Silicon carbide wheel finishing was investigated because this method represented a potential means
of producing crowned roller geometries economically.

Table I. Experimental Finishing Procedures.

Finish #1   Initial rough grinding was with a 100-grit
                 diamond wheel. Final cylindrical grinding
                 was with a 320-grit diamond wheel. Wheel
                 speed, 5500 sfpm; work speed, 600 rpm; tra-
                 verse speed, 0.001"/rev; infeed, 0.001"/pass,
                 decreasing to spark-out. The rms surface
                 roughness was about 8 microinches.

Finish #2   Rods were initially prepared as in Finish #1.
                 A hand lapping operation with a leather strap,
                 impregnated with 6-micron diamond dust, was
                 done with the rods being rotated at 200 sfpm.
                 While stock removal was minimal, the surface
                 roughness was improved to about 3-4 rms
                 microinches.

Finish #3   Rods were initially rough ground with 100-
                 grit diamond wheel. Final finish was by
                 mechanically lapping the rods between two
                 cast iron lap plates charged with 6-8 micron
                 diamond paste. Final surface roughness was
                 2-3 rms microinches.

Finish #4, 5   Rods were initially prepared as in #2 or #3
                 respectively and then reground lightly with a
                 silicon carbide wheel which removed less than
                 0.001 inch from the rod diameter.

Rolling Contact Test Procedure

    To evaluate performance, a rolling contact test
machine[3], marketed by Polymet Corporation, was utilized.

Fig. 1.   Rolling-contact test machine.

The device, shown in Fig. 1, permits rapid evaluation of
bearing materials under nearly pure rolling-contact conditions. A rod specimen is placed on a spindle between
two seven-inch diameter crowned steel wheels.   Loading
of the specimen is accomplished by drawing together, with
a turnbuckle, the two cantilevered loading discs. By
varying the applied load, as measured by a force transducer, the Hertzian contact stress on the specimen may be
varied.   The sample rod is driven at 10,000 rpm and re-
ceives two stress cycles per revolution, resulting in 1.2 X
10^6 stress cycles per hour. The sample is drip-lubricated
with a Type II turbo oil (MIL-L-23699) during test. A
vibration-sensitive transducer monitors the vibration of
the test rig and automatically causes the spindle motor to
be shut down at a preset vibration level; termination defines the sample life, expressed in number of stress cycles.   Shut-downs may be caused by failure in either the
sample or the loading discs. Suspended tests occur when
the loading discs fail or the test is stopped after an arbitrary number of stress cycles. (Failure of the loading
discs required that they be refinished.) In order for a
specimen failure to be considered a valid life test, the
loading wheels were requalified by being run against a

standard steel (M-50, CVM) specimen. The silicon nitride test was considered valid only if the loading discs did not fail in the requalification test and the steel standard failed within a previously established confidence band.

After testing, the specimen surfaces were examined with a low-power optical microscope and with a scanning electron microscope.

EXPERIMENTAL RESULTS AND OBSERVATIONS

Effect of Finishing Procedures

The effects of rod finishing procedures upon fatigue life are summarized in Table II. All tests were conducted at a Hertzian stress level of 800,000 psi (calculated for unlubricated conditions).

Table II. Rolling Contact Life vs Finishing.

Finishing Variant	Number of Rods Tested	Number of Tests	Life (Millions of cycles)	
			Range	Median
#1	1	2	34-55	44
#2	4	25	1 2-80*	31
#3	8	23	0.040-34	0.065
#4 Before SiC refinishing				
	2	8	1.8-28	4.5
After SiC refinishing				
	2	9	0.42-41**	25
#5 Before SiC refinishing				
	2	8	0.04-29	2.2
After SiC refinishing				
	2	4	0.082-0.20	0.14

*   Includes 11 tests suspended after 9.4×10^6 cycles
** Includes 4 tests suspended after 25×10^6 cycles

Of the finishing procedures investigated, Finish #1 produced the longest median life. However, this life is based upon only two tests. Of greater statistical significance is the median life of 31 million stress cycles obtained on #2 specimens. The life of this series is to be contrasted with the median life of 650,000 stress cycles obtained on Finish #3 rods. The lives in the #3 series were highly variable, with many values clustered at the lower end of the range. There were no test suspensions in the #3 series, as all tests ended in true failures.

In order to evaluate the effect of the silicon carbide wheel regrind on fatigue life, the same rods were tested before and after regrinding so as to minimize possible life variations among specimens. Two types of initial finishes were used. The median life of the #2 rods was increased while that of the #3 was decreased as a result of the regrind.

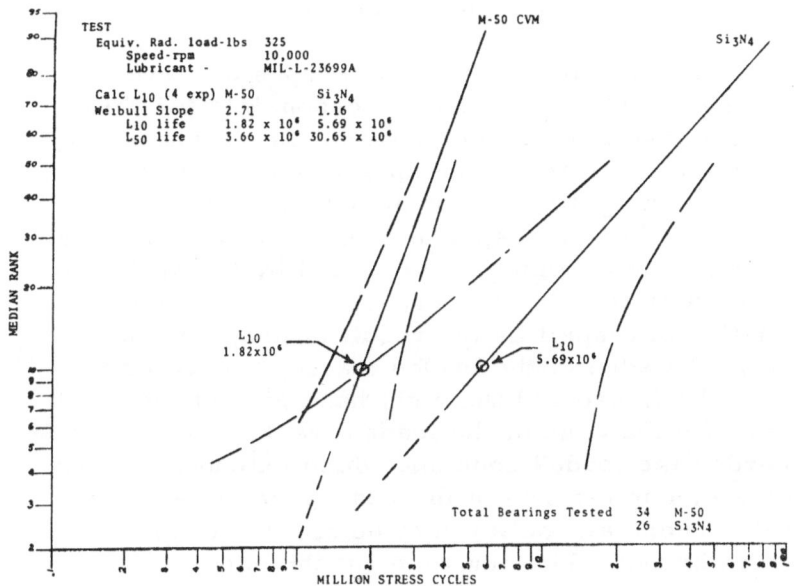

Fig. 2. Weibull Plot of fatigue life.

A Weibull plot of fatigue lives of the #2 silicon nitride samples, tested at a calculated Hertz stress level of 800,000 psi, is shown in Fig. 2. For strictly an orientation comparison, data for M-50 CVM steel are also shown, tested at the same load of 325 pounds. Because of the difference in elastic moduli, the calculated Hertzian stress for the steel specimens was 700,000 psi. The plot shows the silicon nitride to have an L_{10} fatigue life, the number of cycles after which ten percent of the components will fail, of approximately three times that of the bearing steel.

Microscopy of Surfaces

Unlike other ceramics, silicon nitride fails in a rolling contact test by the formation of spalls, that is, by the separation of small volumes of material from the surface. It is spalling which produces the vibration levels which cause automatic test termination. The spalling failure mode is also found in bearing steels.

Fig. 3-6 illustrate various spall types observed on silicon nitride. The mechanisms responsible for three of the spalls have tentatively been assigned. These are based upon observations of the spalls themselves and a comparison with spalling mechanisms known to occur in metal bearings.[4,5] The spall in Fig. 4 initiated from a subsurface inclusion which was exposed by the spall and was located with the electron probe. Fig. 5 illustrates the formation of a spall as the result of propagation of a crack from the edge of the loaded track. The track is vertical in the figure and the right hand side of the spall coincides with the edge of the loaded track. Cracks, which border the loaded zone and which extend partially around the circumference of the rod, were sometimes observed. These are believed to be similar to cracks which form in metal bearing races at the edges of the load application as a result of high geometric stress concentration.[5] This type of cracking does not usually pre-

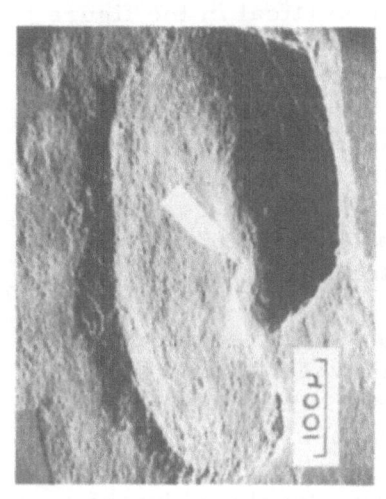

Fig. 4. Inclusion-initiated fatigue spall

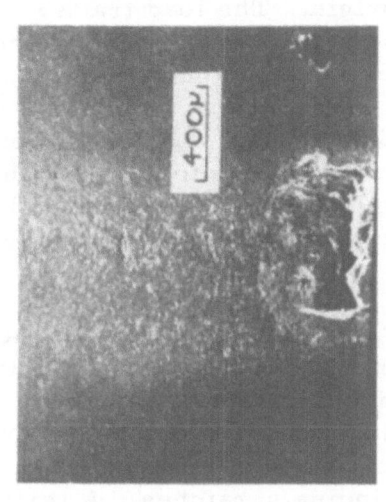

Fig. 6. Porosity-induced fatigue spall

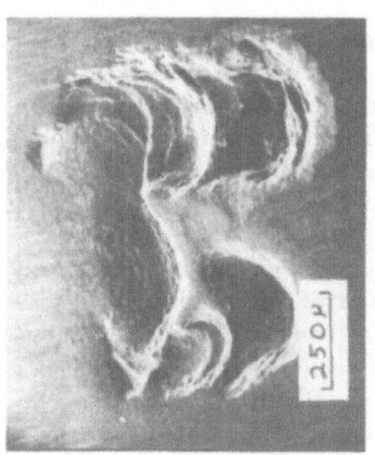

Fig. 3. Fatigue spall

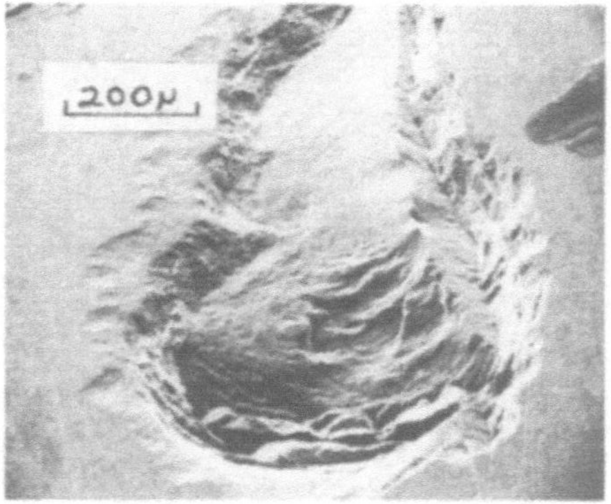

Fig. 5. Crack-initiated fatigue spall

cipitate test termination. However, in Fig. 5 a spall has
formed by the sideways propagation from a circumferential crack. Fig. 6 shows a spall which formed from a surface origin. The load track is again vertical in the figure
and the approximate width of the track is delineated by a
greater surface roughness which resulted from pit formation. This rod was anomalous in that it possessed high
porosity; density was 3.09 g/cm^3 as compared to the normal range of 3.16 to 3.20 for the other rods. As a result,
strength was reduced, permitting pits to form by crushing.
These pits nucleated the fatigue spall, an incipient example of which is seen just above center in Fig. 6.

Fig. 7 shows the as-finished and tested surfaces of a
rod finished with a 320-grit diamond wheel (Finish #1).
The circumferential grinding scratches may be seen running diagonally across both photographs. After rolling
contact, wear in the load track has partially obliterated
the grinding scratches. A texture has developed by the
formation of elongated depressions perpendicular to the
grinding scratches (and to the rolling direction). A concentration of iron was detected in the track, indicating
asperity contact between the rod and the loading wheel.

Fig. 8 shows as-finished and tested surfaces prepared
by Finish #2. Residual grinding scratches were sometimes visible on the as-finished surfaces. Many very
small pits are present. The appearance of the loaded
track is quite similar to the as-finished. The size of the
pits was enlarged little, if at all, during testing.

Fig. 9 shows as-finished and loaded surfaces prepared
by Finish #3. Residual grinding scratches were no longer
evident on the as-finished surfaces. Instead, randomly
oriented short lapping scratches were seen. The background roughness in Fig. 9A appears slightly less than
that in Fig. 8A. Fig. 9B shows a distinctive and typical
feature of loaded tracks finished without an intermediate
grind, i.e., the development of relatively large pits.

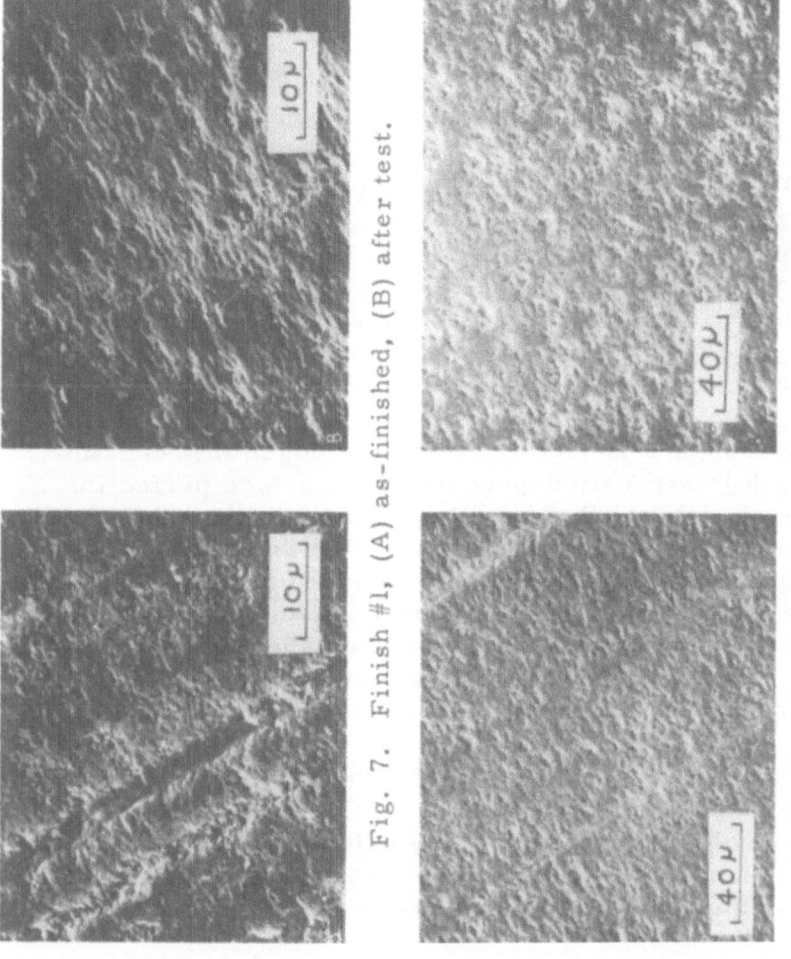

Fig. 7. Finish #1, (A) as-finished, (B) after test.

Fig. 8. Finish #2, (A) as-finished, (B) after test.

Fig. 10 shows areas of a #4 rod. Grinding with a
silicon carbide wheel produced a burnished surface having
relatively smooth areas, spearated by regions containing
pits. Upon being loaded, a thin surface skin has been non-
uniformly removed, leaving islands of smooth regions
against a background of fine-scale surface roughness.

DISCUSSION

The origins of fatigue spalls in metal bearings have
been found to be subsurface or surface.[4,5] A stress
concentrator, for example a grinding scratch or an inclusion, was invariably found at the origin. Since stress
concentrations adversely affect the strength of materials,
a strong correlation may be expected to exist between
material strength and resistance to fatigue spalling.

Although it is generally acknowledged that ceramic
strength is strongly dependent upon surface perfection,
relatively few studies have been published[6] on the effect
of finishing procedures. One study on alumina[8] found the
strength of metallographically polished samples to be similar to that of rough-ground surfaces, and microscopic
examination of the polished surfaces showed remnants of
prior damage. Such persistence of damage is a possible
explanation for the strength-roughness insensitivity.

Surface roughness measured by a profilometer does
not correlate well with strength.[7,8] The stylus tip is
large in comparison with many defects; in addition, roughness is expressed in some averaged form, while it takes
only a single defect to cause fracture.

In the present study, three visually distinct surfaces
were generated by four finishing procedures. The rolling
contact fatigue life was found to be dependent upon the
type of finishing operation employed. The most striking
difference in fatigue performance was found between
Finishes #2 and #3, both of similar roughness and visual

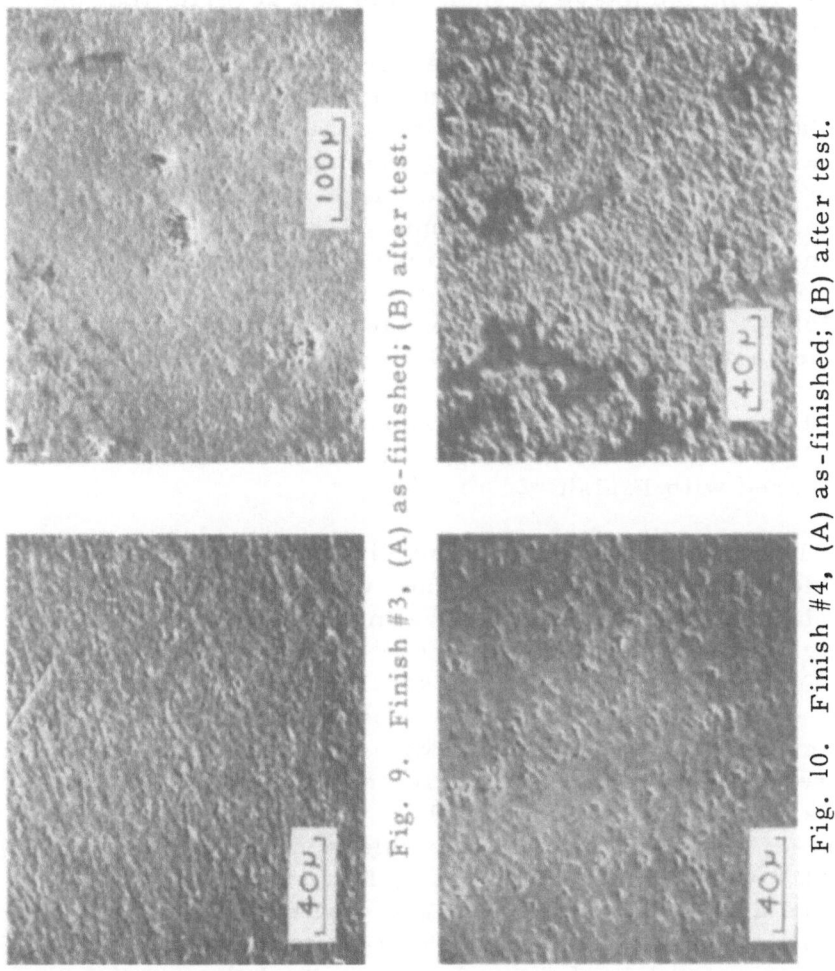

Fig. 9. Finish #3, (A) as-finished; (B) after test.

Fig. 10. Finish #4, (A) as-finished; (B) after test.

appearance. The loaded tracks were different in that
relatively large pits had developed in the tracks of #3
rods. The difference in behavior is explained as follows.
Rough grinding silicon nitride with a 100-grit wheel introduces surface damage of such depth and severity that the
damage is not removed by the standard lapping procedure
used in Finish #3. When these surfaces are subjected to
the high stresses associated with the rolling contact test,
the damage sites act as nuclei for spall cracking. The
large pits which develop during loading are taken to be
sites of the severest grinding damage. Removal of the
100-grit damage by a subsequent 320-grit grind (Finish #2)
removes the large nuclei required for early spalling.
The primary failure mechanism for rods processed by
Finish #2 is not known; both surface origins and inclusion
origins are possibilities.

From the limited results reported, the 320-grit diamond finish (#1) does not adversely affect fatigue life when
compared with Finish #2.

The effect of refinishing specimens with a silicon
carbide wheel apparently depends upon the quality of the
finish prior to refinishing. The median life of rods initially prepared with Finish #2 was increased, while that
for Finish #3 rods was decreased. The burnished appearance of the reground surfaces implies higher grinding
forces and temperatures than with diamond grinding. The
decrease in life of the one series may be explained on the
basis of the severity of the grind, which has caused pre-
existent grinding damage to be enlarged and propagated
into the workpiece. (A similar effect, wherein grinding
cracks are continuously propagated with stock removal,
has been observed in the grinding of hardened steels.)
The increased life of the rods having the better initial
finish is somewhat surprising, since bend tests showed
that even small smounts of silicon nitride stock removal
with a silicon carbide wheel from a 320-grit diamond
finish can result in slight strength reduction. A possible
explanation may be that the burnishing action has produced

a work-hardened or compressive surface layer of incomplete coverage, from which it is more difficult to initiate
a nucleus for a crack. If so, the flaking of the thin skin
from the reground rods during loading could be a manifestation of induced residual surface stresses.

Aided by the finishing studies described in this paper,
a 50-mm bore aircraft-type of roller bearing was designed
and prototypes fabricated. The bearings consisted of
silicon mitride rollers with steel races and cage. Final
shaping of the roller crowns was accomplished by plunge
grinding with a silicon carbide wheel. Of two bearings
tested to date, the first was run for 75 hours at 10,000
rpm under a 2500 pound load (equivalent to a 310,000 psi
Hertz stress) before the test was suspended. The second
bearing is now undergoing test and has already survived
600 hours at the same conditions, a record for ceramic
roller bearings. For a similar all-steel bearing, the
calculated life at which ten percent of the steel bearings
would fail under identical conditions, is 80 hours.

CONCLUSIONS

Hot-pressed silicon nitride is a highly promising
bearing material. It fails by spalling during simulated
tests in a manner similar to bearing steels. Pores, inclusions and residual grinding damage nucleate fatigue
cracks. Their reduction should result in increased
performance. The surface condition of silicon nitride is
critical to rolling contact fatigue life. In surface preparation, damage incurred during rough grinding operations
must be removed by subsequent finishing.

ACKNOWLEDGEMENTS

The authors thank Mr. David V. Sundberg of the
Federal-Mogul Corporation for helpful discussions and
his supervision and performance of the simulated bearing

tests. Special thanks are due Mr. Charles Bersch of the Naval Air Systems Command, Technical Monitor of Contract N00019-72-C-0299, for his helpful discussions.

REFERENCES

1.   Wheildon, W. M. , Baumgartner, H. R. , Sundberg, D. V. and Torti, M. L. , "Ceramic Materials in Rolling Contact Bearings", Final Report on Contract N00019-72-C-0299, 1973.

2.   Alliegro, R. A. , Richerson, D. W. , Torti, M. L. , Washburn, M. E. , and Weaver, G. Q. , "Silicon Nitride and Silicon Carbide for High Temperature Engineering Applications", to be published in Proceedings of the British Ceramic Society Convention, July 1972.

3.   Bamberger, E. N. and Baughman, R. A. , "Development of Fatigue Data for Bearing Materials, " General Electric Report No. R57AGT712, FPLD, November, 1957.

4.   Martin, J. A. and Eberhardt, A. D. , "Identification of Potential Failure Nuclei in Rolling Contact Fatigue", Trans. ASME, Series D, $\underline{89}$, 932 (1967).

5.   Littmann, W. E. and Widner, R. L. , "Propogation of Contact Fatigue from Surface and Subsurface Origins", Trans. ASME, Series D, $\underline{88}$, 624 (1966).

6.   Schneider, Jr. , S. J. and Rice, R. W. eds. , The Science of Ceramic Machining and Surface Finishing, U. S. Nat. Bur. Stds. Spec. Pub. 348, (1972).

7.   Sedlacek, R. and Jorgensen, P. J. , "Processing of Ceramics - Surface Finishing Studies", Final Technical Report of Contract No. N00019-70-C-0179, September 1971.

8.   Pears, C. D. , Starret, H. S. , Bickelhaupt, R. E. and Braswell, D. W. , "A Quantitative Evaluation of Test Methods for Brittle Materials", Technical Report AFML-TR-69-244, Part I, March 1970.

DISCUSSION

P.J. Gielisse (Univ. of Rhode Island):   The relatively
poor surfaces prepared by your method present themselves
in sliding contact (wear mode).   If the surfaces were prepared in the frictional mode, the already proven promise
of Si_3N_4 in these applications could be even more improved.
A step on the road toward this type of surface (not just
roughness) has already been taken in the case where you
grind with SiC.

Author: The present work was concerned with the response
of silicon nitride under cyclical application of high compressive stresses in almost pure rolling contact.   Under
these conditions, failures occur as a result of spall formation, not accumulated wear.   Under sliding conditions
wear may be an important consideration.   It is conceivable
that wear could be of greater relative importance in the
failure of actual bearings (as opposed to the rolling contact experiments), where the stresses are lower and there
are greater surface traction forces.

It was not the intent of the authors to imply that the
surface preparation methods used in this study are the
best possible.   We agree that surface preparation influences bearing behavior (both sliding and rolling) and that
finishing procedures can be optimized.

STRUCTURE AND PROPERTIES OF SOLID SURFACES

Harry C. Gatos

Department of Metallurgy and Materials
Science, Massachusetts Institute of Technology,
Cambridge, Massachusetts

Solid surfaces are giant lattice defects, generated through the abrupt termination of the crystalline lattice. Their structure and properties are a manifestation of their defect nature and not necessarily of the intrinsic characteristics of the solid. Although, in terms of number of atoms, surfaces represent a very small fraction of macroscopic solids (of the order of 10^{-8}) they are more often than not crucial in the characterization, understanding, and overall behavior. Mechanical, electrical, optical or other types of communications with a solid are realized through the surface. Furthermore, all chemical and physical interactions of solids with gases, liquids or other solids take place through the surface atoms. Indeed, surface phenomena bear directly on all facets of basic and applied technology; in advanced electronics, in friction and wear, in catalysis, in the chemical stability or instability of solid systems, on chemical processing, in environmental (pollution) processes, and others; life sciences rest to a large extent on surface phenomena. Like other lattice defects, surfaces often play a very important role in the mechanical behavior of solids, and many types of mechanical failures are initiated at the surfaces.

It is the present purpose to present an overview of the
physical and chemical behavior of the surfaces of crystalline solids. The basic difficulties associated with experimental study, theoretical understanding and control of
solid surfaces will be underscored. The intent is to high-
light certain aspects of the structure and properties of
solid surfaces. For rigorous treatments, literature references will be provided.

The present discussion will be confined to "free"
crystalline surfaces; solid-solid interfaces will not be
included although they are of paramount importance in
friction and wear, sintering, and other fields of science
and technology; similarly metal-oxide and semiconductor-
oxide interfaces, which have become recently a major
factor in solid state science and engineering, will not
be discussed.

GENERAL REMARKS

The experimental and theoretical study and understanding of solid surfaces had not advanced beyond an empirical
or qualitative level until recently. Today, one can say that
our understanding of surfaces rests on a semi-quantitative
framework as a result of striking developments during the
last five to ten years in experimental techniques and theoretical and computational tools.

Two basic problems need to be overcome in experimental approaches to surface studies. One stems from the
fact that the surface atoms are chemically unsaturated and
thus have an intrinsic tendency to react chemically and/or
physically with their environment. All surfaces are con-
taminated in real environments or even in moderately high
vacuum (of the order of 10^{-7} mm Hg) as soon as they are
generated. The second problem is related to the effective
thickness of surfaces, which ranges from one to several
atom layers; thus, techniques are necessary with a resolution on an atomic scale yielding data which can be meaning-
fully related to first principles or engineering parameters.

From the theoretical point of view solid surfaces do not lend themselves readily to the classical or quantum mechanical approaches which have been applied for the last 50 years to the understanding of the structure and properties of crystalline solids. Such approaches are essentially based on the infinite three-dimensional periodicity of the crystalline lattice; such periodicity is lacking in surfaces.

The turning point in the quantitative study and understanding of solid surfaces was the discovery of surface states in semiconductors in 1947[1, 2] (energy levels or electric charge traps within the forbidden energy gap). This discovery led directly to the discovery of the transistor in 1948 and ushered in the era of solid-state electronics. The fundamental role of semiconductor surfaces in solid-state electronics motivated unprecedented efforts on surface studies. Because semiconductors have highly directional bonds (covalent bonding), low coordination number (four) as compared to that of the metals (up to 12), and could be prepared at extremely high purity levels, their surfaces proved to be better suited for quantitative experimental and theoretical studies than any other types of surfaces.

Beginning in the fifties ultra-high vacuum technology advances paved the way for the development of a number of techniques involving the interaction of electric fields, photons, electrons, and ions with "clean" surfaces. The results of such interactions not only yielded basic information on the structure, composition, and properties of surfaces, but also began to provide theoreticians with essential data for guiding the developments of their theoretical models.

Thus, in considering the study and understanding of solid surfaces in a historical framework one can distinguish three major periods: 1. the pre-transistor years, 2. the post-transistor years (1948 to middle sixties), and 3. the recent years (late sixties and early seventies). This paper is not strictly based on such a historical framework.

PHENOMENOLOGY OF SURFACE INTERACTIONS

Interactions of surfaces only with gases will be discussed here. Interactions with liquids are certainly of
basic importance to electrochemistry, corrosion, to a
major segment of industrial processes, and to life sciences;
yet they remain, generally, at a low level of understanding,
and so will not be discussed here.

General Considerations

It has been recognized long ago that surface atoms are
in a state which can be referred to as "unsaturated", "reactive" or "excited". Such a state reflects the fact that
work must be done to generate a free surface corresponding to the energy of the bonds which are broken in the process; the surface energy (work per unit area) at absolute
zero and for one-component systems is the thermodynamic
quantity surface tension (a tensile force, normal to the
surface having a constant magnitude per unit area). The
dependence of the surface tension on temperature and on
additional components (impurities) was introduced by Gibbs[3]
and represents the first quantitative approach to the energetics of surfaces; it has been directly applied with great
success to liquid surfaces. In the case of solid surfaces
additional forces are acting, referred to as surface stresses,
which originate in the long-range interaction of the surface
atoms with the bulk. Such stresses can be compressive or
tensile in nature[4].

There are no experimental or theoretical means to
determine directly the surface energy of solids (surface
tension and the effects of temperature, composition and
stresses). Estimates, however, show that the magnitude
of surface energy is quite high. Thus, the surface energy
in metals is of the order of 1000 ergs/cm^2 (at room temperature)[5]. In the case of copper it is estimated that the
(110) surface energy (32 kcal/mole) is about 30% of copper's
cohesive energy (heat of sublimation, 82 kcal/mole). The
surface energy increases with the bonding energy of the
solid. For diamond, the energy of the (100) surface is

estimated to be about 9,000 erg/cm^2 at 0°K and 5,000 erg/cm^2 at room temperature.

Solid surface-gas interactions are customarily distinguished as physical adsorption, chemisorption and oxidation.

Physical Adsorption

In physical adsorption the heat of the solid-gas interaction is usually less than 10 kcal/mole. Bonds of the van der Waals type, resulting from permanent dipole forces, induced dipole forces or dispersion forces (rather than from direct electron exchange)[6], are involved in this case. Since the van der Waals forces are of the same nature as those responsible for the liquefaction of gases, the adsorbability of gases increases with increasing ease of liquefaction. He, being the most difficult gas to liquefy, has the least tendency to absorb and its heat of adsorption on various surfaces is very small.

Actually, the theoretical model treating the forces acting among inert gases (Lennard-Jones approximation) can be applied reasonably well to covalent and ionic solids in which the electronic configuration of the atoms can be approximated. The attractive forces between two approaching atoms vary with distance, r, as $1/r^6$ and are due to the valence-electron interactions between the two atoms. Bonding reaches the maximum strength at a critical internuclear separation, r_o (Fig. 1); at smaller distances repulsive forces become significant since the closed electron shells of the atoms are approaching interpenetration which is prohibited according to the Pauli exclusion principle.

In the case of metals the valence electron configuration is not well defined and the description of the adsorption forces is far more complex; such description has been pursued on the basis of a number of assumptions and approximations. It turns out that the electronic characteristics of the adatom predominate over those of the metal surface atoms so that adsorption on metals is of a similar nature to that on non-metals[6].

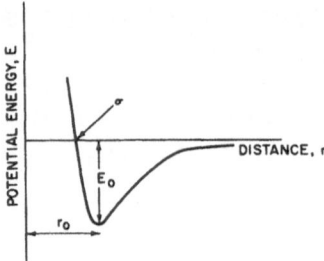

Fig. 1. Schematic representation of atom-atom interactions as a function of interatomic distance according to
Lennard Jones approximation.

In view of the nature of the forces acting in adsorption,
it is apparent that the interaction energies must vary with
crystallographic orientation. The greater the density of
lattice atoms surrounding the atoms, the greater the interaction energy and the stronger the adsorption. Similarly,
adsorption must be most pronounced in lattice holes or
edges.

The dependence of adsorption on orientation has been
demonstrated many years ago[7]. More recent studies have
confirmed such a dependence on quantitative grounds[8].
Thus, it has been found that at room temperature nitrogen
is strongly adsorbed with high initial sticking coefficient
(~ 0.25) on the (100) surfaces of W but on the (111) surfaces
the initial sticking coefficient is small (<0.04), consistent
with the fact that the number of first and second order
nearest neighbor metal atoms surrounding the adatoms is
greater on (100) planes than on the (111) planes of the b.c.c.
metal; nitrogen is not strongly adsorbed on the (110) surfaces of W, in agreement with the above reasoning. It has
also been shown in the potassium chloride-argon systems[9]
that the heat of adsorption of argon is significantly greater
in the (111) than in the (100) surfaces, since in the rock-salt
structure the diatomic configuration of the (111) surfaces
allows the penetration of the argon atoms between surface
atoms, leading to stronger adsorption.

Over and above the orientation effects, solid surfaces exhibit non-uniform behavior in their interaction with gases which is not understood on quantitative bases. Thus, the heat of adsorption on solid surfaces in general decreases sharply with increasing surface coverage, as shown in Fig. 2. For metals such as Ni or W the heat of adsorption of hydrogen decreases by about 40 kcal/mole between near zero coverage and a coverage of about one monolayer[10]. Electrostatic repulsion among the adsorbed atoms is expected to cause a decrease in the heat of adsorption with surface coverage. Calculations on the above system, however, show that such a decrease should be approximately 2.5 kcal/mole, i.e., a factor nearly 17 less than that experimentally determined.

Non-uniform behavior of solid surfaces has been recognized for some time as being responsible for the pronounced variations in catalytic activity on a given catalyst[11]. In many instances only a small fraction of the surface ("active centers") is catalytically active. Catalytic activity may be arrested by amounts of inhibitors far smaller than those required to form a monolayer on the surface of the catalyst.

Chemisorption

In chemisorption the heats of solid-gas interactions are comparable to those of chemical reactions. Thus, the

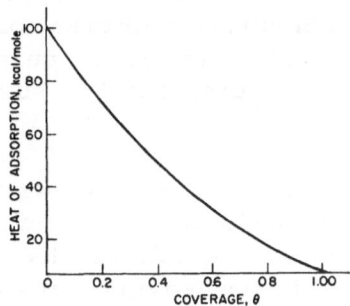

Fig. 2. Typical dependence of the heat of adsorption on surface coverage θ in monolayers.

heats of chemisorption of O_2 and N_2 on W are 194 and 85
kcal/mole, respectively[6]. The exact nature of the interaction forces in chemisorption has not been established,
although extensive study has been devoted to this subject[12].
There is evidence that one or more types of a broad spectrum of interatomic forces may be involved[12, 13].

In the case of chemisorption of gases on metal, experimental evidence (very small surface dipole moment) favors
the view that covalent bonding is involved. In transition
metals such bonding takes place between their partially
filled d-orbitals and the chemisorbed atoms. Extensive
measurements have shown that the magnetic susceptibility
of transition metals decreases upon chemisorption. The
type of d-bond formation must play an import role in
catalytic processes since transition metals are very widely
used as catalysts. In non-transition metals covalent or
partially ionic bonding appears to take place with chemisorbed oxygen or other gases. Oxygen chemisorbs on all
metals except gold.

Ionic bonding appears to prevail in chemisorbed alkali
metals (Na, K, and Cs) on W. Experimentally determined
dipole moments of these chemisorbed alkali metals at low
surface coverage (mutual depolarization may decrease the
dipole moment at high coverage) are in reasonable agreement with those calculated for monovalent ions on the surface[10].

The above types of bonds are not necessarily the only
ones prevailing in chemisorption interactions. A broad
spectrum of bonding exists between the purely covalent
and purely ionic ones (coordinate links, surface radicals,
etc.)[10].

In a recent paper Schrieffer[12] reviews the basic concepts of chemisorption of simple atoms on metals. Even
in these systems, current quantum mechanical treatments
are at best semi-quantitative leaving many basic questions
unanswered. Apparently the alkali adsorbates are the most
amenable to a quantitative theoretical treatment. However,

progress is currently being made in refining the approximations involved in approaching the various aspects of chemisorption interactions and apparently the theory of chemisorption is a fertile area for many years to come.

Oxidation of Surfaces

Interaction of metal surfaces with gases does not necessarily cease with the formation of an adsorbed monolayer (for an extensive treatment, see ref. 14). If a mechanism is available for the continuous exposure of new surface (or interface) atoms, the interaction with the ambient leads to the formation of a three-dimensional phase. Thick surface films have been most extensively studied in the case of oxide formation, although certainly not to the exclusion of other films such as sulfides.

Film formation is complex from a scientific point of view and is important technologically. The phenomenology of film formation on a macroscopic basis has been well formulated; the detailed mechanisms involved are still on a speculative basis. Phenomenelogically, film formation in general can be described in terms of film thickness as a function of time by one of three general relationships: linear, logarithmic and parabolic[14].

It is apparent that in the case of linear growth, the film itself presents no barrier to its further growth; the film remains highly porous and populated with cracks so that essentially the same amount of substrate surface is exposed to the gas at all times. In the other two instances (logarithmic and parabolic growth), the film initially formed, slows down the solid-gas interaction; communication of the substrate with the gas takes place by migration of the gas through the film to the metal-film interface, and/or by migration of metal species, through the film, to the film-gas interface. These mechanisms have been treated in great detail either in general terms or as applied to specific systems (see for example ref. 14). In some instances, agreement has been found between theory and experiment;

however, many questions remain unsettled. It is of
particular interest, for example, to know the role of the
initial stages of film formation in the subsequent film
growth. The study of these stages including rearrangements of the surface atoms upon the initial interaction
has been recently pursued by new experimental techniques
including low energy electron diffraction[15, 16]; the results
obtained are promising in terms of resolving the mechanisms of oxidation.

STRUCTURE OF SURFACES

As pointed out earlier, there is no theoretical means
available for arriving at the atom configuration of solid
surfaces (crystalline structure). The solution of this problem must rest on experimental bases and ranks among the
great challenges in physics and chemistry. Meaningful
structural studies need to be carried out on surfaces containing no adsorbed or chemisorbed contaminant atoms.
Such surfaces can only be obtained in very high vacuum
(of the order of 10^{-10} mm Hg). The rapid development of
techniques for the study of solid surfaces in the last
fifteen years has been possible to a great extent through
the remarkable advances in very high vacuum technology.

The main experimental techniques for the study of
surfaces are based on the interaction of electron fields,
photons, electrons, or ions or with the surface atoms.
Although the techniques to be highlighted here became
widely available in very recent years, the principles,
phenomena or concepts on which they are based have been
known for many years. No attempt will be made to discuss all of the recent techniques, or to discuss extensively any specific technique. The intent is to convey a
general impression of the principles involved in these
techniques, their scope and their limitations. Field-
emission microscopy, although it is not a recent technique
and is rather limited in resolution, will be included as it
led directly to field-ion microscopy and the atom probe.

Field Emission Microscopy

The principles of field emission and the main features of the field emission microscope, invented by Müller[17], are shown schematically in Fig. 3.   The material under study is in the form of a needle tip.   A high electric field is applied between the sample and a positive electrode which is usually a transparent film of tin oxide with a display phosphor layer.   The electric field (of the order of 5×10^6 V/cm) causes electrons to be emitted from the atoms at the tip and form a magnified image of the tip on the phosphor layer.   No lenses (magnetic or electrostatic) are required for the production of the image.

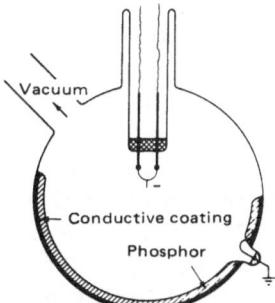

Fig. 3.   Schematic representation of the field-emission microscope.

A field-emission photograph of a tungsten tip is shown in Fig. 4.   The light areas in the figure correspond to crystallographic planes of the tip (high index planes) with relatively low work function (energy required to remove electrons from the atoms into the vacuum); the dark areas correspond to low index planes with relatively low work function.

The sensitivity of the field-emission microscope to changes in work function makes possible the detection of adsorption of a small fraction of a monatomic layer[18] since the work function of a material is very sensitive to adsorbed atoms.

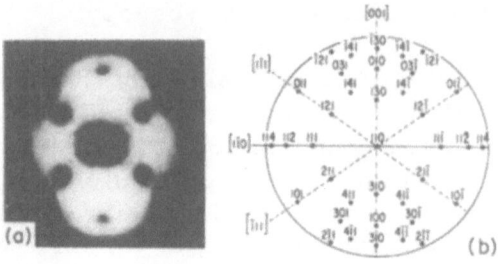

Fig. 4. Field emission photograph of an (110)-oriented
tungsten tip and the corresponding orthographic projection
of the cubic lattice.

Emission microscopy has been used successfully to
study in some detail the interaction of inert gases such
as Xe with metals such as tungsten and molybdenum, and
has helped to resolve some critical aspects of physical
adsorption[6]. It was found, for example, that the binding
energy of xenon on the (111) planes of molybdenum is lower
than that on the (100) plane, in accord with previous
quantitative estimates. For the case of tungsten and
molybdenum, assuming no surface structural rearrangements, it was shown that the binding energy of an adsorbed
atom increases with the number of first and second nearest neighbors surrounding that atom. It was also demonstrated that interaction of surface atoms with gases (specifically, oxygen on nickel) can lead to pronounced structural
rearrangements at the surface.

The main limitation of the field-emission microscopy
is its relatively low resolution (about 20Å). In view of the
advantages of the techniques discussed below, it appears
that field-emission will continue to be used within a rather
limited scope.

The field emission of electrons from the surface is a
quantum mechanical tunneling process, i.e., the electrons
need not overcome the potential barrier of the work function.
In the presence of a high external field the barrier assumes
a triangular shape with a width comparable to the wavelength

of the electrons (as defined by de Broglie's expression relating the momentum of a particle to a corresponding wavelength) near the Fermi level. (Fermi level or Fermi
energy is, in thermodynamic terms, the electrochemical
potential of free electrons or their partial molar free
energy; it represents the energy of electrons at the top
of the energy distribution; accordingly, the work function
is defined as the energy to remove an electron from the
Fermi level out of the material.) Thus, the work function
barrier becomes transparent to the electrons; this tunneling process is quantitatively related to the work function
and the applied electric field (for further discussion see
ref. 19).

Field-Ion Microscopy

In field-ion microscopy, imaging of the surface is
achieved by ions formed very close to the surface atoms[20].
Here also a field is applied between the metal tip and the
phosphor-covered electrode in the schematically shown
arrangement of Fig. 3. However, the chamber now contains an imaging gas, usually helium, at a pressure of a
few mm Hg, and the polarity of the electrodes is reversed.
Fields of the order of 5×10^8 V/cm are required and the
desirable radius of the metal tip is approximately 1000Å.

Field-ion formation at the tip takes place by quantum-
mechanical tunneling as in the case of field-emission[21, 22].
Here, of course, the electrons tunnel from the gas atom
into the metal. There is an optimum distance (of the order
of a few Å) for tunneling. At smaller distances, the energy
of the electrons of the imaging gas is below the Fermi
level of the metal and tunneling is prohibited; at greater
distances the surface barrier is broad and tunneling becomes improbable.

A typical field-ion image photograph is shown in Fig.
5. The bright spots correspond to individual atoms which
can readily be resolved at the edges of the individual planes.
A considerably larger number of planes appears on the ion
image than on the field-emission image. The planes are

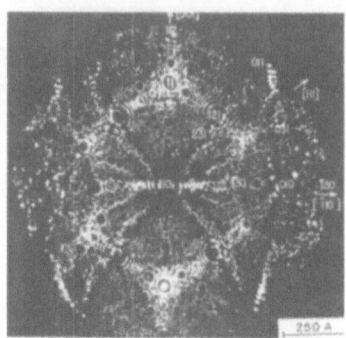

Fig. 5.   Field-ion micrograph of a tungsten tip at 27° K.
Numbers refer to indices of planes.

outlined by the concentric rings.   In low-index planes,
individual atoms are resolved within the planes as well
as at the edges.   The smaller the radius of the metal tip,
the greater the number of atoms contributing to the formation of the image.   For a radius of curvature of 350Å,
about 30 per cent of the surface atoms can be observed.

Although the theory of the ion image formation is still
being developed, field-ion microscopy has already contributed importantly to understanding the structure of surfaces.   Field-ion microscopy made it possible to observe
directly, for the first time, the atomic structure of surfaces.   In some instances, such as in platinum, the overall arrangement of the surface atoms was found to be the
same as that of the bulk.

By combining field-ion imaging and the field-ion
evaporation from the metal tip, it has been possible to see
individual vacancies and even study their behavior (see for
example ref. 21).   In the case of platinum, employing the
(012) planes, it was found that no particularly high strain
is present around the vacancy.   The energy of formation of
vacancies was also determined (approximately 1.2 eV) and
found to be consistent with the value determined by other
experimental means.   Individual interstitial atoms have

been observed in several instances including the platinum and aluminum surfaces and were found to cause structural distortion in their immediate vicinity. No vacancy-interstitial pairs have been observed. Other lattice defects such as screw and edge type dislocations, grain boundaries, and slip have also been directly observed.

The field-ion microscope is now being employed quite extensively in the study of surface reactions such as adsorption, desorption, surface diffusion, and others. It is in these types of studies that field-ion microscopy is likely to make its most lasting contribution. The technique does present some basic limitations: thus, it is readily applicable only to a few metals in view of the severe experimental requirements: the large stresses associated with the high fields necessary for field-ion imaging or field-ion evaporation cause, in most instances, pronounced surface distortion. In fact, even with refreactory metals, it is necessary to carry out the experiments below room temperature to insure surface stability and, thus, the necessary resolution.

Atom-Probe FIM. This technique combines field-emission microscopy with a mass spectrometer of single particle sensitivity[23, 24]. The tip can be moved so that even a single atom in the image can be positioned before a probe hole in the phosphor covered electrode behind which there is a mass spectrometer (Fig. 6). A 2- to 20-nanosecond high-voltage pulse causes field evaporation of the target atom or atoms which are then identified by the mass spectrometer. The atom probe with virtually a single atom resolution is still in its early stages; it is extremely promising although it has the inherent limitations of the field-ion microscopy. A number of interesting results have been obtained including the discovery that adsorption of the image gases He and Ne occurs at temperatures as high as $78°$ K[21].

Electron Diffraction[25, 26]

Electron diffraction is at present the most powerful technique for investigating the relative position of the sur-

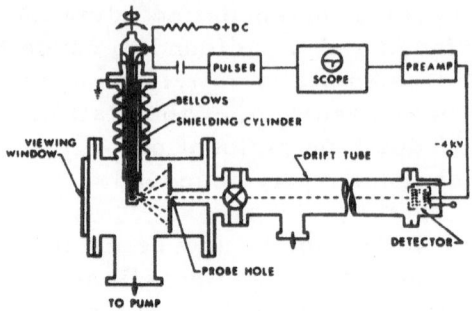

Fig. 6.   Schematic diagram of the atom-probe FIM.

face atoms, although a number of experimental and theoretical problems associated with this technique are not
entirely solved.   Low energy electrons are usually employed (LEED) ranging in energy from 10 to 500 eV.   However, high energy electrons are also used (HEED) to
achieve elastic backscattering from solid surfaces at
glancing incidence of the primary beam (angles near 0°).
HEED is often called RHEED to distinguish the use of
reflected (R) electrons from transmission electron diffraction (THEED) used for the study of bulk structure of
thin films.   LEED is much more widely used then RHEED.

In recent years LEED has been used in combination
with other measurements such as work function measurements, adsorption kinetics and energetics, and Auger
spectroscopy (yielding compositional and bonding information).   Such combined techniques are responsible for the
remarkable strides being presently made towards understanding the relationships between the basic nature of
solid surfaces, their physical properties, and their interactions with gaseous ambients.

Two types of arrangements are used in LEED studies:
a movable Faraday collector for measuring the diffracted
intensity over the angular range in front of the crystal surface, or a display system for observing directly or recording the diffracted patterns (Fig. 7).   In the case of the

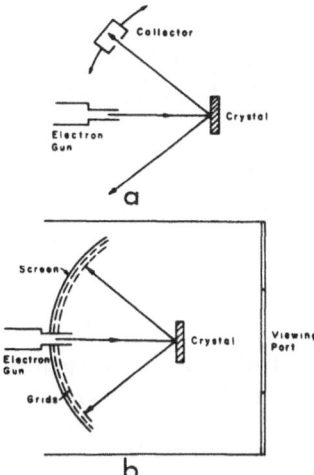

Fig. 7.   Schematic diagram of LEED arrangements   (a)
Faraday collector arrangement,   (b) display arrangement.

display arrangement the electrons impinge on the crystal
surface and those reflected pass through two grids; the
first grid rejects the inelastic electrons while the second
accelerates the elastically scattered electrons which contribute to diffraction.   Relative values of the diffracted
intensities can be obtained from the brightness of the corresponding spots.   More accurate relative intensity values
are obtained with the Faraday collector arrangement.

The LEED patterns can frequently be interpreted by
considering one or two dimensional periodic arrays of atoms
(scattering centers).   Diffraction for a one-dimensional
array of atoms is illustrated in Fig. 8.   The Bragg diffraction relationship applies, except that the interplanar
distance becomes the distance, a, between the atoms:
$n\lambda = a\sin\Phi$, where n is an integer, λ is the wavelength of
the electrons (according to de Broglie's relationship $\lambda =
(150/v)^{1/2}$, where v is the voltage applied), and Φ is the
angle between the incident and the diffracted beam.   The
wavelength at typical Leed energies is of the order of 1Å.
The penetration of the primary electrons at these energies

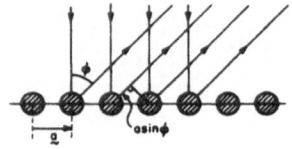

Fig. 8.   Diffraction from a linear array of atoms; the
incident beam is normal to the array (from ref. 25).

is 3-10A, so that diffracted beams originate from only a
few atom layers.   A LEED pattern for a simple structure
is shown in Fig. 9.   Results from a LEED study employing a Faraday collector are shown in Fig. 10.

   The experimental studies have been pursued along
three main objectives: a) the determination of the structure of clean surfaces, b) the study of structural changes
as a result of physical or chemical interactions with gaseous ambients, and c) the concurrent study of the surface
structure and electrical or other properties.   This latter
objective is, perhaps, the most promising, as it could lead
to fundamental structure-property relationships.

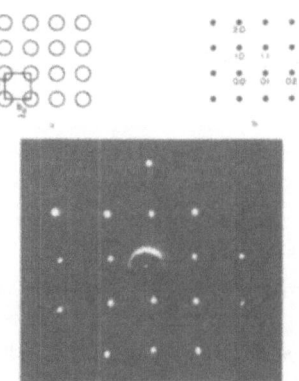

Fig. 9.   (a) Model of (100) plane of a cubic crystal; (b)
diagram of a LEED pattern; (c) photograph of a LEED
pattern from a (100) surface of W (from ref. 25).

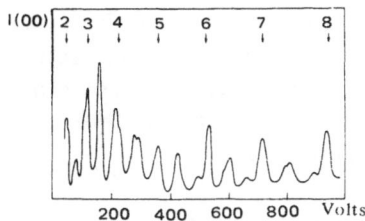

Fig. 10. I-V curve from a (100) surface of W; the arrows indicate the positions of the primary Bragg maxima (from ref. 25).

LEED has been extensively used for the study of surface structure of metallic, ionic and covalent materials. The preparation of "clean" surfaces on an atomic scale is no longer the formidable problem that it used to be even a decade ago. Means for cleaving, ion bombarding, evaporating, etc., in ultra-high vacuum are incorporated in readily available commercial systems.

In the case of metals, it has been found that the surfaces generally exhibit the same lateral symmetry as the bulk. Thus, in the case of (111), (110), and (100), surfaces of Ni no structural rearrangements or superstructures were observed[27]. Similarly, no evidence of reconstruction has been found in ionic crystals[25]. However, there is evidence of vertical displacement of the surface atoms (or shifts of the entire surface plane) as, for example, in the (111) and (110) surfaces of Ni[27] and the (100) surfaces of LiF[28] and KCl[29]. In the case of semiconductors (Ge, Si, GaAs, and others) pronounced surface superstructures have been reported[30]. In this type of covalent (or partially ionic) materials surface reconstruction tends to minimize the number of dangling bonds and thus the surface free energy.

Changes in the LEED patterns resulting from surface interactions with gases can provide an insight into the fundamental nature of adsorption, chemisorption, or oxidation processes. Thus, the initial stages of oxidation, for many years a matter of speculation, are now being

directly established. In the case of the (110) surfaces of
Ni, for example, it was shown that oxygen enters the substrate (110) surface[15]; i.e., strong ordering forces operate
even in the initial stage of oxidation leading to well defined
stoichiometry and orientation.

Combination of LEED studies and electrical or other
properties are of significant promise; thus, an investigation
on germanium has already shown[31] that a direct correlation
exists between the structure of the surface and the electronic
characteristics (surface states).

The theoretical interpretation of the LEED patterns and
intensities has not been resolved with general acceptance.
However, striking progress has been made in the last few
years and the intensity of theoretical efforts continues at a
high level.

Auger Spectroscopy[32]

Auger spectroscopy is based on the analysis of the
energy distribution of secondary electrons emitted by bombarding the surface with an electron beam. Although the
secondary electron emission was discovered in 1925 by
Auger, it was not until 1953[33] that this emission was formulated as a surface analytical tool.

Auger electron emission is schematically shown in Fig.
11. The electronic energy levels for single ionization are
indicated on the left-hand side and correspond to the X-ray
levels, K, L and M. In the present case the valence shell
M (not designated as such in the figure) broadens into a
continuous energy band (valence band) when atoms are
brought together (the density of states in the valence band
is indicated as a shaded area). In this figure a primary
electron has ionized an $L_{2,3}$ level into which the transition
of an outer electron can take place. The energy released
in this transition can be emitted as an X-ray photon or taken
up by another electron; if this energy is sufficient, the
electron can leave the surface, as shown in this case for
an electron at energy V_1. The kinetic energy of the emitted

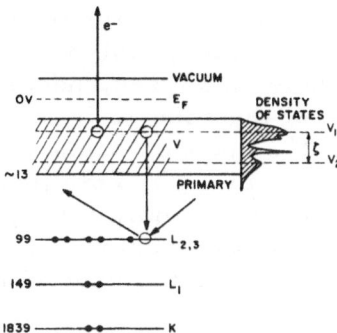

Fig. 11. Schematic representation of the Auger process in silicon; X-ray energy levels are indicated on the diagram; the density of states is drawn in the valence band (from ref. 32).

electron (Auger electron) is characteristic for each element. Thus, with Auger spectroscopy each element can be uniquely identified. As the emitted electrons originate from about the top five atom layers (for an energy of incident electrons of the order of 500 eV) Auger spectroscopy is essentially a surface probe capable of identifying one atom in 10^6.

The analysis of the energy distribution of the electrons provides information on the chemical configuration of the atom. For example, it is possible in some instances to deduce the valence state of the atom. The regions of the energy spectrum originating from the valence band provide information on the band structure and the density of states within the band and even the symmetry of the molecular orbitals of valence electrons (location of the atom in the unit cell).

Numerous investigations employing Auger spectroscopy have been reported in the last five years dealing with surface characterization (chemical composition and nature of bonding) and processes taking place at the surface (type of bonding resulting from surface interactions, diffusion at the surface, segregation of impurities at the surface from the bulk, etc.)[32]

The detailed interpretation of Auger spectra is still
in the development stage. The problem of determining
precisely the volume being analyzed, necessary for the
correct interpretation of the spectra is not entirely solved.
It is essential to know the ejection depth of the escaping
electron; this energy, however, is only indirectly related
(and can be unrelated) to the primary energy of the incident electrons. The situation becomes even more complex
when more than one element is involved in the detected
volume.

Auger spectroscopy is readily incorporated into LEED
experimental arrangements by the addition of relatively
simple electronic instrumentation. LEED-Auger combination is now extensively used. It is an extremely powerful
research tool (perhaps the most powerful now available)
for the study of the structure of surfaces and their interactions. It holds great promise of providing critical information on the basic steps involved during the early
stages of surface interactions which should help resolve
some of the most perplexing questions regarding catalytic
and other surface phenomena.

Scanning Electron Microscopy[34, 35, 36]

The concept of the scanning electron microscope (SEM)
and its demonstration dates back to the thirties, but it was
not commercially introduced until 1965. The SEM is based
on the interaction of an electron beam with the solid surface; such an interaction leads to the generation of secondary
electrons, Auger electrons, backscattered electrons, elastically scattered electrons, X-ray radiation or light. Any
of these electrons or photons can be collected and displayed
on a cathode ray tube (crt) after appropriate amplification.
A schematic representation of the SEM in its emissive mode
(which is most commonly used) is shown in Fig. 12. Electrons from a filament (F) are focused on the sample (S).
The electron beam is scanned by electromagnetic coils in
the form of a scan-raster (TV-type raster). The secondary
electrons emitted from the sample are collected and the
resulting signal is amplified and displayed on the crt. The

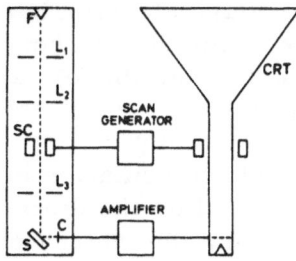

Fig. 12.   Schematic diagram of the SEM (from ref. 35).

crt is synchronized with the electron beam scan so that
there is one-to-one correspondence between the electrons
collected from a given point of the sample and the brightness of a corresponding part on the crt; thus, a signal
map of the scanned area is displayed.   The image magnification is determined by the ratio of sizes of the rasters
on the crt screen and the scanned area.   The crt screen
is usually 10 x 10 cm so that for a magnification of 10X an
area of 1 x 1 cm is scanned; for a magnification of 10,000X,
the scanned area is 10 x 10 μm (an example of a SEM microphotograph is shown in Fig. 13).

The SEM can be operated in different modes depending
on the type of "signal" collected among those generated from

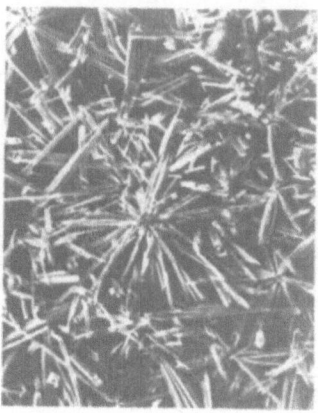

Fig. 13.   Scanning electron microphotograph of an etched
surface of partially crystallized $0.6As_2Se_3.0.4Sb_2Se_3$[38]
(200X).

the interaction of the primary electron beam with the surface. These modes are: emissive (involving any type of electrons or photons pointed out above), absorptive, conductive, and transmissive.

Each mode can provide information on various aspects of surface characteristics. For the emissive mode employing secondary electrons, the energy of which does not usually exceed 300 eV, a surface layer less than 50Å thick can be studied. In the (emissive) X-ray mode the same principle is involved as in the electron microprobe[37]. The emissive mode employing the Auger electrons corresponds to Auger spectroscopy. The resolution of the SEM depends on a number of parameters, including the mode, the size and energy of the primary beam, the nature of the solid, etc. In the emissive (secondary electron mode) a resolution of about 250Å can be obtained routinely. A typical microphotograph obtained with the SEM, in the emissive mode, is shown in Fig. 13.

The discussion of the various operating modes is beyond the scope of this paper. It should be pointed out, however, that the versatility of the SEM is striking indeed (see, for example, references 34, 35, 36) and its potential in the study of solid surfaces will not probably be exhausted for many years to come.

Other Techniques

The above techniques, individually or in combination, provide information on the structure, bonding and composition of solid surfaces, on their physical properties and their interactions with gaseous ambients. The techniques outlined below are primarily analytical tools for determining the chemical composition of solid surfaces on a microscale.

Electron Probe Microanalysis[37]. This powerful analytical tool was introduced approximately twenty years ago and is widely used. It has been under constant development and refinement over that period. An electron beam (typically with an energy of about 30 kV) is focused on the

solid surface; inelastic scattering of the electrons leads to
the emission of X-rays characteristic of the elemental
composition of the surface. The electron beam of the order
of 1 μm in size is scanned over the surface. The penetration depth ranges from 200 to 20,000Å depending on the
operating conditions and the material used. An absolute
sensitivity of approximately 10^{-14}g and a relative sensitivity of about 0.05 wt% can be achieved, usually with an
accuracy in the range of $\pm$ 5%. Elements lighter than sodium
cannot be quantitatively determined in surface consisting of
heavy elements.

Ion Microprobe[39]. This technique combines an ion
sputtering source and mass spectrometry. The ion sputtering source (an ion beam with an energy in the keV range
and a diameter of the order of 10 μm) causes controlled
sputtering and ionization of surface atoms which are then
analyzed in a mass spectrometer. The rate of sputtering
can vary from fractions of a monolayer to several hundred
Å per second. Areas as small as 3 μm in diameter can be
quantitatively analyzed in the part-per-billion range. Spatial images of the sputtered ions can be displayed on a
fluorescent screen with a linear resolution better than 1
μm. The reliable use of this technique became possible
only recently with the development of controlled sputter
ion sources; although destructive, it is powerful indeed,
since it utilizes the extreme sensitivity of mass-spectro-
metry.

Low-Energy Backscattering Spectroscopy[40]. This recent technique is based on the fact that ions with relatively
low energy (a few keV) impinging on a solid surface undergo scattering which can be accurately described by a two-
body collision model. Accordingly, the energy of the back-
scattered ions is a simple function of the mass of the target
atoms. Thus, the necessary apparatus is relatively simple,
requiring a gas-ion beam, an ion energy analyzer, and an
ion detector. This promising technique yields the composition of the outermost atom layer of the surface. It has also
the capability of removing (sputtering) one atom layer at a
time making possible the atom-layer by layer analysis of a
solid surface.

It should be pointed out that no single technique is likely to make striking contributions to the understanding of the nature of solid surfaces. The potential of each will be fully realized only in combination with the others.

SEMICONDUCTOR SURFACES

Atomistic Approach

In semiconductor materials where the chemical bonds are highly directional and the number of near neighbors is relatively small, the chemical or atomistic approach (whereby the atoms are considered as individual entities and the interatomic forces are viewed as electronic interactions, localized between individual atoms) leads to qualitative but most useful models for correlating and predicting certain chemical and physical properties.[41] This approach is also successfully applied to the surfaces of these materials.

The Group IV semiconductors (such as Ge and Si) have the diamond structure (tetrahedral or four-fold coordination - sp^3 bonding). In considering their low index planes it is assumed that the surface atoms undergo no substantial structural rearrangements and that these atoms have one-electron dangling bonds or orbitals, i.e., the formation of these surface atoms results from breaking one (or two) Ge bonds as follows: $Ge:Ge \rightarrow Ge. + Ge.$; a two-dimensional projection of the low index surfaces is shown schematically in Fig. 14. (Note that the atoms of the (111) and (110) planes have one dangling bond each, whereas those of the (100) surfaces have two.) The chemical reactivity and other characteristics of such surfaces can be reasonably related to the tendency of the dangling bonds to acquire a second electron. In the case of germanium, for example, it has been found[42] that the chemical reactivity of the various surfaces in aqueous solutions increases with increasing density of the dangling bonds: (100) > (110) > (111). The dangling bond model has also been used to explain the origin and properties of the surface states[43] (electron energy levels at the surface present within the forbidden energy gap). In fact the density

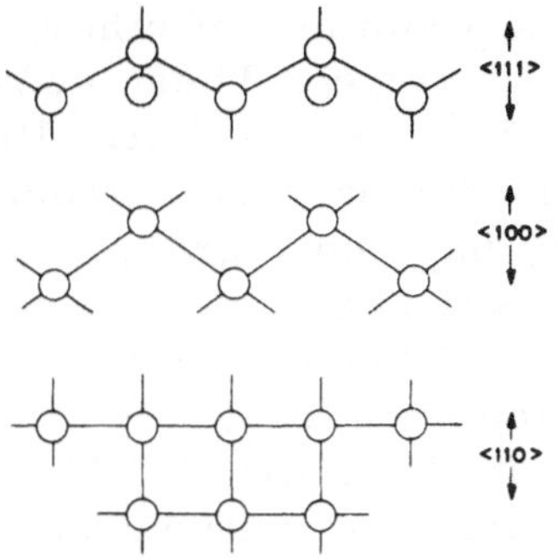

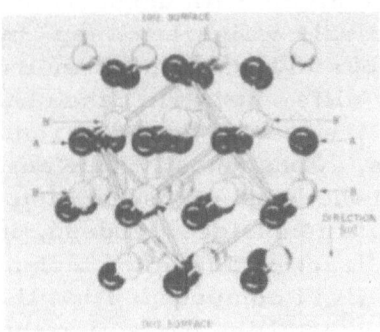

Fig. 14. Two-dimensional projection of the outermost surface layer and the second layer of atoms of the low index planes; diamond structure.

of surfaces in clean Ge surfaces is of the same order of magnitude as the density of the dangling bonds ($\sim 10^{15}/cm^2$).

The chemical approach to surfaces has been particularly fruitful for semiconductors with the zinc-blende and wurtzite structures[44] (in both structures the atoms have a four-fold, sp^3, coordination as in the diamond structure). These structures, being noncentrosymmetric, have two types of {111} surfaces each having different types of atoms as shown in Fig. 15 for the zinc-blende structure. Group III-Group V semiconductor compounds (InSb, GaSb, GaAs et al), some II-Vi compounds (e.g., ZnS, ZnTe, CdTe,

Fig. 15. Zinc blende structure.

HgSe, HgTe) and a small number of others (e.g., beta-SiC) have the zinc blende structure. These compounds are usually designated as AB compounds; their {111} surfaces are designated as A, when they have only group III or group II atoms and B when they have only group V or group III atoms. The formation of such surfaces can be visualized as resulting from breaking A:B bonds. Unlike the case of Ge, however, and for obvious reasons, here the two electrons, A:B → A + :B end up with B atoms[45]. Thus, the A surface atoms have "dangling orbitals" without electrons whereas the B surface atoms have dangling orbitals with two electrons. On the basis of this model one would expect striking differences in the chemical and physical behavior between these surfaces:

The A surfaces with no dangling electrons are expected to be less reactive towards oxidizing agents than the B surfaces. Similarly, adsorption of electron donor molecules or negative ions should be more pronounced on the A surfaces than on the B surfaces. Indeed, in the case of InSb, for example, it has been found that the A (In) surfaces react faster in oxidizing media (by an order of magnitude) than the corresponding B (Sb) surfaces[45]. On the basis of this difference in reactivity it was successfully explained why in various etchants dislocation etch pits develop only on the A surfaces and not on the B[45, 46].

Regarding the structural characteristics of these surfaces, the B surface atoms with unshared electrons can retain their tetrahedral configuration (sp^3 hybridization) much as nitrogen does in the ammonia molecule. The situation is altogether different in the case of the A surface atoms where there are insufficient electrons to fill the tetrahedral orbitals. Consequently, the angles of the bonds of these atoms with the lattice must be distorted, thus, introducing strain at the surface. Indeed, it has been confirmed by X-ray diffraction and other techniques (for InSb and other III-V and II-VI compounds) that the B surfaces exhibit far greater crystalline perfection than the A surfaces[47, 48]. In fact, thin wafers (of the order of 10 μm) with the (111) orientation are spontaneously bent because

of the elastic strain present in the A surface[49]. Furthermore, it has been demonstrated that crystal growth along
the < 111 > direction meets with much greater difficulties on
the A than on the B surfaces[50].

Finally the B surface atoms with two unshared electrons
are expected to act as electron donors, whereas the A atoms
as electron acceptors. Such differences in electrical behavior have been indirectly observed[51].

Electronic Characteristics

In the late thirties it was recognized that the rectifying
action of a semiconductor metal contact was due to the difference in work function between the two solids[52]: electrons
from the solid with the lower work function flow in the solid
with the higher work function until a high enough potential
barrier is set up to prevent further flow of electrons. As
a result, a space-charge region is formed in the semiconductor side of the interface and another of equal and opposite
sign in the metal side. The former extends much deeper
into the bulk because of the much smaller carrier concentration in the semiconductor. The magnitude of the barrier
is equal to the difference in work function and is referred
to as the contact potential.

Accordingly, for a given semiconductor, the degree of
rectification was expected to increase with increasing work
function of the metal. In the late forties it was found that
in the case of Ge and Si the work function of the metal had,
essentially, no effect on the degree of rectification[52].
These results were accounted for by Bardeen's theory[1]
according to which the potential barrier at the semiconductor surface was produced by the presence of surface states
(which trap charges from the semiconductor bulk) at energies within the forbidden energy gap of the semiconductor.
Accordingly, if this potential barrier is equal to or greater
than the difference in work function between the semiconductor and the metal, the contacting metal cannot affect the
rectifying characteristics.

The presence of surface states was experimentally verified by Shockley and Pearson's[2] field-effect experiment. A semiconductor slice was used as one plate of a parallel plate capacitor and a metal as the other. By applying voltage across the capacitor and measuring the change in the conductance of the semiconductor parallel to its surface, it was found that only a small fraction of the induced charges were mobile (i.e., contributed to the conductance); the majority of these charges were trapped in the surface states. By appropriate analysis of the change of the surface conductance as a function of applied field, the energy position and the density of the surface states were determined. (Schematic energy diagrams of semiconductor surfaces are discussed below.) Following their discovery the characteristics of the surface states and their effects on the rectification process were extensively studied, particularly at the Bell Telephone Laboratories. These studies led directly to the discovery and development of the transistor by Bardeen, Brattain and Shockley in 1948.

Surface Photovoltage Spectroscopy. Surface states have been also studied employing the photovoltaic effect, i.e., the appearance of a voltage upon illuminating a semiconductor-metal rectifier[52]. These studies were primarily focused on the role of surface states in trapping and recombination of the excess free carriers generated through band-to-band transitions by illumination[53]. Various optical methods have been developed for the study of semiconductor surfaces, particularly the surface of high energy gap semiconductors, e.g., CdS, ZnO. In this type of material the surface states are generally in poor communication with the bulk and their occupancy does not necessarily obey equilibrium Fermi-Dirac statistics. Thus, field-effect experiments, which have been very successful in the study of germanium and silicon surfaces, are not effectively applicable on these materials.

Recently a new method was developed, "Surface Photovoltage Spectroscopy"[54], which is based on the photovoltaic effect (surface photovoltage) caused by sub-bandgap illumination.   The surface photovoltage is measured as the change in the steady state contact potential difference (cpd) with respect to a vibrating reference electrode by means of an off-null method.   Changes in cpd are a direct measure of the changes in the surface potential barrier as shown in Fig. 16 for a semiconductor-reference electrode system.

It was shown that sub-bandgap illumination leads to discrete electron transitions from the surface states into the conduction band (surface state depopulation) and from the valence band into the surface states (surface state population) while the over-all number of bulk carriers remains essentially unchanged[54].   These types of transitions and the associated changes in the surface potential barrier are illustrated schematically in Fig. 17.   The analysis of the photovoltage as a function of the energy of the incident light leads to the direct determination of the energy positions of the surface states as shown in Fig. 18 for the CdS surfaces[54].   The energy positions of the surface states observed in Fig. 18 and the corresponding electron transitions are

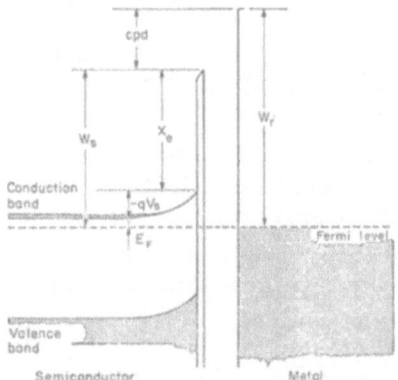

Fig. 16.   Semiconductor-reference electrode system. W_r and W_s are the work functions of the reference electrode and the semiconductor, respectively; X_e is the semiconductor electron affinity; E_f is the Fermi energy in the bulk with reference to the conduction band and V_s is the surface barrier.   Upon illumination only the value of V_s changes; this change is directly determined by changes in cpd, since: cpd = W_r - W_s = W_r - (X_e + E_f - qV_s).

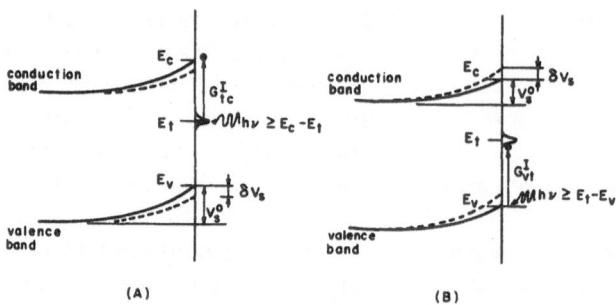

Fig. 17.   Schematic representation of origin of photovoltage:
(a) surface state depopulation (decrease in surface barrier);
(b) surface state population (increase in surface barrier).

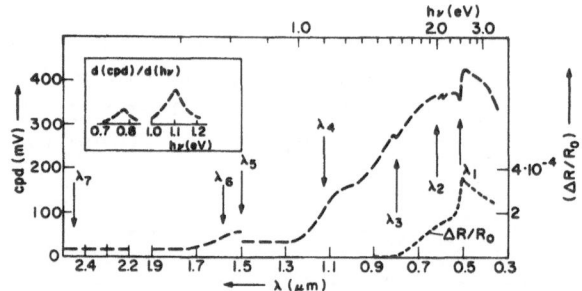

Fig. 18.   Surface photovoltage spectrum and photoconductivity of the prismatic surfaces of CdS in room atmosphere. The insert shows the derivative of the surface photovoltage with respect to photon energy, giving unambiguously the energy position of the surface states (from ref. 54).

schematically summarized in Fig. 19.   The analysis of the transients of the associated transitions leads to the determination of the dynamic[55] parameters of the surface states.

Surface photovoltage has been successfully applied to the study of the electronic characteristics and the effects of gaseous ambients of the surfaces of CdS[54], ZnO[56], and GaAs[57].   It is the only method among all methods based on optical transitions which allows the determination of the

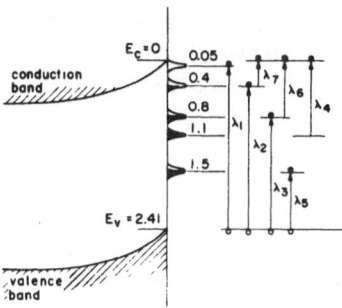

Fig. 19. Schematic representation of the surface states in the prismatic surfaces of CdS and the associated electron transitions (from ref. 54).

energy positions of the surface states, their density and occupancy as well as their dynamic parameters such as the capture cross-sections for electrons and photons[58].

<u>Photomechanical and Surface Piezoelectric Effects.</u>
It has been found[59] that thin (001) wafers or CdS (10 x 3 x 0.015 mm), fastened at one end, exhibited pronounced bending upon illumination with white light. This effect, "photomechanical effect", was also observed on GaAs[57]. It was shown that it results from the light-induced changes in the electric field at the two (001) parallel surfaces (surface barriers). These changes in the electric field induce in turn strains leading to bending, since both surfaces have a depletion layer and consequently the electric field (and thus the strain components) in the two surfaces is in opposite directions.

By modulating the surface barrier with chopped light the wafers are excited to their fundamental frequency vibration. The amplitude of the vibration was found to be a function of the wavelength of the incident radiation, and specifically a function of the changes in the surface barrier. In fact, the amplitude versus wavelength spectra are similar to the surface photovoltage spectra. Accordingly, the photomechanical effect provides a relatively simple means for determining the energy position of surface states.

It is of interest to note that when ambient molecules introduce surface states through adsorption, the amplitude of vibration is very sensitive to even trace amounts of these molecules at the energy of illumination correspond- ing to the energy of the introduced surface states. Accord- ingly, the photomechanical effect can serve as the basis for extremely sensitive detectors of gaseous (or liquid) species. Employing a variety of noncentrosymmetric semi- conductors it should be possible to encompass a number of molecular species which, through adsorption, introduce distinct surface states.

When a stress is applied through mechanical bending on an (001) wafer of a noncentrosymmetric semiconductor, a pronounced change in cpd occurs[60]. This effect, "sur- face piezoelectric effect", was attributed to the polarization induced in the depletion layer by the mechanical stress. The surface piezoelectric effect is consistent with the photo- mechanical effect. It is associated with the modulation of the surface barrier by mechanical stress and thus provides a means analogous to the field effect for the study of surface states. The surface barrier in high energy gap materials usually prevents equilibrium between the surface states and the bulk. Such equilibrium can be conveniently established by lowering the surface barrier through bending, and con- sequently the equilibrium parameters of the surface states can be readily studied[61].

Thus, the surface piezoelectric effect constitutes a very important complement to surface photovoltage spectroscopy. Furthermore, it could serve as an effective means for sen- sing and determining forces (weight for example) or changes in the magnitude or position of forces, provided that such forces are appropriately coupled to the noncentrosymmetric materials so that they induce corresponding strains and thus readily measurable changes in the surface barrier.

Surface Barrier and Catalytic Processes. It was recog- nized many years ago that in chemisorption of gases on solids there is charge transfer between the chemisorbed molecules and the adsorbate. As the understanding of the rectification

phenomena on semiconductor-metal contacts developed,
theoretical models were proposed relating chemisorption
to rectification processes. Since many catalysts (particularly oxides and sulfides) are semiconductors and since
catalytic processes involve charge transfer through chemisorption, it was attempted to explain some aspects of
catalysis on the basis of semiconductor surface characteristics including surface states, the surface potential barrier
and the associated space charge region[62].

A number of experimental investigations[63] were carried
out following the publication of these theoretical models,
employing various semiconductor-gaseous systems in an
attempt to relate the electrical characteristics of semiconductors to catalytic processes. These studies have been
by and large inconclusive because semiconductors in powder
form were used to provide the necessary surface area for
determining the extent of chemisorption and catalytic yields;
no meaningful direct electrical characterization of these
surfaces was possible although interesting findings have
been reported and some qualitative relationships between
the electrical characteristics of the surfaces and their catalytic activity have been reported.

The surface piezoelectric effect together with surface
photovoltage spectroscopy provides a unique approach to the
study of adsorption and catalytic phenomena as related to
charge exchange at the surface and the surface potential
barrier. Employing a non-centrosymmetric material (say
a single crystal of ZnO) the adsorption of a gas, such as CO,
can be determined, down to a fraction of a monolayer, by
measuring the corresponding changes of the surface barrier.
The energy position (within the energy gap) of the adsorbed
gas can be determined by surface photovoltage spectroscopy.
Through appropriate bending of the crystal (in the elastic
range) the surface barrier can be increased or decreased
at will and its effect on the amount of gas adsorbed can be
quantitatively established. In a similar fashion the adsorption of other gases can be individually studied (e.g., O_2
and CO_2). Provided that each gas introduces its own surface states at specific energy levels, then the catalytic

reaction $CO + 1/2\ O_2 \xrightarrow{\text{ZnO}} CO_2$ can be quantitatively studied
by allowing CO and O_2 to adsorb on ZnO and then deter-
mining, through photovoltage spectroscopy (changes in the
surface barrier), the amount of CO_2 formed and the associated time constants of the reaction. The effect of the
initial surface barrier on the reaction can be determined
on the same catalyst by changing the magnitude of the barrier through bending and allowing the reaction to take place
again. This type of experiment is currently being designed.

THEORETICAL STATUS

In a theoretical approach to solid surfaces one is concerned with the wave functions and energy levels of an
electron in the presence of a finite array of ion cores exposed to vacuum. Suitable approximations of the potential
functions at the surface must be developed so that the resulting Schrödinger wave equation can be solved. However,
the limited periodicity restricts the application of Bloch's
theorem (the potential is taken to be the same in each unit
cell and periodic boundary conditions are imposed) which
has been applied successfully in provided solutions to
Schrödinger's equation for the bulk of solids where infinite
periodicity prevails.

As early as 1932 Tamm[64] used the Kroning-Penney
model (one-dimensional array of delta-function potential
barriers) but considered a semi-infinite array terminated
at a lattice point by a constant finite potential. He found
that one discrete level corresponding to a wave function
was localized near the surface for each energy gap. Goodwin[65] confirmed the presence of surface states in the forbidden gaps by using the nearly-free electron approximation
as well as the tight-binding approximation in one as well as
in three dimensions. For every surface state in the forbidden gap a state was absent from the bulk energy bands.
At about the same time Shockley[66] considered a three-dimensional perfectly periodic potential including the surface
region. Employing the tight-binding approximation he found
that for small interatomic spacings one surface state per

surface atom exists with energies near the middle of the
forbidden gap. Shockley's results showed that both in the
free-electron as well as in the tight-binding approximation
the surface states were associated with the change in potential in the surface cells.

As pointed out earlier the experimental behavior of
semiconductor-metal contacts was unambiguously explained
by Bardeen's hypothesis that localized surface states, with
energies lying within the energy gap, are present on semiconductor surfaces leading to a potential surface barrier
which depends on the density and occupancy of the surface
states. More recently Koutecky and Tomasek[67] using the
tight-binding approximation and assuming that the overlap
integrals for the surface atoms are the same as those of
the bulk atoms, found that surface states exist, independent
of the values of the integrals, with energies near the middle
of the gap. This result is similar to that found by Shockley.

Attempts have been made by Tomasek[68] to consider surfaces of semiconductors with specific lattices such as those
with the diamond lattice. He used a tight-binding approximation, considering overlap integrals not extending beyond
the first neighbors. He found the presence of Shockley states
but with energies just below the bottom of the conduction band.
By considering distortions in the outermost two layers he
specified conditions under which no surface states are present.

Heine[69] has reintroduced the free-electron approach to
surface states and treated the bulk states in the Bloch-wave
approximation but with complex wave vectors. In this case
Bloch wave functions characterizing the bulk are matched at
the surface to evanescent waves, i.e., waves which have an
exponentially decaying character away from the surface due
to the complex wave vectors. These waves lead to surface
states.

In the case of ionic crystals Mark and Levine[70] have
approached the problem of surface states by calculating the
difference in electrostatic energy (Madelung) between the
bulk and surface ions.

The above theoretical approaches (based on covalent or ionic structures) are still of a qualitative nature as they do not take into consideration the actual positions of the surface atoms, which are not necessarily those expected from the bulk lattice structure. They do show the presence of surface states which in semiconductor surfaces have been experimentally found and studied. At best semi-quantitative agreement between theory and experiment can be claimed.

In the case of metals theoretical approaches meet with additional difficulty stemming from the fact that both the bulk and surface electron structure is characterized by delocalized orbitals and overlapping bands. The simplest model considers non-interacting electrons in a potential well bounded by potential walls of infinite height[71]. The core ionic charge is assumed to be uniform within each unit cell. Because this surface model does not correspond to a real metal with the same electron density, but to an artificial one, the name "Jellium" was given to the model.

The approach to metal surfaces has been refined in recent years by considering local distributions of the electron gas[72]; calculations of the work function of metals have been carried out based on the refined Jellium model[73]. The results are certainly of interest although still not in very good agreement with experiment (discrepancies range from 5 to 30%). Wave matching techniques like those used in semiconductor surfaces have been also applied[74,75]. These approaches have shown the presence of surface states in metals; however, experimental evidence for the presence of surface states based on observed photoemission peaks in Ni[76] is still a matter of controversy.

More recently a molecular cluster approach has been developed for solid surfaces[77]. This approach considers bonding at surfaces in the form of localized clusters corresponding to the properties of disordered bulk. The distinct advantage of this approach is that it can be applied to thin films as well as to chemisorption and catalytic reactions, even in the case of transition metals which present additional difficulties to theoretical treatments.

During the last few years, progress in theoretical treatments has been guided by and coupled significantly with fundamental experimental results obtained from the interaction of solid surfaces with photons, electrons and ions, and the prognosis for important future theoretical developments is very encouraging.

CONCLUDING REMARKS

In the last five years unprecedented strides have been made in the understanding of the structure, bonding and interactions of solid surfaces. The basic and applied work during that period can be clearly characterized as a cascade-type or self-feeding process. It is difficult to associate the main stimulus with a specific factor such as: (a) the importance of surfaces in numerous aspects of technology and life sciences, (b) the confrontation of the experimental and theoretical scientists with the unique challenges that solid surfaces present, and (c) the inevitable (and earlier unattainable) development of powerful experimental techniques. It is easy to see, however, that there is an intimate interplay of all these factors.

The physicist or chemist, having resolved some basic questions on the atom structure of clean surfaces, turned to relationships of structure and electronic properties; then to the nature of the initial stages of surface interactions with ambients. These studies pointed directly to the fundamentals of technological processes such as solid state electronics, oxidation, catalysis, corrosion, friction and others. On the other hand, the experimental results based on the direct or indirect interaction of surfaces with electric fields, electrons, ions, and photons provided, for the first time, factual guidelines to the theoretical chemist and physicist; a number of classical applied and basic disciplines began to converge on a new framework: the structure-bonding-property relationships of surfaces. Chemical engineering, metallurgy, electrical engineering, chemistry, physics, biology and perhaps others are participating, interacting, and contributing to the study of sur-

faces. A new trend in "surface science and engineering"
is emerging and perhaps it will, if properly nourished,
follow in the footsteps of materials science and engineering.

Lest these remarks be misleading, it must be pointed
out that only a beginning has been made in understanding
the nature of solid surfaces and their interactions. Significant progress has been essentially confined to "clean"
surfaces. "Real" surfaces remain, by and large, on an
empirical or qualitative basis. However, one can look
with great optimism into the future as the activities on
"Surface Science and Engineering" are further consolidated.

ACKNOWLEDGEMENTS

The author is indebted to the National Aeronautics and
Space Administration and to the National Science Foundation
for supporting his work on semiconductor surfaces.

REFERENCES

1. J. Bardeen, Phys. Rev. 71, 717 (1947).
2. W. Shockley and G. L. Pearson, Phys. Rev. 74, 232 (1948).
3. J. W. Gibbs, "Collected Works," Yale University Press, 1948, Vol. 1.
4. See for example W. W. Mullin in "Metal Surfaces," ASM, 1963, p. 17.
5. For a review see A. D. Adamson "Physical Chemistry of Surfaces," Interscience Publishers, Inc. 1960.
6. See for example G. Ehrlich in "Metal Surfaces," ASM, 1963, p. 221.
7. See for example T. N. Rhodin, J. Am. Chem. Soc. 72, 4343 (1950).
8. T. A. Delchor and G. Ehrlich, J. Chem. Phys. 42, 2686 (1965).
9. D. M. Young, Trans. Faraday Soc. 48, 548 (1952).
10. B. M. W. Trapnell, "Chemisorption," Butterworths Scientific Publications, London, 1955.

11. See for example O. Beeck and A. W. Ritchie, Disc. Faraday Soc. $\underline{8}$, 159 (1950).

12. J. R. Schrieffer, J. Vac. Sci. Technol. $\underline{9}$, 561 (1972).

13. T. B. Grimley, J. Vac. Sci. Technol. $\underline{8}$, 31 (1971).

14. O. Kubaschewski and B. Hopkins, "Oxidation of Metals and Alloys," in 2nd edition, Academic Press, 1962.

15. R. L. Park and H. E. Farnsworth, J. Appl. Phys. $\underline{35}$, 2220 (1964).

16. J. W. May and L. H. Germer, Surface Sci. $\underline{11}$, 443 (1968).

17. E. W. Müller, Zeit. Physik $\underline{106}$, 541 (1937).

18. E. W. Müller, Zeit. Physik $\overline{108}$, 668 (1938).

19. "Modern Diffraction and Imaging Techniques in Materials Science," edited by S. Amelinckx, R. Gevers, R. Remaut, and J. Van Landuyt, North-Holland Publishing Company, 1970, E. W. Müller, p. 683.

20. E. W. Müller, Zeit. Physik $\underline{131}$, 136 (1951).

21. E. W. Müller, ref. 19, p. 701 and 717.

22. S. S. Brenner, in "Metal Surfaces," ASM, 1963, p. 305.

23. E. W. Müller, J. A. Panitz and S. B. McLane, Rev. Sci. Instr. $\underline{39}$, 83 (1968).

24. S. S. Brenner and J. T. McKinney, Appl. Phys. Letters $\underline{13}$, 29 (1968).

25. For a general review see P. J. Estrup, ref. 19, p. 377.

26. For a general review see P. J. Estrup and E. G. McRae, Surface Sci. $\underline{25}$, 1 (1971).

27. A. U. McRae and L. H. Germer, Ann. New York Acad. Sci. $\underline{101}$, 627 (1963).

28. E. G. McRae and C. W. Caldwell, Surface Sci. $\underline{2}$, 509 (1964).

29. I. Marklund and S. Andersson, Surface Sci. $\underline{5}$, 197 (1966).

30. J. J. Lander, Progr. Solid State Chem. $\underline{2}$, 26 (1965).

31. M. Henzler, Surface Sci. $\underline{9}$, 31 (1968); J. Appl. Phys. $\underline{40}$, 3758 (1969).

32. For a general review see C. C. Chang, Surface Sci. $\underline{25}$, 53 (1971); numerous papers on Auguer spectroscopy studies have appeared in the recent volumes of Surface Sci.

33. J.J. Lander, Phys. Rev. 91, 1382 (1953).

34. P.R. Thornton, "Scanning Electron Microscopy,"
    Chapman and Hall, Ltd., London, 1968.

35. For a general review see C.R. Booker, ref. 19, p.
    553.

36. For a general review see G.W. Kammlot, Surface
    Sci. 25, 120 (1971).

37. For a review on electron probe microanalysis see
    W. Reuter, Surface Sci. 25, 80 (1971).

38. N.S. Platakis and H.C. Gatos, J. Electrochem. Soc.
    119, 914 (1972).

39. For a general review see A.J. Socha, Surface Sci.,
    25, 147 (1971).

40. For a general review see D.P. Smith, Surface Sci.,
    25, 171 (1971).

41. H.C. Gatos and A.J. Rosenberg, in "The Physics
    and Chemistry of Ceramics," edited by C. Klingsberg,
    Gordon and Breach Science Publishers, Inc., 1963,
    p. 196.

42. W.W. Harvey and H.C. Gatos, J. Electrochem. Soc.
    105, 654 (1958).

43. P. Handler and W.M. Portnoy, Phys. Rev. 116, 516
    (1959).

44. H.C. Gatos in "Interdisciplinary Approach to Friction
    and Wear," edited by P.M. Ku, National Aeronautics
    and Space Adminstration, Washington, D.C., 1969.

45. H.C. Gatos and M.C. Lavine, J. Electrochem. Soc.
    107, 427 (1960).

46. M.C. Lavine, H.C. Gatos and M.C. Finn, J. Electro-
    chem. Soc. 108, 974 (1961).

47. H.C. Gatos, M.C. Lavine and E.P. Warekois, J.
    Electrochem. Soc. 108, 645 (1961).

48. E.P. Warekois, M.C. Lavine, A.N. Mariano and
    H.C. Gatos, J. Appl. Phys. 33, 690 (1962).

49. R.E. Hanneman, M.C. Finn and H.C. Gatos, J.
    Phys. Chem. Solids 23, 1553 (1962).

50. H.C. Gatos, P.L. Moody and M.C. Lavine, J. Appl.
    Phys. 31, 212 (1960).

51. H.C. Gatos, M.C. Finn and M.C. Lavine, J. Appl.
    Phys. 32, 1174 (1961).

52. For a detailed treatment of semiconductor surfaces

see: A. Many, Y. Goldstein and N. B. Grover, "Semi-conductor Surfaces," North-Holland Publishing Company, 1965; D. R. Frankl, "Electrical Properties of Semiconductor Surfaces," Pergamon Press, 1967.

53. W. H. Brattain, Phys. Rev. 72, 345 (1948).

54. J. Lagowski, C. L. Balestra and H. C. Gatos, Surface Sci. 29, 213 (1972).

55. J. Lagowski, C. L. Balestra and H. C. Gatos, Surface Sci. 29, 203 (1972).

56. J. Lagowski, E. S. Sproles and H. C. Gatos, Surface Sci. 30, 653 (1972).

57. J. Lagowski, I. Baltov and H. C. Gatos, Surface Sci., in press.

58. H. C. Gatos and J. Lagowski, J. Vac. Sci. Technol. 10, 130 (1973).

59. J. Lagowski and H. C. Gatos, J. Appl. Phys. Lett. 20, 14 (1972).

60. J. Lagowski and H. C. Gatos, Surface Sci. 30, 491 (1972).

61. J. Lagowski and H. C. Gatos, Proc. Int. Conf. Semi-conductor Physics, Warsaw 1972, p. 1462.

62. P. B. Weisz, J. Chem. Phys. 20, 1483 (1952); P. Aigrain and C. Dugas, Z. Elektrochem. 56, 363 (1952); K. Hauffe and H. J. Engell, Z. Elektrochem. 56, 366 (1952).

63. For a general review and further discussion see R. F. Baddour and C. W. Selvidge, in "Progress in Solid State Chemistry," edited by H. Reiss, Pergamon Press Vol. 3, 1967, p. 45; T. A. Goodwin and P. Mark in "Progress in Surface Science," edited by S. G. Davison, Pergamon Press, Vol. 1, 1972, p. 1; see also paper by T. J. Gray in present volume.

64. I. E. Tamm, Physik Z. Sowjet 1, 733 (1932); Z. Phys. 76, 849 (1932).

65. E. T. Goodwin, Proc. Camb. Phil. Soc. 35, 205, 221, 232 (1939).

66. W. Shockley, Phys. Rev. 56, 317 (1939).

67. J. Koutecky and M. Tomasek, J. Phys. Chem. Solids 14, 241 (1960).

68. M. Tomasek, Surface Sci. 2, 8 (1964).

69. V. Heine, Surface Sci. 2, 1 (1964).

70. J.D. Levine and P. Mark, Phys. Rev. 144, 751 (1966).
71. G. Herring in "Metal Surfaces," ASM, 1952, p. 1;
     R.P. Ewald and H. Juretschke in "Structure and Prop-
     erties of Solid Surfaces," edited by R. Gomer and C.S.
     Smith, University of Chicago Press, 1953, p. 82.
72. N.D. Lang and W. Kohn, Phys. Rev. B1, 4555 (1970).
73. N.D. Lang and W. Kohn, Phys. Rev. B3, 1215 (1971).
74. R. Haydock, V. Heine, M.J. Kelley and J.B. Pendry,
     Phys. Rev. Lett. 29, 868 (1972).
75. S.J. Gurman and J.B. Pendry, Phys. Rev. Lett. 31,
     637 (1973).
76. D. Eastman in "Experimental Electron Spectroscopy,"
     edited by E.A. Sherley, North-Holland Publishing Com-
     pany, 1972.
77. K.H. Johnson and R.P. Messmer, J. Vac. Sci. Technol.,
     in press; R.P. Messmer, C.W. Tucker and K.H. John-
     son, Surface Sci., in press.

DISCUSSION

K.K. Verma (Alfred): In amorphous semiconductors, including chalcogenide glasses, the energy band model is
modified by the tailing of the extended energy states into
the forbidden gap and by the presence of localized energy
states within the gap. These modifications are related to
the lack of long-range order in the material, local stoichiometric and structural deviations and other kinds of possible
traps. Apparently, these are the phenomena do to the
bulk material. Strictly surface studies of related material
too have evidently shown the presence of energy states in
the forbidden gap.

Since the task of investigation of the bulk and the surface phenomena is similar, e.g., temperature and frequency effects on electrical conductivity, photoconductivity,
thermally stimulated currents, SCLC, then how can one
distinguish and isolate the surface effects and other sources
of localized energy states in the forbidden gap of the energy
band model?

Author: In general, localized states at the surface and in
the bulk can be assigned by various methods - direct or

indirect - to bulk or surface phenomena. For example, when a method (based on conductance, photovoltage, etc.) detects pronounced changes resulting from various surface treatments, various ambient atmospheres, or surface orientation, then one can be quite certain that this particular method is primarily measuring surface properties. When there is no variation in localized states between the bulk and the surface, then surfaces are not of particular significance to the electronic characteristics of the material.

D. Dove (Univ. Florida): Could you please comment on the measurement of electronic states at the interface between two materials.

Author: Interface states are very important in Metal-Oxide-Semiconductor (MOS) and Semiconductor-Oxide-Semiconductor (SOS) electronic devices. Their study presents far greater difficulties than the study of free-surface states. Methods have been developed and measurements carried out with varying degree of success. These methods are based on the study of the capacitance-voltage or current-voltage characteristics of these devices. For a discussion of the general subject an appropriate reference is: S. M. Sze, "Physics of Semiconductor Devices," Wiley-Interscience, 1969.

L. L. Hench (Univ. Florida): Are the property changes observed in the new technique described due entirely to electronic shifts or is an alteration of local stoichiometry, e. g., anion coordination configurations, also involved?

Author: The effects observed in surface photovoltage spectroscopy are strictly due to electron transition from the surface states into the conduction band or from the valence band into the surface states. I do not believe that stoichiometry in the real sense plays a role. However, there is evidence (see text) that rearrangements of the surface atoms take place in these types of materials. It is reasonable to expect that such atom rearrangements must affect the electronic behavior of the surfaces. However, no direct quantitative information is available.

A. Choudry (Univ. Rhode Island): Could you comment on the new technique you referred to for measuring the surface

traps, particularly how it differs from the rather well
established optical and electrical methods?
Author: The main advantage of Surface Photovoltage
Spectroscopy is that it allows the study of surface states
in high energy gap semiconductors, e. g., CdS, ZnO.   In
these materials the surface states are not in equilibrium
with the bulk and the standard techniques (Field-effect for
example) are inapplicable.   It differs from the other
optical methods in that it is based on sub-band gap illumination; consequently there is no interference from band-to-
band transitions (photoconductivity); such interference is
one of the disadvantages of the other methods employing
illumination with energies greater than the energy gap.

CATALYSIS IN CORRELATION WITH THE DETAILED ELECTRONIC STRUCTURE OF THE SURFACE

Thomas J. Gray

Director, Atlantic Industrial Research Institute
Halifax, Nova Scotia

The earliest suggesions of an inter-relationship between the electronic constitution of solids and the adsorption, desorption, and reaction processes which occur at
their surfaces came in the work of Pisarzheveskii[1] and
Langmuir[2]. These relationships are also implicit in the
work of Roginski[3] and were in some measure amplified
by Wagner[4] and Nyrop[5]. Formulation of a correlation
between electronic structure and catalysis was simultaneously developed by Garner and coworkers[6] and by Volkenshtein[7] which, in association with the rapid advances
made in solid state theory, particularly as applied to
semiconductors, has led to the almost universal acceptance
of a general correlation between the electronic constitution
of a catalyst and the adsorption and reaction processes
which occur at its surface. Nevertheless, the oversim-
plified model adopted by many authors mitigates against
a wider utilization of the general concepts in endeavouring
to tailor catalysts for specific applications. There is a
marked reluctance to accept the complexity of a realistic
model deriving from modern concepts of the solid state.
It has been repeatedly emphasized that in many particulars
the minority electronic defects have greater significance
than the majority defects as the rate-controlling species.
Drawing on classical terminology they may be regarded as
the "promoters".

241

In many instances the electronic theory of catalysis differs from older concepts only on the basis of semantics. Furthermore, analysis of the electronic constitution of the solid is an adjunct to a more general consideration of electron transfer processes occurring at the surface with the generation of ionized species and quasi-free radicals which may in many instances constitute the "active site". The "Multiplet Theory" of Balandin[8] can also be interpreted as a function of the electronic constitution of the solid. The electronic theory of catalysis, of itself, should not be construed as an exclusive theory of catalysis but rather as the common denominator coordinating the viable portions of many other theories.

It is essential to establish a reasonable, albeit complex, model for the electronic constitution of a solid catalyst both in respect of the surface and the bulk. On the basis of solid state concepts it is permissible to represent the surface of a catalyst electronically as a region with many localized electronic levels existing in the forbidden zone between the uppermost filled valence band and the conduction band. This representation must include Tamm[9] or localized surface states and reflects the defect constitution of the solid. While Tamm states exist even for an ideal crystal lattice due to the asymmetry of lattice termination, the more important levels are those due to surface structural defects, impurities, adsorbed species and the like. Shockley[10] considered a more generalized variant of surface states. On either basis the density of intrinsic surface states is about $10^{15} cm^{-2}$. Although the concept of intrinsic surface states is widely used, there is as yet no direct experimental proof of their existence. Inferential proof is, however, derived from investigations of extremely clean surfaces of super-pure semiconductors. It is possible to identify two types of surface state, i.e., "fast" states for which the residence time is 10^{-7} sec or less, and "slow" states where the residence time is 10^{-2} sec or longer as first established by Kingston[11]. It would, however, be naive to assume that these could arbitrarily be divided into intrinsic and extrinsic states. The concept of surface states enabled Bardeen[12] to explain the experimental observations on silicon and germanium that the contact potential is independent of bulk doping and that the rectifying characteristics of point contact diodes are independent of the work function of the metal.

The presence of lattice imperfections in the bulk as well as in the surface, such as Schottky or Frenkel defects, impurities, including deviations from stoichiometry, dislocations and general disorder, all contribute localized electronic impurity levels situated between the uppermost valence band and the conduction band. These and other electronic characteristics have been considered in considerable detail in the many standard texts on solid state physics.

The correlation between electronic structure and catalysis is extensively treated in the publications of Gray[6, 13], Stone[14], Volkenshtein[7] and others. The surface electronic energy states are represented diagrammatically in Fig. 1. It will be observed that these representations are significantly more complicated than those advanced by Hauffe[15], Volkenshtein[7], and others. Particular attention is drawn to the minority electronic levels and surface states which are capable of acting as trapping levels since they appear to play a dominant role in catalysis. This is particularly the case with finely divided materials of high surface area where the surface characteristics predominate and may be very different from those of the bulk. It is considered essential that this complexity be fully appreciated in order that the significance of promoters and other catalyst modifiers can be better understood. It has been emphasized by Gray[13] that, contrary to the views expressed by Volkenshtein, it is frequently the minority components (promoters) that have the greater significance. If the significant factor were only the surface "depletion" or "enhancement" zone it would be difficult, if not impossible, to account for the activity of complex systems such as typical transition metal oxides illustrated in Fig. 1c. However, the most telling argument against correlations based on these zones

is the realization that relatively few sites on a catalyst
surface are active.

The model applies not only to semiconductors but also
to relatively insulated solids such as are frequently used
as catalysts or catalyst substrates. Investigations after
the suggestion of Joffe have established that the band model
description of electronic states, originally developed on the
basis of the periodic lattice in a crystalline solid, is in

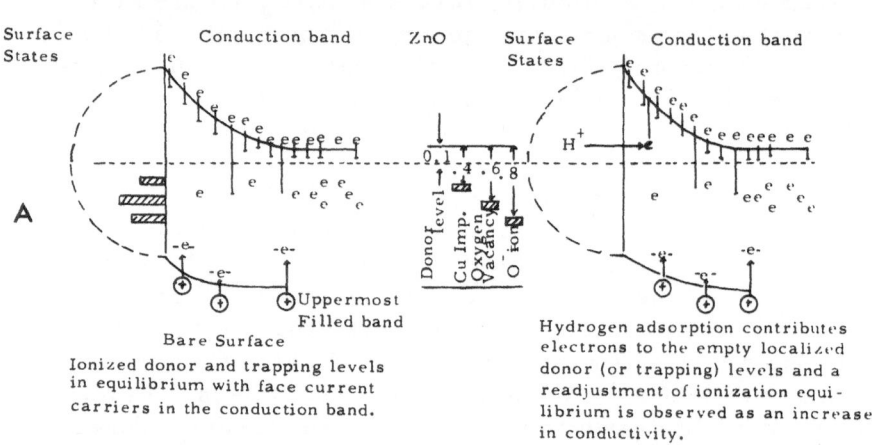

A

Bare Surface

Ionized donor and trapping levels
in equilibrium with face current
carriers in the conduction band.

Hydrogen adsorption contributes
electrons to the empty localized
donor (or trapping) levels and a
readjustment of ionization equilibrium is observed as an increase
in conductivity.

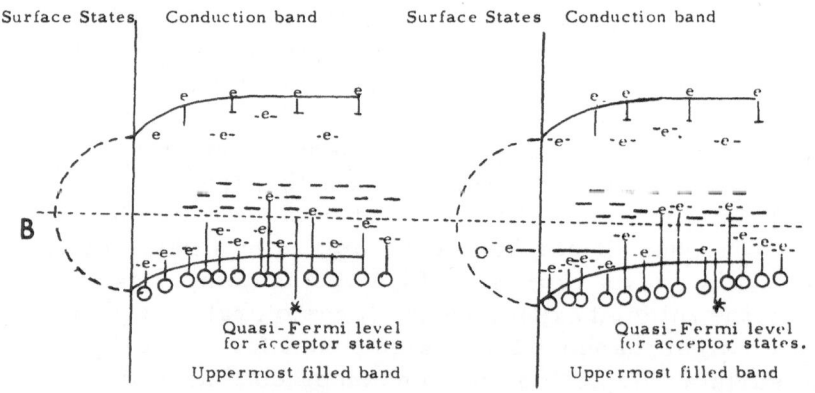

B

Quasi-Fermi level
for acceptor states

Uppermost filled band

Bare Surface

Ionized acceptor states in equilibrium
with unionized states withdraw electrons
from the uppermost filled band generating
positive holes occasioning p-type conductivity by counter migration of electrons in
the filled band.

Quasi-Fermi level
for acceptor states.

Uppermost filled band

Oxygen Adsorption

Electron transfer occurs between the
occupied acceptor levels and the adsorbed oxygen promoting further ionization and a reestablishment of equilibrium with an increase in the number
of positive holes in the filled band.

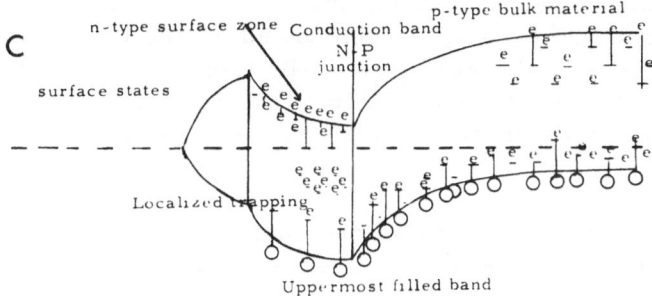

Fig. 1. Surface electronic energy states. (a) predominantly n-type semiconductor, (b) predominantly p-type
semiconductor and (c) complex catalyst surface.

most respects equally applicable to noncrystalline solids
with only short range periodicity. While certain special
attributes pertain to these systems, they can still be considered on a unified basis. As the specific surface area
increases so does the significance of surface electronic
constitution and energy states and in all probability these
characteristics dominate for high surface area catalyst
materials.

Adsorption on a catalyst surface with the establishment of a fugitive and weak bond involves electron transfer to a localized electronic state or "active site", which
experience indicates as being few in number. The adsorption of a reactive species on the surface of a catalyst results in electron transfer involving localized impurity
levels with the deeper-lying levels being predominant in
the earlier stages of adsorption, with a higher heat of
adsorption, longer residence time for the adsorbed species
and greater propensity towards extensive degradation of
the catalyst. In contrast, the low energy sites are amenable to rapid reaction and desorption. This concept must
be appreciated in the light of a dynamic system with the
probability that the deepest lying traps may become perma-
nently blocked but still not interfere with the net reaction
process. Inherent in the concept is the feature that quasi-

free radicals or other ionized species at the surface, of
moderate life-time, may be construed as the catalytic
species, whereas the associated electron transfer may
be that aspect amenable to study by measurements of
semiconductivity, photoconductivity, and associated properties.

Typical correlations can be visualized by considering
oxidation reactions over metal oxides. The adsorption of
oxygen on p-type semiconductors occurs with the transfer
of an electron between the occupied acceptor levels and
the adsorbed oxygen species, most probably according to
the reaction

$$O_2 + e^- \longrightarrow O + O^-$$

originally proposed by Gray[16] and confirmed in EPR studies by Mikheikin, Mashchenki and Kazanskii[17] and by
Wong[18].

The activity of the O^- ion has been widely established
and characterized in the work of Kiwan[19], Tench[20],
Kazanskii[21], and others. The presence of O_2^- molecular
ions has also been substantiated by EPR on many catalyst
surfaces, but their contribution to catalytic activity is
certainly far less than that of O^-. However, the electronic model is not significantly changed if the O_2^- molecular ion is involved as advocated by Cope and Campbell[22]
or the various other species of oxygen identified in the
work of Tench[20]. The electron transfer occasions a
shift in the electronic equilibrium with an increase in the
number of positive holes in the uppermost filled band,
reflected by an increase in electrical conductivity. Conversely, hydrogenation catalysts are frequently predominantly n-type semiconductors, for which the adsorption
of hydrogen or a hydrocarbon results in an electron transfer to a localized ionized donor level. The adsorption and
electron transfer to the donor states occasions a readjustment in the ionization equilibrium and is reflected as an

increase in electrical conductivity resulting from more electrons in the conduction band. Such catalysts are active in a wide variety of hydrocarbon reactions and the model also correlates directly with the concept of carbonium ion formation.

A more complex model is, however, frequently necessary to cover the situation arising when a material which is normally p-type in the bulk develops an n-type surface as a result of pretreatment under reducing conditions. Of significance in this case is the existence of a p-n junction near the surface which significantly modifies electronic mobility.

Advancing from these initial concepts, the presence of a multiplicity of levels agrees with the concept of energetic heterogeneity of the adsorption sites on a catalyst's surface. The deeper-lying sites present on a freshly prepared catalyst frequently result in the adsorption and retention of hydrocarbon species with excessive cracking, diminishing to a somewhat lower but more constant level of activity after the adsorption of only a few millimoles of reactant. Clarke has shown that both high- and low-energy sites are important in polymerization of ethylene, with the probability that the higher energy sites initiate the chain reaction which propagates through the low-energy sites; in the same context it is observed by ESR that Cr^{4+} and Cr^{5+} are both necessary ingredients in the polymerization of ethylene and can be associated with the deeper-lying electronic levels.

It is probable that in many instances the actual catalytic species is in fact the quasi-free radical existing at the discrete electron trapping site. This may be in the form of a pseudo-organometallic complex and can frequently be identified as such by infrared spectroscopy. Nevertheless, the electron transfer process between the catalyst and the adsorbed species represents a common unifying feature.

EXPERIMENTAL

Techniques for the establishment of the electronic constitution of catalysts before, during and after catalytic reaction have been reviewed in considerable detail by Gray[13, 26]. A working catalyst presents many conflicting electronic aspects, the analysis of which can only be derived by simultaneous observations of many parameters in correlation with actual catalytic processes. Unfortunately, the majority of information available on the correlation between electronic properties and catalysis is inadequate to provide that information essential to defining a representative model. These data should establish:

a) Number and type of the majority and minority current carriers and variations with temperature.

b) Band gap, distribution and occupancy of all energy levels as a function of temperature.

c) Mobility of the current carriers under all conditions.

d) Nature, concentration, and characteristics of surface states.

This information should be determined during all stages of catalysis and typically would include data on (a) the original catalyst material, (b) the catalyst after pretreatment, (c) the variations occasioned by the simple adsorption and desorption of the individual reactants and products, and by any intermediates, (d) the variations during the progress of a number of "type reactions" selected for simplicity, and (e) the variation of electronic properties as a function of the "ageing" of the catalyst. To this information regarding the electronic structure of the solid surface should be added information regarding the constitution of the ionized adsorbed species. This can be derived by ESR measurements and by infrared spectroscopy. These techniques are reviewed in detail by Kokes[23], Hair[24], and Amberg[25].

In the majority of investigations endeavouring to establish a correlation between the electronic constitution of a catalyst and its activity, electrical conductivity measurements have been the sole characterization and then not always under operating conditions. These measurements, particularly when utilizing D.C. procedures, are at best inaccurate and frequently invalid. Even A.C. measurements at a single and usually low frequency do not provide satisfactory data. A catalyst can be visualized as resembling "Swiss cheese", diagrammatically represented in Fig. 2. Under certain specific conditions this can be analyzed as the electrical analogue and a separation made between the characteristics of the surface and those of the "bulk" material. Utilizing A.C. bridge measurements over a wide range of frequencies and following the analysis by Gray[26], unique values of surface conductivity correlating with adsorption, reaction and desorption can frequently be established. Even so caution must be applied to the interpretation of these relationships, since severe interfacial polarization can develop at intergranular contacts. These effects can frequently be identified by the behaviour of the relaxation time derived from the temperature coefficient of the maximum for the dielectric loss peak. The typical dependence of electrical properties on frequency is illustrated in Fig. 3.

No single A.C. bridge is capable of providing the accurate data over the very wide frequency range required

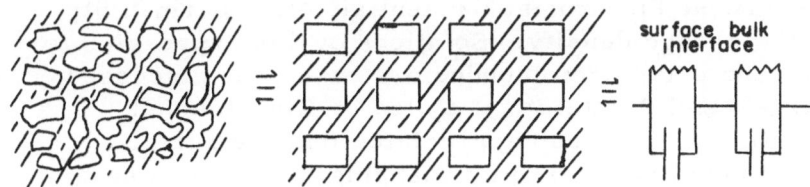

Fig. 2. "Swiss Cheese" model of catalyst and the electrical equivalent.

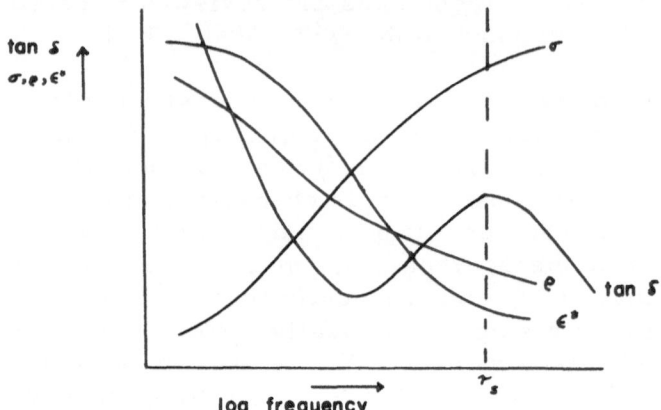

Fig. 3.   The typical dependence of electrical properties
on frequency.

to establish the complex conductivity of a catalyst.   A
combination of conventional commercial A. C. bridges
adequately covers the audio and radio frequency ranges,
but the very low frequency range which is often vitally
important requires special attention[27].

Surface states on silicon and germanium single crystal
semiconductors have been extensively investigated in relation to solid state device technology.   The distribution
of the electrostatic potential in a surface space-charge
layer is given by the Poisson equation,

$$\nabla^2 \Psi = \frac{-4\pi}{\epsilon} \rho \tag{1}$$

where Ψ is the electrostatic potential, ϵ the permitivity,
and ρ the charge density.   Solutions of this expression for
both n- and p-type semiconductors have been discussed
by Garrett and Brattain[28] and also by Kingston and
Neustadter[29].   Bardeen appreciated that the surface
potential could be altered by changing the gaseous environment at the surface, thereby altering the distribution
and density of the slow states which govern the Fermi

level at the surface.   Measurements of the field effect of
surface conductivity with changing gaseous environment
can be used to study surface states.

Surface recombination relates the difference between
the number of recombining minority carriers and the number of generated carriers at the surface as directly proportional to their density, deriving therefrom a surface
recombination velocity.

Surface states may be investigated in relation to surface recombination centers, recombination velocity and
charge capture as a function of the adsorption, reaction
and desorption of foreign species.   Several techniques
exist for the measurement of surface recombination
velocities, one of the most widely used being that of
Many[30] utilizing longitudinal pulses superimposed on a
transverse field in a double-bridge circuit measuring
carrier lifetimes and surface conductivity.   The photo-
conductive decay method is in many ways superior, avoiding many of the disadvantages of the Many technique.   However, neither of these methods is particularly suited to the
investigation of the kinetics of processes occurring at a
surface.   For this purpose the steady-state photoconductivity method can be advantageously employed as proposed
by Novototskii-Vlasov[31].   Unfortunately, these techniques
do not lend themselves to investigations involving small
particle polycrystalline catalyst materials.

The technique of thermally stimulated electron current measurements developed by Gray and Amigues[13] provides direct information on a majority of these aspects, incorporating at the same time measurements of electrical
conductivity, photoconductivity and associated properties.
By this technique it is possible to establish the position,
occupancy, origin and significance of the multiplicity of
levels occurring in either simple or complex catalyst
systems.

Measurements to establish the mobility of current
carriers in catalyst materials are even more difficult than

those of conductivity, which must be made simultaneously.
One common method, Hall coefficient determination, while
relatively simple for metals or single-crystal semiconductors, becomes more difficult for dense polycrystalline
bodies and extremely difficult, almost to the point of im-
possibility, in the case of lightly compressed pellets or
loose powders of catalysts. A very few measurements have
been reported but, unfortunately, the measurement techniques have all been inappropriate. Dual A.C. Hall effect
procedure using both an alternating magnetic field and an
A.C. energizing field of different frequency, making measurements over a range of frequencies, appears to be the
only appropriate approach to this difficult problem which
is further compounded by the significant magneto-resistive
component likely to be observed in catalyst materials.
Even if it were possible to achieve satisfactory Hall effect
measurements, analysis to separate the mobilities of the
majority and minority carriers is still very difficult.

A technique showing greater potential is that of the
photoelectromagnetic effect (PEM) which is observed
when a semiconductor situated in a magnetic field is
illuminated on a surface perpendicular to the magnetic
field and the emf generated mutually at right angles to
these directions is measured. This effect was first reported by Kiikoin and Naskov[32] and analyzed by Frenkel[33]
and was recognized as resulting from the Hall effect due
to photoinduced diffusion currents. The possible advantage in adopting this technique would be to facilitate the
separation of the specific significance of majority and
minority mobilities under static and reaction conditions.

Measurements of Thermally Stimulated Electron Currents

Thermal glow (TG) curves together with thermally
stimulated electron currents (TSC) have received considerable attention by solid state physicists in establishing
the existence and occupancy of electronic trapping levels
first proposed by Urbach[34]. The technique has been discussed in detail and the theoretical considerations analyzed

in the work of Randall and Wilkins[35], Garlick[36], Heijne[37], and others. A sample of material is cooled to a low temperature, conveniently 77° K, and exposed to illumination of energy exceeding the band gap. This results in the saturation of photoconductivity and almost full population of the various electronic levels when the illumination is suddenly removed. After a brief period for dark adaptation, the specimen is warmed at a linearly programmed rate, typically 2-20° C/min., while the conductivity and/or the optical emission is recorded as a function of temperature. Peaks in optical emission and the corresponding peaks in electrical conductivity occur as a result of the thermal emptying of successive electronic trapping levels. From the position and character of these peaks the energy, concentration and occupancy of the individual electronic levels can be deduced. The equipment is illustrated in Fig. 4.

The powdered catalyst is contained in a fused silica cell, open on the upper surface, incorporating electrodes and a thermocouple. The connections are carefully insulated to insure a leakage resistance of not less than 10^{15} ohms since the resistance encountered in measurements is frequently in this range. The collimated illumination is filtered from infrared to avoid the quenching effects frequently encountered. However, provision is made for the supplementary application of quenching or monochromatic radiation intended to stimulate specific energy levels. Stringent precautions are observed to avoid illumination falling on the electrodes lest photovoltaic effects invalidate measurements. Thermal glow curves can be investigated by a fiberoptic light pipe and an external high-gain photomultiplier selected for very low noise and wide spectral response.

The apparatus is coupled with a vacuum/gas handling system for degassing to 10^{-9} Torr with palladium and silver diffusion thimbles for the precision admission of hydrogen and oxygen respectively. A low pressure of high purity helium is generally used during the measurements to ensure satisfactory heat transfer.

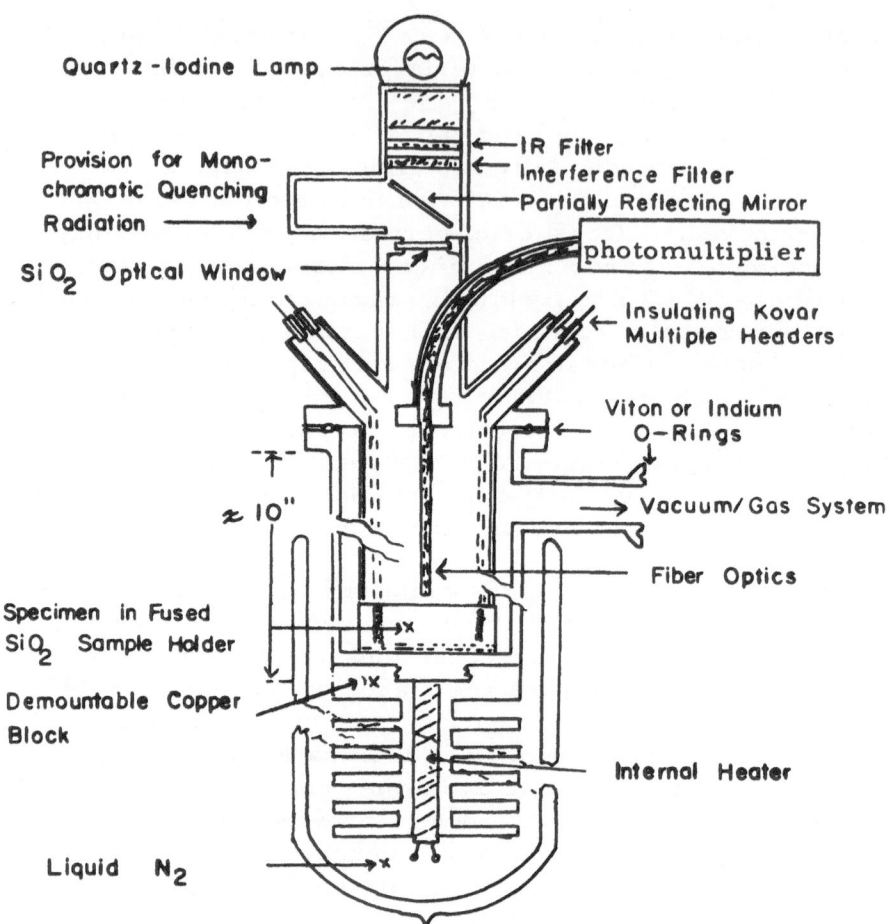

Fig. 4.   Apparatus for the measurement of thermally
stimulated electron current and optical thermal glow curves.

One important variation in the technique, of particular
advantage in the investigation of catalyst materials when
unsatisfactory resolution of a large number of peaks is encountered, is a modification of the photoelectret measurements of Semak, Chepur, and Zolotarev[38]. In the modified
procedure introduced by Gray[39], the specimen is polarized
at low temperature under saturation illumination and, after
the light has been extinguished, the short-circuit current

for the sample, reflecting the discharge of photoelectrets, is measured during the linear rise in temperature. This aids significantly in the resolution of individual emission peaks which are otherwise frequently masked by extensive overlap. Unfortunately, the technique does not lend itself to accurate determination of occupancy of the particular level. All measurements employ either two- or four-electrode orientation feeding through a Keithley electrometer to an X-Y recorder. Standardized pretreatment of the catalyst is adopted, followed by adsorption, reaction and desorption with intermediate determinations of TSC and TG curves. Provision is incorporated for the use of supplementary illumination by monochromatic radiation of less than band gap energy to inhibit adsorption for specific electronic energy levels.

Data derived from these measurements can establish the position, number and occupancy of the various electronic levels between the uppermost filled and conduction bands and can establish preliminary correlation with the adsorption of individual species. However, in order to establish correlation with practical catalytic processes, extensive screening is necessary for a large number of closely related valence-controlled impurity-doped catalysts over well-defined composition ranges for a variety of type reactions. To facilitate such measurements, a rising temperature reactor technique has been developed capable of generating in a few hours data which by classical methods might well occupy weeks or months. The equipment is illustrated diagrammatically in Fig. 5 from the original design by Gray and Oswin[40]. A variant of the system has been described by Bridges and Houghton[41] including an analysis of the kinetic data. A more detailed analysis has been presented by Gray and Lowery[42]. The vessel is a miniature reactor with a preheat section and a catalyst zone for 5-10 cm^3 of catalyst in a 5 cm bed with upward gas flow. The surface area of the catalyst after pretreatment is first determined, using a variation on the flow technique of Nelson and Eggertsen[43]. Subsequently, the reactor temperature is raised linearly at a rate of

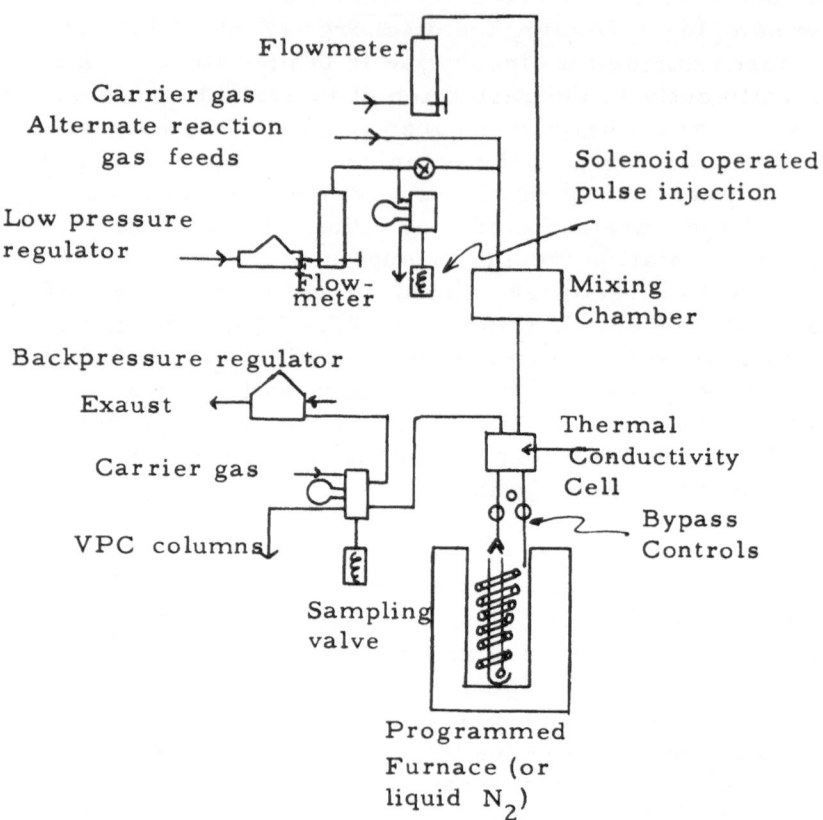

Fig. 5.   Rising temperature catalyst screening reactor system.

1-2°/min. with a feed of carrier gas and either a constant flow of reactant or a succession of small pulses at 50° C or 10° C intervals.   In the former case the flow thermal conductivity gauges give direct conversion data whereas with the latter technique detailed results are available sequentially from the VPC apparatus.   The temperature is programmed to rise to a predetermined maximum, hold for a fixed period and then cool while continuing to record data.

For a simple heterogeneous catalytic reaction for which the reactant can be completely converted to product over a representative temperature range, plotting conversion against temperature results in a sigmoid plot. Analysis readily establishes that the slope of this curve at the point of inflexion is dependent on the apparent activation energy for the catalytic process and is not significantly affected by moderate changes in operating conditions. The actual temperature for the point of inflexion does vary with velocity, turbulence and nonuniformity of temperature distribution throughout the bed. Nevertheless, under carefully selected and standardized conditions, this temperature can be used as a qualitative indication of relative activity. Furthermore, after holding the catalyst under reaction conditions at an elevated temperature followed by a reducing temperature study, the relative rise in temperature of the point of inflexion is a very useful measure of the "ageing" of the catalyst.

When using the pulse technique, data on adsorption and retention on the catalyst can be obtained by the flow-gram for the individual pulses. This also provides a direct indication of materials balance and gives a rapid indication of complicating factors. The technique has been used to establish the retention of polymer on moderately high energy sites for polymerization catalysts during the early stages of adsorption without interference with the main polymerization process taking place on sites of lower energy. The technique lends itself readily to fully automated, computerized screening of catalysts and the analysis of the relevant data.

RESULTS

While several analyses of TSC measurements have been advanced, Dittfeld and Voigt[44] have concluded that the Bube[45] method was the most satisfactory. It relates the temperature for the maximum TSC or glow curve to the trap depth and determines that the initial rise in the TSC is given by the simple exponential relation

$$i = n_o \cdot S \cdot \exp^{(-E/kt)} \tag{2}$$

where i is the current and S a frequency constant related
to the ability of the electron to escape from the trap at
depth E with an initial occupancy of n_o. The analysis assumes negligible retrapping; however, modification to
accommodate retrapping does not alter the exponent, so
that the use of this simple expression is justified for determination of E. If the actual occupancy is to be determined, then more exact analyses are essential and the
degree of retrapping must be assessed.

Attention has so far been concentrated primarily on
zinc oxide and titanium dioxide (anatase). The former is
of considerable importance not only in catalysis but also
in electron photography where adsorption of dye sensitizers,
oxidant modifiers and the resin vehicle all relate directly
to the detailed electronic structure. Data on this system
have already been reported by Gray and co-workers[13, 42].

The most important aspects of these investigations
were the establishment of the dominant significance of O^-
energy level and the oxygen vacancy level, both being more
important in adsorption and catalysis than the majority
donor level. These levels are associated with energies
of 0.8 eV and 0.6 eV respectively and are confirmed by
the thermoluminescent studies of Maenhout-Van der Vorst[46]
and of Seitz, Tinter and Hirthe[47]. Of further interest is
the identification of copper (0.4 eV) and lead (0.25 eV) impurity levels, even at concentrations of parts per billion,
and the existence of an ionic conduction process. Recent
data by Gray and Lowery[42] have shown the effect of particle size, while corresponding data have been observed
related to morphology variations.

The adoption of the photoelectret technique aids mater-
ially in the establishment of the distribution of levels as
illustrated in Fig. 6.

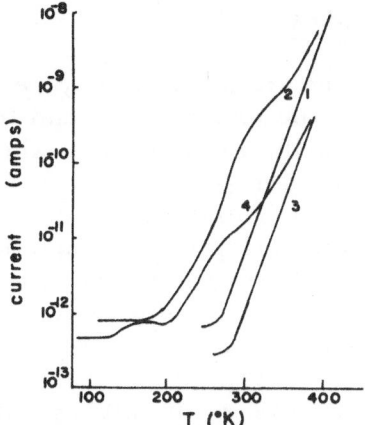

Fig. 6. Distribution of levels obtained in adoption of photoelectret techniques.

Although less well understood, important information can also be obtained by extending the TSC measurements above room temperature. Results for pure anatase illustrating the effect of adsorption of hydrogen, oxygen and propylene are illustrated in Fig. 7.

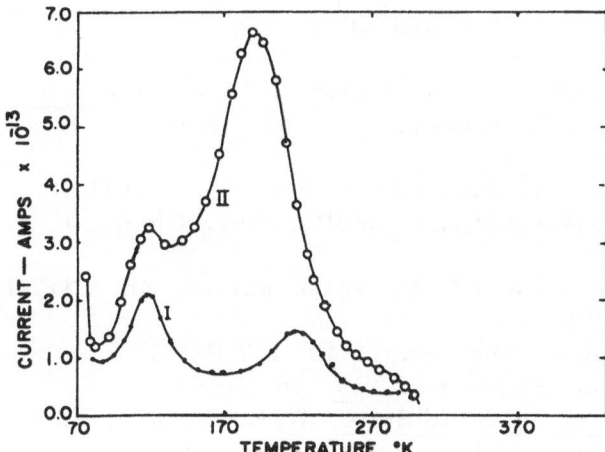

Fig. 7. Effect of adsorption of hydrogen, oxygen and propylene for pure anatase.

CONCLUSION

The extensive information which may be derived from these investigations can be expected to lead to a more accurate appreciation of the inter-relationship between detailed electronic structure and catalysis, including the important aspects of intermediate complex formation and desorption. Direct correlation with "impurity-" or "promoter-" related electronic levels has been established while at the same time there is evidence of protonic and carbonium ion formation.

While the electronic theory of catalysis does not purport to be totally definitive it can be regarded as the unifying basis to co-ordinate many important and often conflicting postulates present in earlier theories of catalysis. It must be regarded as an adjunct to these older concepts rather than as a total replacement. The acceptance of the more complex model is imperative and should lead to a better understanding of the nature of the catalytic process and of the catalyst surface, both of which are essential for satisfactory tailoring of catalysts for specific purposes.

REFERENCES

1. L. V. Pisarzhevskii, in Theories Electronique de la Catalyse, F. F. Volkenshtein, ed., Masson, Paris, 1961.
2. I. Langmuir, J. Am. Chem. Soc. 40, 1361 (1918).
3. S. Z. Roginskii and E. I. Shultz, Ukr. Chem. J. 3, 177 (1928).
   S. Z. Roginskii and F. F. Volkenshtein, Zh. Fiz. Khim. 29, 483 (1955).
4. K. W. Wagner, Ann. Phys. 40, 817 (1913).
5. J. W. Nyrop, Phys. Rev. 39, 967 (1932).
6. Chemistry of the Solid State, W. E. Garner, ed., Academic Press, New York, 1955.
   T. J. Gray, Defect Solid State, Wiley (Interscience, New York, 1957).
7. F. F. Volkenshtein, Advanes in Catalysis, Vol. 12, 189.

Electron Theory of Catalysis on Semiconductors, Fitzmatgiz, 1960.

Kinetika i Kataliz 2, 481 (1961).

Trans. Faraday Soc. 31, 209 (1961).

Surface Properties of Semiconductors, A. N. Frumkin, ed., Consultants Bureau, New York, 1964, p. 279.

8. A. A. Balandin, Modern Status of the Multiplet Theory of Heterogeneous Catalysis (In Russian), Plenum Press.

9. I. E. Tamm, Zh. Eksp. Teor. Fiz., 3, 34 (1933).

10. E. Shockley, Phys. Rev., 56, 317 (1939).

11. R. H. Kingston, J. App. Phys., 27, 101 (1956).

12. J. Bardeen, Phys. Rev., 71, 717 (1947).

13. T. J. Gray, Defect Solid State, Wiley (Interscience) New York, 1957.

    T. J. Gray and S. D. Savage, Disc. Far. Soc., 28, 159 (1959).

    T. J. Gray, N. G. Masse and C. C. McCain, J. Phys. Chem., 63, 472 (1959).

    T. J. Gray, Actes 2nd Congr. Intern. Catalyse, Paris, 1960, Editions Techniq. Paris, 1961, p. 1561.

    T. J. Gray and D. O. Carpenter, Proc. 3rd Intern. Congr. Catalysis, Amsterdam, 1964, Wiley, New York, 1965, p. 463.

    T. J. Gray and P. Amigues, Surface Science, 13 (1969) 209-221 (1968).

    T. J. Gray, Measurements of Semiconductivity, Photoconductivity and Associated Properties of Catalysts, Academic Press, Inc., New York, 1968.

14. F. S. Stone, Disc. Faraday Soc., 29, 211 (1959).

    F. S. Stone, and T. I. Barry, Proc. Roy. Soc., A255, 124 (1960).

15. K. Hauffe and H. J. Engell, Z. Electrochem., 56, 366 (1952); 57, 762 (1953).

16. T. J. Gray, Chemistry of the Solid State, Academic Press, New York, 1955.

17. I. E. Mikheikin, A. I. Maschchenki and U. B. Kazanskii, 8, 1363 (1967).

18. N. B. Wong and J. H. Lunsford, J. Chem. Phys., 55, 3007 (1971).

19. T. Kiwan, Proc. Int. Congr. Catalysis, Amsterdam 1964, p. 493.

20. A.J. Tench, JCS Faraday (1), 68, 1169; 1181 (1972);
    JCS Faraday (1), 69, 857 (1973).
21. V.B Kazanskii, Int. Symp. on Heterogeneous Catalysis,
    Roermonde, 1967.
    V.A. Shvets, V.M. Vorotinsev and V.B. Kazanskii,
    Kinetika i Kataliz, 10, 356 (1969).
22. J.O. Cope and I.D. Campbell, JCS Faraday (1), 69,
    1 (1973).
23. R.J. Kokes, Experimental Methods in Catalytic Re-
    search, R.B. Anderson, ed., Academic Press, New
    York, 1968.
24. M.L. Hair, Infrared Spectroscopy in Surface Chemistry,
    Dekker, New York (1967).
25. C.H. Amberg, The Solid Gas Interface, E. Alison
    Flood, ed., Dekker, New York (1967).
26. T.J. Gray, Experimental Methods in Catalytic Re-
    search, R.B. Anderson, ed., Academic Press, New
    York, 1968.
27. D.J. Scheiber, J. Res. Natl. Bur. Std., C65, 23
    (1961).
28. C.G.B. Garrett and W.H. Brattain, Phys. Rev., 99,
    376 (1955).
29. R.H. Kingston and S.F.J. Neustadter, J. App. Phys.,
    26, 718 (1955).
30. A. Many and D. Gerlich, Phys. Rev., 107, 404 (1957).
31. Y.F. Novototskii-Vlasov, Trudy (Lebedev), 48, 1
    (1969).
32. I.K. Kiikoin and M.N. Naskov, Phys. Z. USSR, 5,
    586 (1934).
33. J. Frenkel, Phys. Z. USSR, 5, 586 (1934); ibid. 8,
    185 (1935).
34. F. Urbach, Zitzber. Akad, Wiss. Wien, Math.-Naturw.
    Klasse, 139, 353 (1930).
35. J.T. Randall and M.F.H. Wilkins, Proc. Roy. Soc.,
    London A184, 390 (1945).
36. G.F.J. Garlick, "Photoconductivity" in Encyclopedia
    of Physics, Springer-Verlag, Berlin (1956).
37. L. Heijne, Philips Res. Repts. Sup. 4, 99 (1961).
38. D.G. Semak, D.V. Chepur and V.F. Zolotarev, Fiz.
    Tverd, Tela 9, 1242 (1967).

39. T.J. Gray and N. Lowery, Disc. Far. Soc., 1972 (In Press).

40. T.J. Gray and H.G. Oswin, Catalysis, State University of New York College of Ceramics, at Alfred University, Alfred, New York, 1958.

41. J.M. Bridges and G. Houghton, J. Am. Chem. Soc., 81, 1334-8 (1959).

42. T.J. Gray and N. Lowery, 4th Can. Congress on Catalysis, Halifax, May 1971.

43. F.M. Nelson and F.T. Eggertsen, Determination of Surface Area Adsorption Measurements by a Continuous Flow Method, Shell Development Co., Emeryville, California.

44. H.J. Dittfeld and J. Voigt, Phys. Stat. Sol., 3, 1941 (1963).

45. R.H. Bube, Photoconductivity in Solids, John Wiley, New York (1960).

46. Maenhout-Van der Vorst and F. Van Craeynest, Phys. Stat. Sol. 9, (1965) 749.

47. M.A. Seitz, W.F. Tinter and W.M. Hirthe, Mats. Res. Bull., 6, 275 (1971).

DISCUSSION

H.C. Gatos (MIT): I believe that it is essentially impossible to refer to surface states (or surface impurity levels) in a "semiconductor sense" on the basis of powder materials and the reported measurements. In fact I doubt that a "Schottky-type" barrier exists in the powder surfaces reported. We have recently found in our laboratory (employing surface photovoltage spectroscopy) that the surface states spectra are entirely different in the basal and the prismatic surfaces of ZnO single crystals. We have also found that both intrinsic and extrinsic states (although they lie below the Fermi level) are in very poor communication with the bulk. (J. Lagowski, E.S. Sproles and H.C. Gatos, Surface Sci. 30 653 (1972); H.C. Gatos and J. Lagowski, J. Vac. Sci. Technol. 10 130 (1973).

PHYSIOLOGICAL FACTORS AT BIOCERAMIC INTERFACES

L. L. Hench

Ceramics Division, Department of Materials
Science and Engineering, University of Florida,
Gainesville, Florida

The objective of this chapter is to review the current
state of understanding of the variables that can influence
the behavior of ceramic materials when exposed to a
physiological environment. The physiologic-ceramics
interface is extraordinarily complex. The interface varies with time, location, species, and age of animal, and
is subject to the presence of external stresses of various
types as well as being influenced by pharmeceutical treatments.

THE NEED FOR BIOCERAMICS

Nearly everywhere that one turns in modern medical
surgery there becomes an awareness of needs for improved
biomaterials. Few of the medical materials which are in
current use, mainly polymers and metals, have been
specifically designed for the particular function that they
are serving. This is because general principles for de-
signing a material to produce a specific physiological
interfacial response are not available. These principles
must eventually come from systematic investigations of
the physiologic-materials interface. However, in the
meantime surgeons must use the best materials available.
Materials selected usually offer a compromise between

biological acceptability and mechanical performance.
Consequently, additional improvement in behavior of
materials for most surgical applications is possible.

Improved methods for mechanical fixation of orthopedic devices is one area of surgical need. Corrosion
of metallic bone plates, screws, nails, and pins left in
the body for appreciable lengths of time often results in
failure and requires surgical removal of the devices.
Non-removeable devices of this type are needed. Replacements for sections of long bone are not currently available.
The high incidence of long-bone damage for automobile
accidents, bone cancer, war wounds, osteomyelitis make
a satisfactory long-bone prosthesis highly important. Repair of joints damaged by severe arthritis is also of extreme importance. Replacement involves the use of polymethylmethacrylate cement which polymerizes in the bony
defect to hold in place various types of metallic and polymer joint prostheses. The attachment is entirely mechanical and there is growing evidence of possible dangers
associated with the temperature rise accompanying polymerization and by the presence of the monomer in the body.
An improved method of fixation of joint prostheses that
would eliminate or minimize use of polymethylmethacrylate
would be a major surgical advance.

Materials used in the cardiovascular system must have
surface characteristics to prevent build-up of blood clots
which will occlude the prostheses or break off and travel
in the blood stream to occlude other parts of the cardiovascular network. Sufficient material development has
occurred that replacements for arteries and for vein and
heart valves are now commonly employed in surgery. However, additional advances in materials performance must
be obtained in order for artifical hearts to become a reality.

Occlusion of tubes for shunting neural fluids from head
injuries in neurosurgery is presently a limitation on the
neurosurgical procedures that can be employed and on the
monitoring of patients after surgery. Nonoccluding catheters for drainage of cerebral spinal fluid from hydro-

encephalic children is also an urgent need in neurosurgery.
Again the materials interface is the limiting factor. Electrical stimulation of the nervous system has been shown to
be the practical way for achieving relief of chronic pain
and producing bladder control, and it has potential for producing artificially generated vision and hearing. Long-
term stability of the electrode interface with neural tissues
must be assured for each of these applications. Such information is only now being obtained.

It has been routinely demonstrated that the use of
external dialysis units can keep kidney-failure patients
alive. Improvements in size, efficiency, and cost of such
units require a better understanding of the materials-
dialysis fluid interface. The ultimate goal of an internal
dialysis unit, an artificial kidney, will require major material advances. Even use of simpler urinary devices
such as catheters is currently limited by chronic occlus-
ions, resulting in painful and recurrent surgical replacement. Development of nonoccluding materials surfaces
again will provide a major surgical advance.

Prosthetic materials for repair of bony surgical defects in the maxillofacial area are also needed. Such defects are occasioned by accidents and cancer. Repair is
essential not only for normal eating function but also for
psychological purposes. One of the most difficult aspects
of prosthetic materials for such applications is that they
must not only be compatible with hard tissues but a variety
of soft tissues that have different and important functions.
Surgical augmentation of the mandibular ridge is such an
example. A large population exists with resorption of the
bone around tooth root sockets resulting from advanced
periodontal disease. The resorption of bone makes the
fitting and wearing of dentures extremely difficult. Augmentation of this bony ridge by a prosthetic material makes
the wearing of a normal denture plate possible.

Tracheal prostheses represent a similar type of problem, i.e., combination of compatibility with both soft and
hard tissues that have very specific biological functions.

Another biointerface problem that is important to nearly
every individual is that of the compatibility of dental implants.   Nearly every individual loses one or more permanent teeth in his lifetime. If the tooth can be replaced
by an artificial one that would remain stable in the tooth
socket without producing bony resorption or interfacial
infection, we could retain a normal oral environment for
a considerably lengthened period of time. However, the
materials variables that are necessary for prevention of
the migration of the epithelial layer of the gingival tissues
are not known. Extensive migration of the epithelial layer
eventually results in the loosening and removal of the tooth
implant. Resorption of bone around tooth implants also
occurs and the factors responsible are also not known.
Such bony resorption can lead to the same problem as
advanced periodontal disease mentioned above. Although
there are indications from a number of clinical investigators that dental implants may be successful, until the
basic research results of materials-physiological interface factors are identified and understood, such applications
will always retain an aura of suspicion.

    We now live in an era where many medical and dental
researchers are most willing to discuss advanced materials approaches to their problems. This open channel of
communication is evident from the large number of recent
symposia in this area[1-4]. As discussed above, a need for
new materials in medicine is high. However the level of
understanding the materials variables that are influencing
the acceptance or rejection of materials in these areas is
low. This is in no small part due to the difficulties of
understanding the nature of the surface and interface of
materials even under stable and well controlled environments. Adding the complexities of understanding the material interface under a dynamic, rapidly changing, and
multi-cellular living system makes the problem exceeding-
ly difficult. However if the principles for understanding
surfaces and interfaces of ceramics discussed in this symposium can be applied to the problem over the next decade,
an important human and social payoff in medical-dental
applications will be sure to be realized.

NATURE OF THE PHYSIOLOGIC ENVIRONMENT

It is essential to recognize that the behavior of a ceramic material in the body is controlled by the interface between the material and the body tissues or fluids. The interface is dynamic; changing locally with function, fluctuations in body chemistry or alterations in histological features. The materials interface can interact passively with the various physiological factors or the material may undergo chemical or microstructural alteration at the interface resulting in a changed interface. Likewise the presence of the material may alter the physiological constituents perturbing them to produce a different physiological environment than would be present without the materials there.

The cells in each portion of the body differ in specificity, form, function and reproducibility. The general morphological configuration of many of the cells that may be encountered at the interface of the bioceramics material is indicated in Fig. 1. The cells include undifferentiated primitive or stem cells, a holdover from the embryonic state of the individual and with the potential for rapid reproduction. It also has potential of developing into one of the more specialized type of cells which perform specific functions. However in order for this change, called differentiation, to occur, the primitive cell must be exposed to appropriate local environmental conditions and appropriate stimulus. The presence of a synthetic material may alter the local conditions such that the primitive cell can differentiate in only one direction. Limitation of primitive cell differentiation may be one of the major negative factors involved in biomaterials interfacial behavior.

Fibroblasts (Fig. 1b) are especially critical cells from the standpoint of the biomaterials interface since they generate constituents such as collagen (Fig. 1c) that comprise major portions of the connective tissues of the body.

One of the many different types of epithelial cells is shown in Fig. 1d. They provide the internal lining of our body cavities and our external surface and thus are in effect

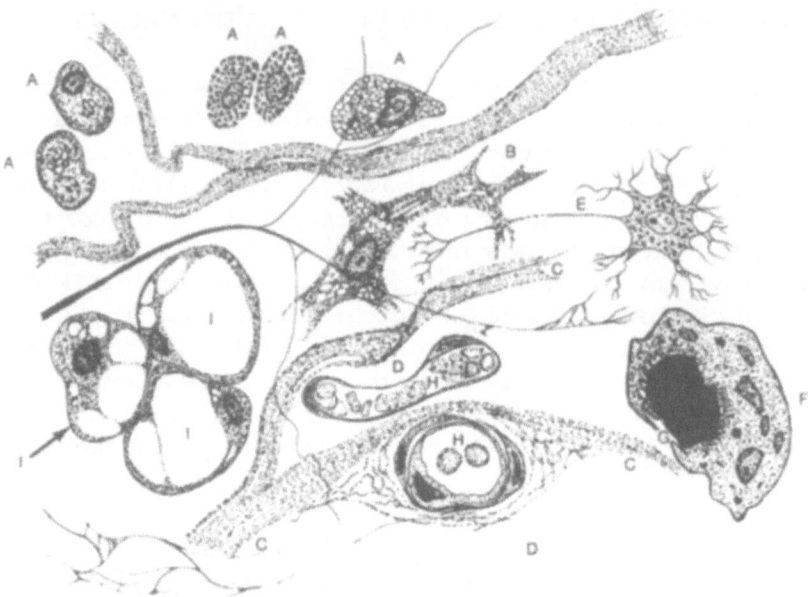

Fig. 1. Schematic of physiological constituents that may
be in contact with bioceramic implants: (A) various primitive or stem cells, (B) fibroblast cell, (C) collagen
fibrils, (D) epithelial or endothelial, (E) neuron, (F)
macrophage, (G) foreign particle, (H) blood cells, and
(I) fat cells.

the packaging material for the body. Any prosthetic device
that is in contact with the outside world must be compatible
with epithelial cells. The neuron shown in Fig. 1e is one
example of the type of cells that comprise the nervous
system. The long appendages along the neuron, called
dendrites and axons respectively, interact with each other
carrying electrical signals by interfacial chemical dis-
charges. Bioelectrode materials must be compatible with
such neural tissues in order to function as long term implants.

Large cells with many nuclei, termed macrophages,
(Fig. 1f) are produced in tissues by differentiation when
foreign particles (Fig. 1g) need to be destroyed. The cells

are provided with oxygen for respiration by blood cells
(Fig. 1h). Long-term energy storage is accomplished by
fat cells (Fig. 1i).

Osteoblasts (Fig. 2a) are encountered in many prosthetic applications because these are the cells that produce
the mineralizing tissue that comprises bone. Osteoblastic
activity is favored by an alkaline local environment.
Osteoclasts, Fig. 2b, are responsible for bone resorption
by dissolving the hydroxyapatite bone mineral phase with
a local acidic environment. Prosthesis repair must be
accompanied by osteoclast activity during the first four
weeks in order for dead bone to be removed.

Understanding the origin, structure and function of the
cells that comprise the body is essential for understanding
interactions of the physiological environment with materials. The reader is referred to Ham[5], Stevenson[6], and
Lehninger [7] as starting points for such an understanding.

Although each cell is a separate identity in the body,
the cells interact with each other to form physiological
structures. This interaction occurs predominately at the
cell membrance, i. e. , the external wall of the cell which

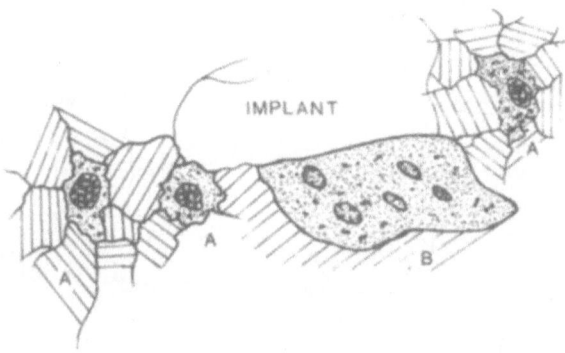

Fig. 2. Osteoblasts (A) producing newly mineralizing bone
(cross hatched area) while an osteoclast (B) is resorbing a
portion of old bone at an implant surface.

contains within it the various substructures that make up
the living organism[8]. The presence or absence of adhesion of cell membrances with bioceramics is one of the
primary factors influencing implant behavior. Consequently, understanding the nature of the cell membrance and
cellular adhesion is important in understanding the mate-
rials-cell interface. The cell membrance can be considered in its simplest form to be constructed of two protein
monolayers covering both sides of a lipid core, as depicted
in Fig. 3a[9]. The protein layer results in a hydrophillic
surface for the membrane whereas the lipid core gives
the internal portion of the membrane a hydrophobic character. More recent theories suggest a dynamic cell membrane structure which includes fluctuating interactions of
water and mucopolysaccharides (MPS) with the protein-
lipid bilayer[10, 11]. The MPS on cell surfaces may provide
histocompatibility antigens, regulate cellular specificity
and perhaps provide bonding sites between the cell membrance and other surfaces. Mucopolysaccharides (Fig. 3b)
are made up of sugars that are covalently bonded to poly-
peptide chains. These macromolecules form varied configurations with each and are linked by glycosidic bonds.
A cell membrane with MPS interactions is shown in Fig.
3c. Hydrophobic portions of the amino acids extend into

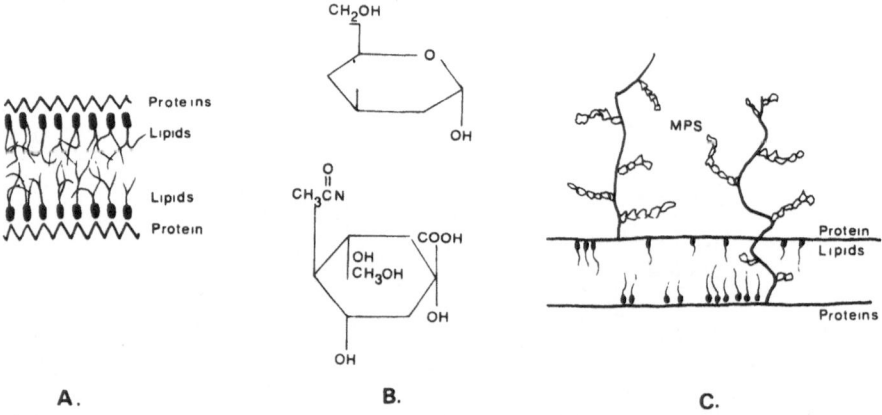

Fig. 3. Features of cell membranes. (A) Bilayer protein-
lipid model, (B) two typical sugar groups in mucopolysaccharides (MPS), (C) membrane model with MPS.

the lipid sections of the membrane. Hydrophillic portions extend into the cell gap, while hydrocarbon chains of phospholipids extend inward from both sides of the outer walls.

Thus, bonding between charged sites on the hydro-phillic portions of the MPS or the membrane proteins with a ceramic surface is possible[11]. However, sufficient type and concentration of charge on the ceramic is necessary to displace the water bound to the membrane. Because of the disordered character of the membrane surface and the amorphous ground substance (MPS) coating, often termed cement, it is very difficult to characterize the nature of intercellular or cell-implant bonds.

Adhesion between cells appears to fall within four types: 1) A few cells exhibit close adhesion with mem-branes no further than 20 Å apart. 2) Most cells show approximately a 100-200 Å gap between the respective membranes. 3) Some cells show gaps between neighbor-ing membranes up to several microns in dimension, and 4) certain cells exhibit fan-shaped intercellular bonding features across the membrances called desmosomes[12].

The presence of intercellular material such as MPS tend to prevent adhesions closer than about 100 Å from forming because of an increase of the effective viscosity of the fluctuating membrane-water layer[12]. The increased viscosity results in a large decrease in rate of fluid drain-age from the membrane interface and therefore the time required for forming a close adhesion is greatly increased. Such intercellular substances likewise prevent detachment of cells for similar reasons.

A membrane feature, termed a nexus, has also been described as a region of specific bonding[8, 13]. The feature appears in photomicrographs of epithelia and smooth and striated muscle as an array of knoblike particles thought to be ATP-ase covering about 5% of the cell area. The nexi are associated with regions of 20 Å close adhesion and are the sites of action potential spikes and rythmic oscil-lations. Such regions of specific charged groups may be

responsible for the presence or absence of cellular adhesion with implant surfaces as well.

SURFACE FEATURES OF BIOCERAMICS

All four of the major classes of surface features, e. g.,
chemical, crystallographic, microstructural, and topographical, may be important in influencing the physiolog-
ical-ceramic interface.   These features and some possible
physiological interactions are illustrated in Figs. 4 and 5.

Fig. 4 shows a magnified section of a possible two-
phase ceramic body consisting of Al_2O_3 grains with a glassy
phase in the grain boundaries.   Studies of Al_2O_3 implants
show that alumina is relatively inert in the physiological
environment[14, 15].   Thus Al_2O_3 does not alter the local
chemical composition or pH of body fluids.   In contrast,
our previous studies have shown that many different glass
compositions react in the aqueous environment comprising
the body in a way that does alter the local chemical composition and pH[16-19].   Consequently, a two-phase ceramic

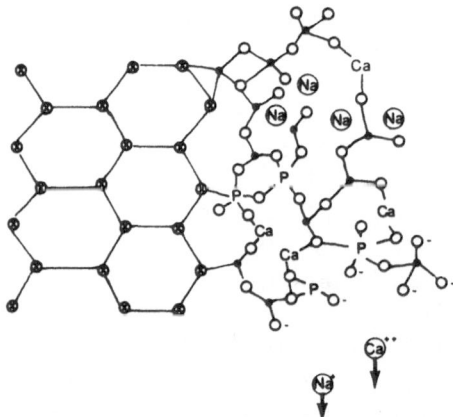

Fig. 4.   Atomistic model of 45S5 bioglass (see Table I) for
composition interface with Al_2O_3 after surface dissolution
in the physiological environment.

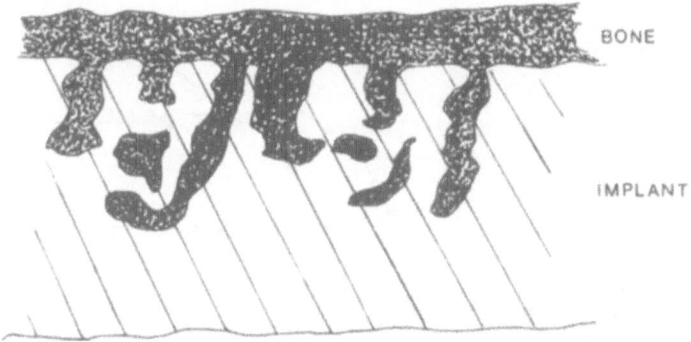

Fig. 5. Mechanical stabilization of bioceramic implant surface by means of tissue ingrowth through pores.

material may be designed, such as shown in Fig. 4, that incorporates the behavior of both inert and reactive phases in a way that produces a desired dynamic response at the material-physiological interface. Such "tailor-making" of ceramic materials for specific time-dependent physiological responses is one of the primary advantages offered by ceramics as biomaterials.

An example of the use of multiphase ceramic technology to control surface pH is shown in Fig. 6. Samples of variable percentages of a surface reactive bioglass (see Table I for composition) and $0.3\mu m$ Al_2O_3 were cold pressed and sintered to $\sim 0\%$ porosity. Specimens 1 x 0.5 x 0.1 cm were suspended in triply distilled deionized water, which had an initially acidic pH, at room temperature under static conditions. As shown in Fig. 6, changes in the solution pH are dependent upon the percentage of Al_2O_3 present. The rate of pH change has been shown to correlate with the formation of a chemical bond of implant materials in bone through other studies[16-19]. The variation in surface pH rate change in the previous studies was accomplished through alteration of glass composition, e.g., addition of B_2O_3 or CaF to the composition. Thus the results shown in Fig. 6 illustrate a new approach towards modifying the time dependent surface reactivity of bioceramics without alteration of bioglass composition.

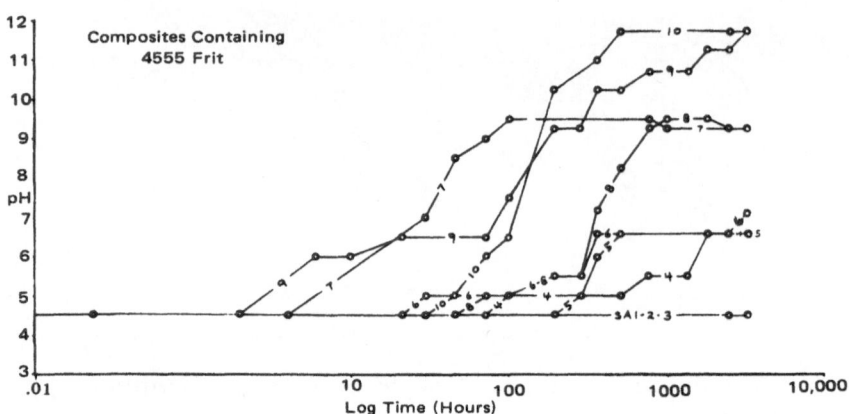

Fig. 6. Time dependent change of pH of static deionized H_2O at $25°C$ exposed to 0% porosity 45S5 bioglass-Al_2O_3 composites. (See Table I for compositions).

Table I. Composition of 45S5 Bioglass-0.3μm Al_2O_3 Composites.

Composite Code	Weight % of 45S5 Frit*	Weight % of Linde A Al_2O_3
SA-1	10	90
SA-2	20	80
SA-3	30	70
SA-4	40	60
SA-5	50	50
SA-6	60	40
SA-7	70	30
SA-8	80	20
SA-9	90	10
SA-10	100	0

* Composition of 45S5 Bioglass: 45 w/o SiO_2, 6 w/o P_2O_5, 24.5 w/o CaO, 24.5 w/o Na_2O.

There are several advantages in having this alternative method for control of surface reactivity. First, the total concentration of soluble surface ions can be varied without other ions being added to the composition. Second, fabrication by powder processing is possible. Third, the inert Al_2O_3 phase serves as a second-phase strengthening agent as well as a modifer of surface reactivity.

Fig. 5 illustrates another application of microstructural control to physiological processes. As developed by Hulbert, et al[14, 15] and discussed elsewhere in this proceedings, if the pore size of a biologically unreactive ceramic is sufficiently large, tissue ingrowth will occur. The tissue will be infiltrated with blood vessels in a normal manner producing a physiologically and mechanically stable interface with the bioceramic.

Investigations have also shown that high-porosity ceramic implants that are chemically reactive in the physiological environment, such as $Ca_3(PO_4)_2$, dissolve during a 6-12 weeks period of time[20]. The porous reactive implants are replaced by bone during the dissolution process if they are implanted in a bone cavity. Thus ceramic materials exhibit a complete spectrum of physiological reactivity depending upon composition and porosity.

Previous studies in our laboratory have shown that microstructural variation of fully dense bioglasses and bioglass-ceramics do not result in a change in physiological response if the surface reactivity of the material is the same[17, 18]. In other words, if the physiological response is controlled by surface chemical reactions, microstructure is important only as it affects the chemical reactions. This is an important result since it eliminates crystallography, point and line defects, and grain boundaries as critical physiological variables independent of their influence on surface reactivity. Our studies on glass corrosion have shown that surface topography, and consequently surface area, can have an appreciable effect on surface solubility rates[21-23].

In summary, the surface features of ceramics which have been shown to alter the living physiological environment include: 1) composition of glassy phase, 2) volume fraction of glassy phase, 3) composition of crystal phase, 4) surface area exposed to physiological fluids, and 5) cross-sectional area of pores at the physiological interface.

Recent detailed information has been generated regarding the influence of the composition of the surface glassy phase on physiological processes[19, 24]. If a soda-lime-silica glass with P_2O_5 additions (Table I) has sufficient CaO and P_2O_5 content the glass will form a stable surface gel when exposed to an aqueous solution. The gel formation is accompanied by Na^+, Ca^{2+}, Si^{4+}, and P^{5+} loss from the glass, but at rates that produce a silica- and phosphate-rich film. Local pH becomes alkaline as a result of the cation exchange with H^+. When the combined surface-ion release and alkaline pH occurs during the normal 4-6 weeks period when osteoclasts and osteoblasts are differentiating in bone repair, bone growth occurs at the implant site. Collagen fibrils generated by the osteoblasts as well as the osteoblasts themselves become trapped in the amorphous gel phase on the glass surface. Therefore, adhesion of the cells and organic constituents to the glass surface is an inorganic analog to the cell membrane-MPS structure described above (Fig. 3c).

The presence of sufficient calcium and phosphate groups from intercellular fluids, local alkaline pH, and an environment compatible with alkaline phosphatase enzymes result in precipitation of hydroxyapatite bone mineral, with the collagen-cellular structure producing fully mineralized bone. Bone thus grows in intimate contact with the bioglass surface, being cemented into a chemical bond through the formation of a stable amorphous inorganic gel that polymerizes, entrapping the physiological constituents. A schematic of this transition zone between the bioglass surface and bone is shown in Fig. 7. Recent publications[16-19] provide transmission electron micrographic evidence of this interface structure, obtained in living animals.

MATURE BONE

CEMENT LAYER FROM IMPLANT SURFACE GEL ($\sim$0.1 μm)

UNDISSOLVED IMPLANT SURFACE

NORMAL BONE MINERAL

COLLAGEN FIBERS

ECTOPIC CRYSTALS

Fig. 7. Schematic of mechanism of interfacial bonding of bioglass implants to newly mineralizing bone.

The results discussed above for bioglass behavior in the physiological environment of bone are quite specific for bone. When the same glasses, or glass-ceramics of the same composition, are implanted in muscle, tendon, or epithelial tissues the implants are rejected[18]. The physiological environment of soft tissues does not favor the presence of an alkaline pH or excess concentrations of calcium, phosphorous, and silicate ions as does bone. Consequently macrophages are produced in proximity to the implant. The macrophages engulf and dissolve portions of the ceramic surface, gradually breaking up the material until it disappears. The ingestion process is very similar to what the body does to the corrosion products of metals.

Consequently, studies conducted to date have shown that microstructural and chemical features of the surface of ceramics and glasses strongly influence physiological responses to the material. A specific combination of physiological responses are required for each part of the body. Therefore, a ceramic material may be designed to achieve a specific response by employing modifications and various combinations of surface chemistry and microstructural features. It is for this reason that ceramic and glass surface science holds great promise for significant advances in medical science in the years to come.

ACKNOWLEDGMENT

The author gratefully acknowledges the financial support of NSF Grant GH-33008 and the assistance of research colleague, A. W. Smith, for the alumina composite studies.

REFERENCES

1. The Chemistry of Biosurfaces, Vols. 1 and 2, M. L. Hair, editor, M. Dekker, Inc., New York, 1972.

2. Symposium on Materials and Design Considerations for the Attachment of Prostheses to the Musculo-Skeletal System, S. Hulbert, editor, Interscience Publishers, J. Wiley & Sons, New York, 1973.

3. C. W. Hall, S. F. Hulbert, S. N. Levine, and F. A. Young, Bioceramics-Engineering in Medicine, Interscience Publishers, J. Wiley & Sons, New York, 1972.

4. Biomaterials, L. Stark and G. Agarwal, editors, Plenum Press, New York, 1969.

5. A. W. Ham, Histology, Lippencott, Inc., Philadelphia, 1969, 6th edition.

6. W. K. Stephenson, Concepts in Biochemistry: A Programmed Test, J. Wiley & Sons, New York, 1967.

7. A. L. Lehninger, Bioenergetics, W. A. Benjamin, Inc., Menlo Park, California, 1971.

8. E. L. Benett and D. Delbauffe, "Plasma Membrane as a Model of Complex Organization," in Cell Membrane, Academic Press, New York, 1968.

9. J. A. Lucy, "Experimental Models for Plasma Membrane," in Biological Membranes, D. Chapman, editor, Academic Press, New York, 1968.

10. F. Wold, Macromolecules: Structure and Function, Chap. 9, Prentice Hall, Englewood Cliffs, New Jersey, 1971.

11. Richter, Scarpelli, "Biological and Pathological Aspects," in The Cell Membrane, D. Champman, ed., Academic Press, New York, 1968.

12. A. S. G. Curtis, Biol. Rev., 37, 82-129 (1962).

13. J. P. Revel and I. Susumma, "Surface Components of Cells," in Specificity of Cell Surfaces, W. Davis, ed., 1967.

14. J. J. Klawitter and S. F. Hulbert, J. Biomed. Res. Symposium, No. 2 (Part 1), Interscience, New York, 1972, pp. 161-229.

15. S. F. Hulbert, J. J. Klawitter and R. B. Leonard, "Compatibility of Bioceramics with the Physiological Environment," in Ceramics in Severe Environments, W. W. Kriegel and H. Palmour, III, editors, Plenum Press, New York, 1971.

16. L. L. Hench, "Ceramics, Glasses and Composites in Medicine," Medical Instrumentation, 7 (2) 136 (1973).

17. L. L. Hench, R. J. Splinter, W. C. Allen and T. K. Greenlee, Jr., J. Biomed. Res. Symposium, No. 2, Interscience, New York, 1972, pp. 117-143.

18. L. L. Hench, H. A. Paschall, W. C. Allen and G. Piotrowski, "An Investigation of Bonding Mechanisms at the Interface of a Prosthetic Material," Report #3, August 1972, U. S. Army Med. R & D Contract No. DADA 17-70-C0001.

19. L. L. Hench and H. A. Paschall, "Histo-chemical Effects at a Biomaterials Interface," J. BioMed. Mat. Res. Symposium, "Prostheses and Tissue: The Interface Problem," S. Hulbert, editor, in press.

20. S. N. Bhaskar, J. M. Brady, L. Getter, M. F. Grower and T. D. Driskell, Oral Surg., 32, 336 (1971).

21. D. M. Sanders and L. L. Hench, J. Am. Ceram. Soc., 56 (7) 373-378 (1973).

22. D. M. Sanders and L. L. Hench, "Environmental Effects on Glass Corrosion Kinetics," Am. Ceram. Soc. Bull. 52 (9) 662-65 (1973).

23. D. M. Sanders and L. L. Hench, "Surface Roughness and Glass Corrosion," Amer. Ceram. Soc. Bull. 52 (9) 666-69 (1973).

24. A. E. Clark and L. L. Hench, "Effect of F^-, B^{3+}, and P^{5+} Additions on the Solubility of Soda-Lime-Silica Glasses," to be published.

DISCUSSION

R. Snyder (Alfred): Do you observe muscle attachment to your implants?

Author: During the first several weeks after implantation very tight contiguous cell membranes engulf the surface of the Ca- and P-containing implants. This appears to be attachment. However, beyond 8 weeks the implant materi- al begins to be destroyed by the cells by a dissolution pro- cess. Therefore, these bioglass-ceramics, that form a stable bond with bone, become rejected by muscle owing to lack of tolerance of the tissues for Ca and P ions.

P. J. Gielisse (Univ. Rhode Island): Is it known why the amorphous or pseudo-amorphous state is apparently necessary for a positive relationship between the organic and inorganic interacting materials to develop? Do you feel that it is more structurally rather than energetically related?

Author: Our data indicate that the surface gel must form at the time that the collagen fibrils are being formed by the asteoclasts in order for the fibrils to become attached to the gel structure. We do not know as yet whether chemi- cal bonds are involved in the entrapment or whether there is just a mechanical imbedding of the fibers within the gel. The fibrils and mucopolysaccharides may actually become part of the gel structure.

P. B. Adams (Corning): What is role of P_2O_5 - controlling chemical durability of glasses or adjusting compatibility?

Author: Recent study by A. E. Clark and myself show that the P_2O_5 addition results in a stable SiO_2-rich film form- ing on the bioglass surface during aqueous attack. This amorphous gel film forms within hours and continues to develop in thickness with time. Eventually the Ca and P ions in the glass migrate through the SiO_2-rich surface to form the hydroxyapatite-like crystals which interlock with the naturally mineralizing bone.

CHARACTERIZATION OF TISSUE GROWTH INTO PELLETS AND PARTIAL SECTIONS OF POROUS PORCELAIN AND TITANIA IMPLANTED IN BONE

S. F. Hulbert, L. S. Bowman, J. J. Klawitter, B. W. Sauer and R. B. Leonard

Clemson University
Clemson, South Carolina

Porous titania and porous porcelain were selected for this investigation* because of their chemical inertness and ease of fabrication. Titanium metal implants have demonstrated a high degree of tissue tolerance[1, 2]; these are certain to have maintained some oxide layer on the surface. Also, porcelain was shown to be non-toxic when implanted intramuscularly in rabbits[3].

Klawitter[4] and Talbert[5] produced porous calcium aluminate by reacting pressed powder mixtures of $CaCO_3$ and Al_2O_3. The decomposition of $CaCO_3$ caused the escape of CO_2, resulting in a porous material whose pore interconnections were similar to the pore size itself. Klawitter[4] demonstrated a small degree of hydration of the implanted pellets; however, calcium aluminate was shown to be non-toxic, and it did not elicit inflammatory response. Therefore, this material was used as a qualitative reference material for tissue reaction and degree of osseous ingrowth.

It has been shown that mineralized bone growth into porous calcium aluminate is limited to a minimum interconnecting pore size of 100 microns[4, 5]. For this study, a calcium aluminate pore size of 150-200 microns was used.

─────────────────────
*Work supported by the Office of Naval Research

The two test materials, porcelain and titania, were developed so that their pore size was greater than 100 microns, thus allowing for mineralized osseous ingrowth.

EXPERIMENTAL PROCEDURE

Fabrication of Samples

The dry powder mix[4] was composed of 41 weight percent 150-200 micron $CaCO_3$ and Reynolds RC-152 DBM alumina. It was wet mixed to a thick stable consistency, dried, pressed at 4000 psi into 6. 3 mm diameter cylindrical pellets, fired for 20 hours at 1455° C in a silicon carbide electric resistance furnace, and cooled in the furnace to room temperature. Porcelain of a composition shown in Table I was ball milled and foamed in a manner similar to that described by Ryshkewitch[6]. Hydrogen peroxide is stirred into the aqueous ceramic slurry including a polyvinyl alcohol binder, poured in molds, allowed to dry, and fired at 1200° C for one hour.

Table I. Porcelain Batch

Al_2O_3 (RC-152 DBM)	33g
Nepheline Syenite (Blue Mountain)	33g
Georgia Kaolin	20g
Tennessee Ball Clay	14g

The titania foamed samples were prepared in a similar fashion and fired 12 hours at 1480° C. The surface pores of the fired samples were substantially smaller than those in the interior. The outside layer was removed and shaping was completed by grinding on a 120-grit silicon carbide abrasive disc. The samples were ultrasonically cleaned and were autoclaved at 260° C for 30 minutes.

Specimen Properties

The properties of the implants are summarized in Table II and III. Mercury porosimetry was used to characterize the distribution of the pore size interconnections of the three ceramic materials. The calcium aluminate had approximately 63% of its apparent pore volume enclosed by pore interconnections greater than 100 microns. The porcelain and titania implant materials had 33% and 45%, respectively, of their apparent pore volumes enclosed by pore interconnections greater than 100 microns.

Table II. Archimedes Density in Water, Average Pore Dimensions, and Volume Fraction of Pores.

Material	Archimedes Density	Average Pore Size	Volume Fraction of Pores
	(g/cm^3)	(Microns)	(Point Count)
Calcium Aluminate	2. 03	265	0. 51
Porcelain	1. 73	480	0. 46
Titania	2. 48	295	0. 58

Table III. Fracture Stress, Modulus of Elasticity, and Strain to Fracture.

Material	S_f (psi)	E (psi)	E_f (in. /in.)
$CaO. Al_2O_3$	4100	$0. 37 \times 10^6$	$1. 62 \times 10^{-2}$
Porcelain	4100	$0. 34 \times 10^6$	$1. 65 \times 10^{-2}$
TiO_2	4300	$0. 34 \times 10^6$	$1. 73 \times 10^{-2}$

Implantation of Samples

Twelve mature dogs weighing approximately 40 pounds each were divided into two groups of six animals per group. Porcelain was implanted in one group and titania in the other. In each dog, an approximately 2 1/2-cm long bone section (one-fourth to one-half the bone diameter) was removed from one femur and replaced by a partial section of a test material. Two pellets of test material (titania or porcelain) and two pellets of reference material (calcium aluminate) were implanted in the other femur. Half of the dogs in each group were observed for 6 weeks and half for 12 weeks.

A high-speed air drill, under a constant flush of saline, was used to prepare the implant site for the ceramic. Just enough bone was removed to accommodate the partial section implant which was held in place by three encircling 3-0 chromic gut sutures. In the femur receiving the pellets, a knife blade was used to disturb the periosteum over the area to be drilled by an orthopedic hand drill. The holes, 5 mm in diameter, were drilled through the lateral cortex into the medullary cavity. The pellets (two pellets of test material and two pellets of calcium aluminate) were then press fitted into these holes using finger pressure.

Radiographs were taken immediately following surgery to demonstrate the position and fit of the partial section and pellets. Radiographs were taken at intervals of 2 weeks until the dogs were sacrificed to determine the status of the operated limbs and the degree of incorporation of the implant. To insure a three-dimensional interpretation of the surgically treated femora, ventrodorsal and medio-lateral radiographs were made at each time period. At necrospy, most of the tissue was stripped away and detailed radiographs taken for a further examination of ingrowth of mineralized bone.

Histology

At necropsy, the implant specimens, adjacent bone and approximately 3 mm of the surrounding soft tissue were retrieved and placed in 10% buffered neutral formalin for 168 hours to insure penetration of the fixative. After washing for 15 minutes in running water they were transferred to 70% ethyl alcohol and dehydrated by 30 hour soaks in 80%, 95% and absolute ethyl alcohol. The first 2 hours in each alcohol was carried out in a vacuum of approximately 300 mm of mercury. They were then placed in unpolymerized methyl-methacrylate which had been previously treated with 5% sodium hydroxide to remove the inhibitor and to which 0.5 g of anhydrous benzoyl peroxide per 100 ml of monomer had been added as a catalyst. The specimens were soaked for 48 hours, the first 2 hours under vacuum, in the unpolymerized solution and then transferred to small containers of partially polymerized methyl-methacrylate. This thick monomer had been prepared by heating the unpolymerized monomer from which the inhibitor was removed and the catalyst had been added. Because of the large size of the tissue specimens, there was a possibility of local overheating and bubble trapping in the hardening methyl-methacrylate, causing damage to the tissue. For this reason, the polymerization process was allowed to take place at room temperature in 7 to 21 days.

The blocks were then sectioned longitudinally on a diamond saw, both parallel and perpendicular to the long axis of the bone. Fig. 1 shows the sectioning scheme.

To insure optical microscopic sections with good morphologic detail, a hand grinding was employed. Thick slices (ca. 500 microns) containing the implant were mounted on 55-micron well-slides using Canada balsam, and ground to 40-100 microns on 240, 320, 400, and 600 grit rotating silicon carbide discs.

Contact microradiography was used to confirm the presence of mineralized bone within the pores of the ce-

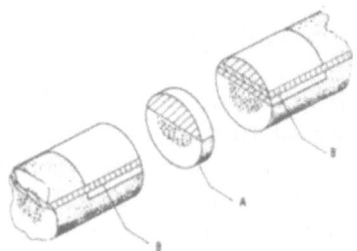

Fig. 1. Histologic sectioning of the partial section implant:
(A) cross section, (B) longitudinal section.

ramic, as described by Jowsey . Since the thin sections
were also stained with Paragon-1301, a direct correlation
between the appearance of osteoid and mineralized bone,
histologically and microradiographically, could be made.
As a result of the staining, nuclei were colored deep blue-
gray, nucleoli, a darker blue. The cytoplasm of most
cells stained a pale blue-gray. Erythrocytes stained
orange to red and the elastica of blood vessels and collagen
fibers stained red. The extracellular constituents of
mineralized bone were unstained, while osteoid (unmineralized bone) stained pink.

RESULTS

Tissue Growth into Pellet Implants

The results of the pellet implants were consistent in
all the dogs at the time periods studied. Gross examination
at necropsy demonstrated normal appearance with no signs
of adverse tissue reaction. Histologic analysis revealed
fibrous and osseous ingrowth into all of the samples (Figs.
2 and 3). Fibrous tissue infiltrating the pellets appeared
normal and vascular. In areas of the titania and porcelain
test pellets situated in the marrow cavity, normal hemo-
poietic tissue was present. Osseous tissue ingrowth was
comprised of mineralized and unmineralized bone (osteoid).
The amount of fibrous and osseous tissue growth into the

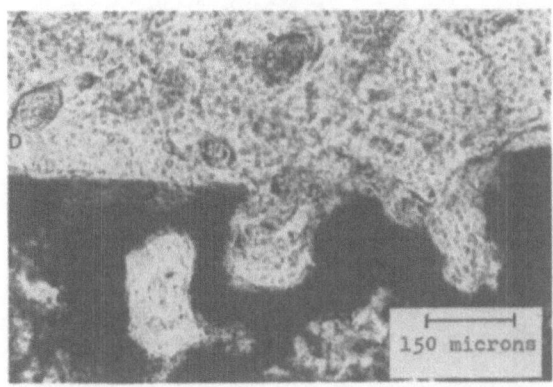

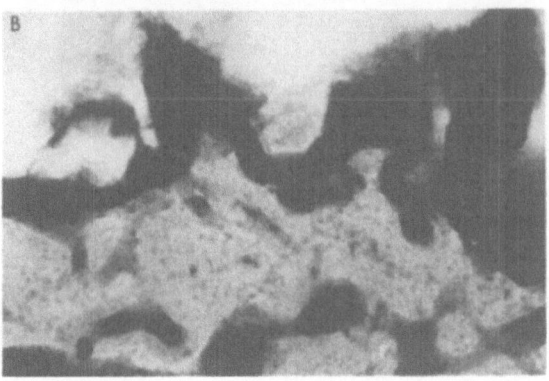

Fig. 2. Osteoid seam at the bone-ceramic interface of a 12-week calcium aluminate pellet implant. (A) Photomicrograph, and (B) Microradiograph.

pellets related directly to the length of time of implantation, pore size, and number of pores.

The maximum depth and the average depth of mineralized bone growth into the pellets are presented in Table IV as a function of implant type and length of implantation.

The amount of mineralized bone growth into the reference material (calcium aluminate) was considerably less than into the test materials (porcelain and titania). This

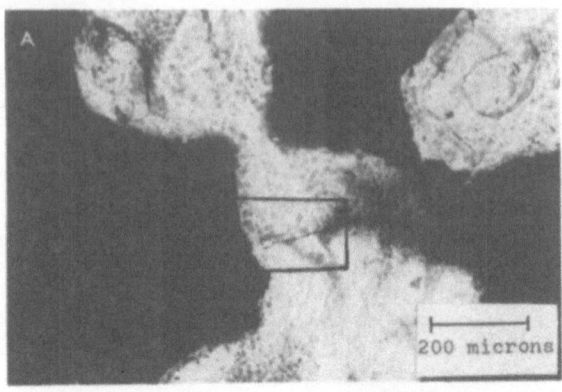

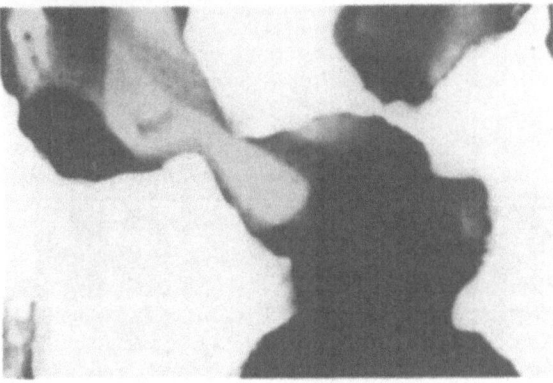

Fig. 3. Edge of mineralized bone lined by osteoblasts
growing 1200 microns into a pore of a 12-week titania
pellet implant. (A) Photomicrograph, and (B) Microradiograph.

can be attributed to 1) the presence of large surface pores
in the test materials (up to 900 microns) as compared to
150-275 micron surface pores on the reference material,
and 2) the apparent inertness of the test materials when
implanted in the skeletal system compared with the hydration tendency of the reference material which raised
the local pH and retarded bone mineralization.

Table IV.   Depth of Mineralized Bone Growth into Pellet
Implants.

| Implant | 6 Weeks | | 12 Weeks | |
	Average (Microns)	Maximum (Microns)	Average (Microns)	Maximum (microns)
Calcium Aluminate	175	400	500	900
Porcelain	750	1100	1300	1800
Titania	850	1150	1400	1750

Tissue Growth into Partial Section Implants

Six partial sections were observed for porcelain and
six for titania.   Three were evaluated 6 weeks following
surgery and the remaining three after 12 weeks.   All the
sites appeared normal, with no evidence of any adverse
tissue reaction.

Microscopic analysis revealed ingrowth of osseous
and fibrous tissue (Figs. 4 and 5).   The amount and depth
of penetration of mineralized bone was dependent upon the
length of implantation, and the position of the partial section in relation to the lines of stress transmitted along the
cortical shaft of the femur.

The results of the partial section implants were not
uniform for the material and time period involved.   The
depth of mineralized bone ingrowth for each implant is
presented in Tables V and VI.

All the partial section implants demonstrated a lack
of adverse tissue reaction with some degree of mineralized
bone ingrowth.   Since both ceramic test materials were
biologically compatible for the implantation periods observed, the success of the partial sections depended on
the incorporation of the implant as a stress transmitting
section of the femoral shaft.   The partial sections demon-

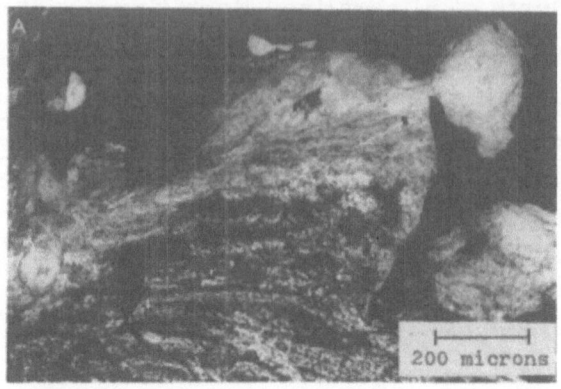

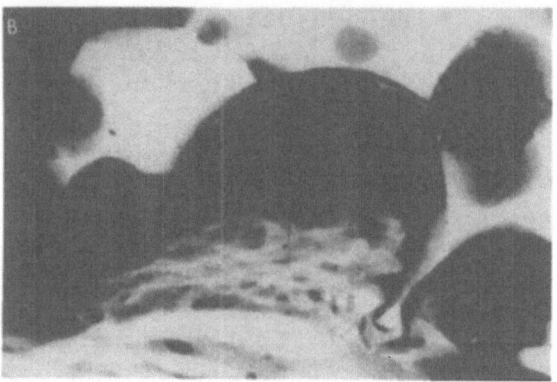

Fig. 4. Mineralized bone growth into the anterior edge
of a 12-week porcelain partial-section implant. (A) Photomicrograph, and (B) Radiograph.

strated some degree of cortical thickening at the proximal
and distal ends of the implant at 2 to 6 weeks following
surgery (Fig. 6). If the cortical thickening continued
along the medial side of the implant and the medial cortex
opposite the implant demonstrated an increased density,
then the bone was remodeling around the implant not allowing stress to be transmitted through the ceramic. The
implants which showed a recession of the cortical thickening at the proximal and distal ends 4 to 8 weeks following

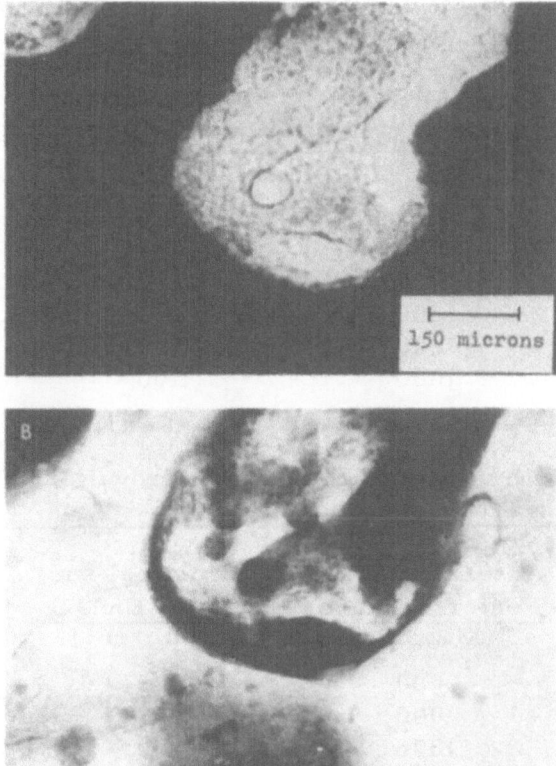

Fig. 5. Mineralized bone growth 800 microns into the posterior edge of a 12-week titania partial-section implant. (A) Photomicrograph, and (B) Radiograph.

surgery and which had no increase in the cortical density opposite the implant, apparently were becoming a functional part of the femoral shaft.

DISCUSSION

The mechanical and physical properties of the reference material (calcium aluminate) paralleled values reported in the literature for percent porosity, apparent

Table V. Depth of Mineralized Bone Growth into Six
Six-Week Partial Section Implants.

Material	Anterior and Posterior Edges (Microns)	Proximal and Dista Ends (Microns)	Medial Side (Microns)
Porcelain	800	0	900
"	650	0	1000
"	700	0	500
Titania	600	0	700
"	800	0	600
"	1000	1100	500

Table VI. Depth of Mineralized Bone Growth into Six
Twelve-Week Partial Section Implants.

Material	Anterior and Posterior Edges (Microns)	Proximal and Distal Ends (Microns)	Medial Side (Microns)
Porcelain	400	0	800
"	900	1400	0
"	1000	0	900
Titania	800	1300	0
"	600	0	1200
"	600	0	900

bulk density, and distribution of pore size interconnections.
The strength determinations for all three ceramic materials
appear to be within the values necessary for a weight-bear-
ing application; however, the influence of sharp notches,
aqueous environments or fatigue-type load-time profiles
on the porous specimens is unknown. In most engineering
design, a safety factor of 5 is acceptable. It appears from
this research that this factor is provided by the materials
tested. This is further evidenced by the absence of any
material failures during the implant period.

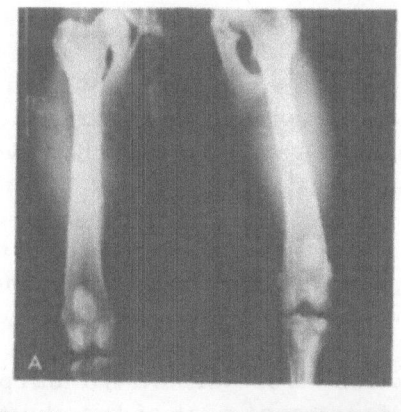

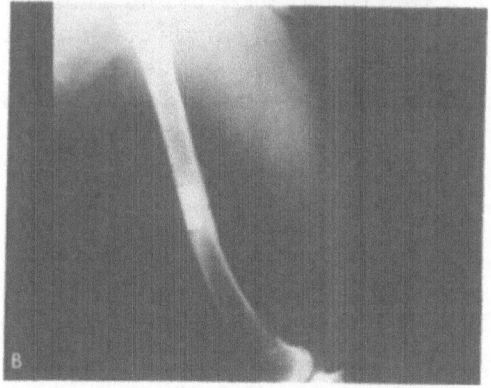

Fig. 6. Radiographs (A) ventro-dorsal and (B) medio-lateral of a porcelain partial-section implant 4 weeks postoperatively, demonstrating increased cortical density at the proximal and distal edges of the ceramic and also the medial cortex opposite the implant. Note the trans-verse lines opposite the implant on the medio-lateral view (B) secondary to the encircling sutures.

CONCLUSIONS

1. Three large-pore ceramic materials were introduced in the shape of pellets into the femurs of mature dogs without detectable signs of any adverse tissue reaction surrounding or within the implants at 6 and 12 weeks.

2. Mineralized bone was observed to penetrate and to be
    in direct apposition to the ceramic, deep within the
    pellets of the porcelain and titania test materials.
3. Mineralized bone growth into the reference material
    (calcium aluminate) was observed; however, minerali-
    zation of the osteoid tissue adjacent to and within pores
    was retarded, possibly due to hydration of the implant.
4. One-inch partial sections of the two test materials were
    implanted in the midshaft of the femur and no chronic
    inflammatory response or foreign body reactions were
    observed.
5. The partial sections were either 1) incorporated into
    the femur as a functional part of the shaft, or 2) separ-
    ated by fibrous tissue at the proximal and distal ends
    with the bone remodeling and transmitting stress a-
    round the implant. The results depended upon the
    initial fit of the implant and its position in relation to
    the lines of stress transmitted along the cortical shaft
    of the bone.

REFERENCES

1. Beder, O. E. and G. Eade. "An investigation of tissue
    tolerance to titanium metal implants in dogs," Surg.
    39, 470-473 (1958).
2. Lueck, R. A., J. Gallante, W. Rostoker and R. D. Ray.
    "Development of an open pore metallic implant to per-
    mit attachment to bone," Surg. Forum. 20, 456-457
    (1969).
3. Leonard, R., Unpublished data, Clemson University,
    Clemson, S. C. (1971).
4. Klawitter, J. J. "A basic investigation of bone growth
    into a porous ceramic material," Ph. D. Thesis,
    Clemson University, Clemson, S. C. (1970).
5. Talbert, C. D. "A basic investigation into the potential
    of ceramic materials as permanently implantable
    skeletal prostheses." M. S. Thesis, Clemson Univer-
    sity, Clemson, S. C. (1969).
6. Ryshkewitch, E. "Compression strength of porous

sintered alumina and zirconia." J. Am. Ceram. Soc.
<u>36</u>, 65-68 (1953).
7. Jowsey, J., P.J. Kelly, B.L. Riggs, A.J. Bianco,
    D.A. Scholz, and J.C. Cohen, "Quantitative micro-
    radiographic studies of normal and osteoporotic bone."
    J. Bone Joint Surg. <u>47A</u>, 785-806, 872 (1965).

DISCUSSION

<u>W.B. Crandall (IITRI)</u>: TiO_2 tends to become unbalanced
in Ti/O ratio upon heating. If it is non-stoichiometric, it
may alter the charge balance at the surface and thus change
its characteristic in the body.
<u>Author</u>: The titania samples were fired at 1480°C for 12
hours and furnace cooled. Earlier studies[1] indicate that
these firing conditions produced stoichiometric titania.
While there is considerable evidence that surface charge
is an important property in determining the compatibility
of implants[2], it is postulated that the physiological compatibility of titania is primarily due to its very limited
solubility and the fact that titanium ions are relatively
non-toxic[3]. No prosthetic material is completely compatible with the physiological environment. The only material which is completely compatible is autogenous tissue.
The criteria used for determining compatibility of implant
materials is minimal adverse tissue response. In the 40
ceramics systems evaluated to date, titania and alumina

---

[1]S.F. Hulbert and M.J. Popowich, "Kinetics and Mechanism
of the Reaction Between TiO_2 and $SrCO_3$," Mat. Sci. Res. <u>4</u>,
422-445, 1969.

[2]S.F. Hulbert, S.N. Levine and D.D. Moyle, <u>Prosthesis
and Tissue - The Interface Problems</u>," John Wiley & Sons,
1974.

[3]P.G. Laing, A.B. Ferguson and E.S. Hodge, "Tissue Reaction in Rabbit Muscle Exposed to Metallic Implants,"
Biomed. Mat. Res. <u>1</u>, 135-149, 1967.

have shown the least amount of adverse tissue response[1].
K. D. Reeve (Australian AEC): How important is it to control pore structure?
Author: The microstructure and chemistry of porous implants determines the type of tissue ingrowth. The minimum interconnection pore size range necessary for mineralized, osteoid, and fibrous tissue is as follows:

Tissue	Interconnection Pore Size Range
Mineralized Bone	100 microns
Osteoid	40 - 100 microns
Fibrous	5 - 15 microns

P. J. Gielisse (Univ. Rhode Island): Should the TiO_2 and Al_2O_3 ceramics have a completely porous structure or just surface porosity?
Author: It is not necessary for a ceramic implant to be completely porous to have tissue ingrowth. Only the surface need be porous. It is important to have a good blood supply for the ingrown tissue, and thus only the outer surface should be porous and porosity should be limited to the outer 2,000 microns. Surface undulations of appropriate dimensions will also result in tissue ingrowth and mechanical interlocking. The surface undulations, however, must be at least 100 microns across and 50 microns deep. Just the presence of surface roughness will not result in mechanical interlocking. In the case of these materials, the stability of the implants is due primarily to mechanical interlocking by tissue ingrowth into surface undulations; there is no evidence that a direct chemical bond will develop between TiO_2 and Al_2O_3 with bone.

---

[1]S. F. Hulbert, F. W. Cooke, J. J. Klawitter, R. B. Leonard, B. W. Sauer and D. D. Moyle, "Attachment of Prostheses to the Musculo-Skeletal System by Tissue Ingrowth and Mechanical Interlocking," Materials and Design Considerations for the Attachment of Prostheses to the Musculoskeletal System, S. F. Hulbert, S. N. Levine and D. D. Moyle, Eds., John Wiley & Sons, 1973.

PHASE DISTRIBUTION IN SOLID-LIQUID-VAPOR SYSTEMS

Ilhan A. Aksay, Carl E. Hoge and
Joseph A. Pask
Lawrence Berkeley Laboratory, University of
California, Berkeley, California

Spatial distribution of phases in a solid-liquid-vapor system are described by the classical Young's equation[1]

$$\gamma_{sv} - \gamma_{s\ell} = \gamma_{\ell v} \cos \theta \quad , \qquad (1)$$

where γ is the interfacial tension between solid-vapor (sv), solid-liquid (sℓ), and liquid-vapor (ℓv) phases, $\gamma_{sv} - \gamma_{s\ell}$ is the driving force for wetting, and θ is the contact angle at a solid-liquid-vapor triple point as measured through the liquid phase. Furthermore, in systems where the solid phase is polycrystalline

$$\gamma_{ss} = 2\gamma_{sf} \cos \frac{\Phi}{2} \quad , \qquad (2)$$

where γ_{ss} is the interfacial tension at the solid-solid grain boundary, γ_{sf} is either $\gamma_{s\ell}$ or γ_{sv}, and Φ is the dihedral angle[2] at a solid-fluid-solid triple point measured through the fluid phase. Both of these equations have been extensively used in various fields to describe the conditions of mechanical equilibrium of a capillary system under chemical non-equilibrium conditions, without explicitly considering the effect of chemical reactions on the interfacial

The work was supported by the United States Atomic Energy Commission.

tensions. Recently, it has been shown[3] that an interfacial
reaction or diffusion of a component from one bulk phase
to the other across an interface results in a transient decrease in the corresponding interfacial tension by an amount
equal to the free energy of the effective chemical reaction
per unit area at that interface. As a consequence of this
transient lowering of interfacial tension values, phenomena
such as spontaneous spreading and permeation of a liquid
phase along a solid-solid interface followed by a pull-back
may be observed.

In this report, the thermodynamics of a solid-liquid-
vapor system and the mechanics of wetting under chemical
non-equilibrium conditions, based on the model of Gibbs[4],
are briefly discussed. These sections are followed with
discussions on the effect of chemical reactions on phase
distribution in liquid phase sintering. In the last sections,
the results of some sessile drop and permeation studies
in the MgO (solid) + $CaO-MgO-SiO_2-R_2O_3$ (Al_2O_3, Cr_2O_3,
or Fe_2O_3) (liquid) system are discussed in order to illustrate the effect of interfacial chemical reactions on microstructure development.

THERMODYNAMICS OF A SOLID-LIQUID-VAPOR SYSTEM

The simple but elegant model of Gibbs[4] for surfaces
has been used in the treatment of the thermodynamics of
various interfacial phenomena. In this section, once again,
this model is utilized to show the effect of chemical reactions on the interfacial tension.

Let us first consider a system with two homogeneous
phases, α and β. The interphase between these two phases
is not a two-dimensional boundary but one of finite thickness
which includes the regions that are influenced by surface
forces and whose properties grade into the bulk. However,
for simplicity in mathematical expressions, Gibbs[4] assigned
any property of the interphase to the interface, a dividing
plane or surface, and assumed the bulk phases to be homogeneous and at equilibrium up to this interface. Using this

approach, it can be shown[3] that the total differential, dG, of a two phase system (α and β), assuming an isothermal and isobaric process, is

$$dG = \sum_i \mu_i^\alpha dn_i^\alpha + \sum_i \mu_i^\beta dn_i^\beta + \left(\frac{\partial G^{\alpha\beta}}{\partial A}\right) dA +$$

$$\sum_i \left(\frac{\partial G^{\alpha\beta}}{\partial n_i^\alpha}\right) dn_i^\alpha + \sum_i \left(\frac{\partial G^{\alpha\beta}}{\partial n_i^\beta}\right) dn_i^\beta + \sum_i \left(\frac{\partial G^{\alpha\beta}}{\partial n_i^{\alpha\beta}}\right) dn_i^{\alpha\beta}, \tag{3}$$

where subscripts of the partial derivatives are omitted for brevity, μ_i^α and μ_i^β are the chemical potentials of component i in the α and β phases, and

$$\left(\partial G^{\alpha\beta}/\partial A\right)_{T,p,n_i} = \left(\partial G/\partial A\right)_{T,p,n_i} = \gamma_{\alpha\beta}$$

is the surface tension of the $\alpha\beta$ interface. The last three terms reflect the effect of compositional variations in the bulk and the interphase regions on the interfacial free energy.

The relationship between the interfacial or surface tension, $\gamma_{\alpha\beta}$, and the specific interfacial free energy, $g^{\alpha\beta}$, can be shown by expressing $\left(\partial G^{\alpha\beta}/\partial A\right)$ in terms of the surface excess Γ_i as

$$\gamma_{\alpha\beta} = \frac{\partial G^{\alpha\beta}}{\partial A} = \sum_i \left(\frac{\partial G^{\alpha\beta}}{\partial \Gamma_i}\right)\left(\frac{\partial \Gamma_i}{\partial A}\right) = \sum_i \left\{ A \left(\frac{\partial g^{\alpha\beta}}{\partial \Gamma_i}\right) + \left(\frac{\partial A}{\partial \Gamma_i}\right) g^{\alpha\beta} \right\} \left(\frac{\partial \Gamma_i}{\partial A}\right),$$

and

$$\gamma_{\alpha\beta} = g^{\alpha\beta} - \sum_i \mu_i^{\alpha\beta} \Gamma_i, \tag{4}$$

where $\mu^{\alpha\beta} = \left(\partial g^{\alpha\beta}/\partial \Gamma_i \right) = \left(\partial G^{\alpha\beta}/\partial n_i \right)$ is the chemical potential of component i at the interface. At the first instant of formation of a surface by mechanical means, the composition of the surface is identical to that of the bulk; and thus, since $\sum_i \mu_i^{\alpha\beta} \Gamma_i = 0$, $\gamma_{\alpha\beta}^0 = g^{0,\alpha\beta}$, where $\gamma_{\alpha\beta}^0$ and $g^{0,\alpha\beta}$ are the pure dynamic tension and specific free energy of the interface, respectively. With time, as adsorption takes place and the interfacial region approaches equilibrium conditions, the interfacial tension decreases[5,6] towards a static value as shown in the upper curve of Fig. 1, and it differs from the corresponding interfacial specific free energy $g^{\alpha\beta}$ as expressed in Eq. (4).

Now, the total differential of the free energy of a solid-liquid-vapor system (at constant temperature and pressure, after neglecting the effect of curvature on the pressure and assuming that the interfacial tensions are independent of orientation) is

$$dG = \sum_i \mu_i^s dn_i^s + \sum_i \mu_i^\ell dn_i^\ell + \sum_i \mu_i^v dn_i^v + \gamma_{s\ell} dA_{s\ell} + \gamma_{sv} dA_{sv} + \gamma_{\ell v} dA_{\ell v} +$$

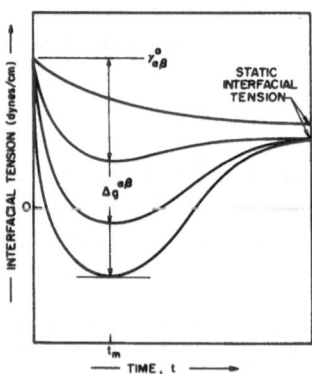

Fig. 1. Variation of dynamic interfacial tension with time during a chemical reaction between two phases. The degree of minimization of the interfacial tension at t_m is proportional to $\Delta g^{\alpha\beta}$. In case of a pure adsorption process, no minimum is observed (top curve)[3].

$$\sum_{\alpha,\beta} \left\{ \sum_i \left(\frac{\partial G^{\alpha\beta}}{\partial n_i^{\alpha}} \right) dn_i^{\alpha} + \sum_i \left(\frac{\partial G^{\alpha\beta}}{\partial n_i^{\beta}} \right) dn_i^{\beta} + \sum_i \left(\frac{\partial G^{\alpha\beta}}{\partial n_i^{\alpha\beta}} \right) dn_i^{\alpha\beta} \right\}, \quad (5)$$

where the summation $\sum\limits_{\alpha, \beta}$ is taken over all three interfaces.

At total thermodynamic equilibrium, $dG = 0$; then since the variations of mass are independent of the variations of area,

$$\gamma_{s\ell} dA_{s\ell} + \gamma_{sv} dA_{sv} + \gamma_{\ell v} dA_{\ell v} = 0, \quad (6)$$

and

$$\sum_i \mu_i^s dn_i^s + \sum_i \mu_i^\ell dn_i^\ell + \sum_i \mu_i^v dn_i^v +$$

$$\sum_{\alpha,\beta} \left\{ \sum_i \left(\frac{\partial G^{\alpha\beta}}{\partial n_i^{\alpha}} \right) dn_i^{\alpha} + \sum_i \left(\frac{\partial G^{\alpha\beta}}{\partial n_i^{\beta}} \right) dn_i^{\beta} + \sum_i \left(\frac{\partial G^{\alpha\beta}}{\partial n_i^{\alpha\beta}} \right) dn_i^{\alpha\beta} \right\} = 0. \quad (7)$$

These two equations outline the conditions for mechanical and chemical equilibrium of the system, respectively; at this point the γ values correspond to those for static interfacial tensions. Equation (6) directly yields[4, 7] Young's equation as presented in Eq. (1), when a flat and rigid solid surface is assumed, and the effect of the gravitational field and the curvature on the pressure in the liquid and the vapor is neglected. Similarly, when Eq. (6) is written in terms of ss and s$_\ell$ interfaces, the dihedral equation, Eq. (2), can be shown[8] to outline the conditions for mechanical equilibrium of a solid-fluid-solid system.

When the conditions of Eq. (7) are not satisfied throughout the system, the phases of the solid-liquid-vapor system will react with each other through the interfaces to achieve a state of chemical equilibrium. During these non-equilibrium dynamic conditions, the γ's change, and the areas change correspondingly in an effort to maintain mechanical equilibrium, as represented by Eq. (6), until dG becomes zero and the system reaches a state of chemical equilibrium. Volume changes occuring during the reactions, if significant, will affect the physical configuration of the system.

Mass transfer across the interfaces must result in a net decrease of the free energy of the system at any time; otherwise, the reaction will not proceed. At the first instant of formation of an interface, however, only the interfacial region is involved in the chemical reaction, and thus the corresponding initial decrease in the free energy of the system is totally attributed to the decrease in the free energy of the interfacial region since the free energies of the bulk phases are not affected. The magnitude of the decrease in the specific interfacial free energy, $(-)\Delta g^{\alpha\beta}$, then is directly equal to $(-\Delta G^{\alpha\beta}/A)$. The corresponding interfacial tension is similarly reduced by an amount equal to $(-)\Delta g^{\alpha\beta}$ (Eq. (4)), as schematically shown in Fig. 1. If it is assumed that the free energy of the reaction between the phases in the interfacial region is comparable but not necessarily equal to that between the bulk phases, the value of $(-\Delta G^{\alpha\beta}/A)$ could be substantially high, and for an approximate interfacial region thickness of 20 Å, a decrease of as much as 1,000 ergs/cm^2 could be realized in the magnitude of the specific interfacial free energy and thus the interfacial tension[3]. Experimentally, negative interfacial tensions are often measured[9-11] during such chemical reactions that result in spontaneous spreading[9-17] or emulsification phenomena[18].

Under chemical equilibrium conditions, however, specific interfacial free energies and static interfacial tensions are always positive since the bulk phases are more stable than the interfaces. Thus, after the completion of the reaction at the interface followed by its continuation into the bulk regions by diffusion, the incremental contributions of the $\sum_{i} \left(\partial G^{\alpha\beta}/\partial n_i^{\alpha} \right) dn_i^{\alpha}$ and $\sum_{i} \left(\partial G^{\alpha\beta}/\partial n_i^{\beta} \right) dn_i^{\beta}$ terms in Eq. (5) must be such that $\gamma_{\alpha\beta}$ increases towards a static interfacial tension value. With time, the contributions of these terms will decrease and become minimal because of the decrease in the chemical potential or composition gradient from the interface into the bulk phases as they approach chemical equilibrium. Therefore, after the initial decrease, $\gamma_{\alpha\beta}$ increases and gradually approaches the static interfacial tension of the reacted bulk phases (Fig. 1), which could be higher or lower than the dynamic interfacial tension, γ^0, of the unreacted phases but should not differ from it drastically. In comparison, the top curve in Fig. 1 also shows the variation of the interfacial tension with time for a pure adsorption process.

MECHANICS OF WETTING UNDER CHEMICAL NON-EQUILIBRIUM CONDITIONS

Let us now consider the specific effects of several types of reactions on the solid-liquid-vapor system, assuming that chemical equilibrium exists between the vapor and the condensed phases but not between the solid and the liquid. The reactions to be considered are those that result because (i) only the solid is not saturated with some or all of the components of the liquid, (ii) only the liquid is not saturated with some or all of the components of the solid, (iii) both phases are unsaturated with respect to the other, and (iv) a compound forms at the interface.

Several dynamic stages associated with the first type of reaction are shown schematically in Fig. 2. At time t_o, Fig. 2a illustrates the instantaneous quasichemical equilibrium involving no interfacial reaction between the liquid and the solid. Young's equation (Eq. (1)) may then be expressed only in terms of the initial dynamic surface tensions. Now, as the solid solution reaction proceeds at the interface, the dynamic specific interfacial free energy, $g^{0,s\ell}$, will change by an amount $\Delta g^{s\ell}$ due to the free energy of the reaction; a corresponding change in $\gamma_{s\ell} = \gamma_{s\ell}^{o} + \Delta g^{s\ell}$ with time occurs, as shown in Fig. 1. When the diffusion rates of the reacting components and thus the growth rate of the reaction product are slow enough relative to the flow rate of the liquid drop, the liquid at the periphery of the drop will remain in contact with unreacted solid that has an unaltered γ_{sv}^{o} as long as $\Delta A_{s\ell}$ is positive; the driving force for wetting $\gamma_{sv}^{o} - \left(\gamma_{s\ell}^{o} + \Delta g^{s\ell}\right)$ which is increased by the amount

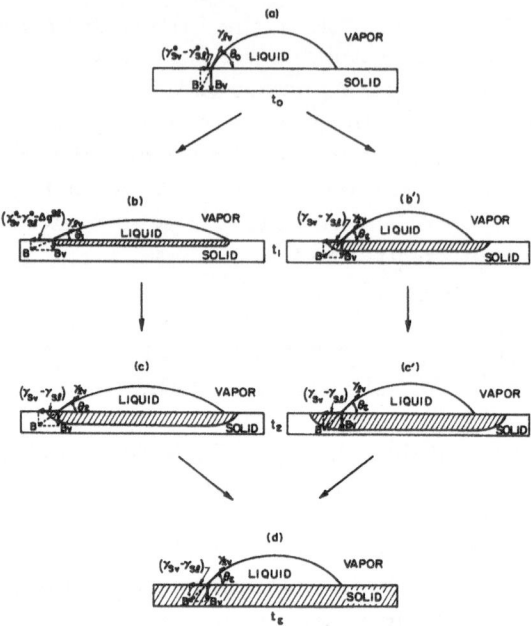

Fig. 2. Schematic representation of the various dynamic stages of a sessile drop when the initial solid is not saturated with some or all of the components of the liquid. The path "abcd" corresponds to the case where the growth rate of the reaction product is slower than the flow rate of the liquid drop; and path "ab'c'd" corresponds to the case where the growth rate of the reaction product is faster than the flow rate of the liquid drop[3].

$(-)\Delta g^{s\ell}$ remains constant. If the maximum driving force at t_m (Fig. 1) exceeds $\gamma_{\ell v}$, then spreading occurs[17]; and if the force does not exceed $\gamma_{\ell v}$, the contact angle continues to decrease until a transient mechanical equilibrium is reached as represented by t_1 in Fig. 2b. At this point, however, diffusion in the solid continues as shown schematically in Fig. 2c; γ_{sv}^{o} ahead of the liquid periphery then is also decreased by an amount $\approx (-) \Delta g^{s\ell}$.[19] The driving force for wetting therefore decreases, and the contact angle increases to a new value of θ_e corresponding to the one for mechanical and chemical equilibria for the system (Fig. 2d). During this pull-back stage, the drop may break into isolated smaller drops if the thickness of the original drop decreases considerably during the transient spreading stage[17].

On the other hand, when the diffusion rates of the reacting components in the solid are fast relative to the flow rate of the liquid drop[3], both γ_{sv}^o and γ_{sl}^o will simultaneously decrease by an amount $(-)\Delta g^{sl}$, and the liquid at the periphery of the drop will remain in contact with reacted solid, as represented schematically in Fig. 2b'. The driving force for wetting in this case does not change drastically from that due to the initial dynamic surface tensions (Fig. 2a) and remains essentially constant while the system moves to chemical equilibrium (Figs. 2b', 2c', and 2d).

Throughout these entire sequences, shown schematically in Fig. 2, the amount of material dissolved by the solid was considered to be small enough to be neglected so that the solid surface remained flat. However, if the specific volume of the solid solution phase at the interface differs appreciably from that of the unreacted solid, analysis by use of Young's equation as applied to experimentally measured contact angles could be misinterpreted because of the resulting non-existence of a flat solid surface.

Several dynamic stages associated with a reaction of type (ii), where only the liquid is not saturated with the solid, are shown schematically in Fig. 3. Figure 3a shows the configuration at t_o when the liquid phase first comes into contact with the solid and Young's equation may be used to express the conditions for mechanical equilibrium in terms of the dynamic interfacial tensions. After the initial reaction the composition of the liquid around the periphery and at the solid-liquid interface rapidly approaches equilibrium compositions relative to the solid; correspondingly, γ_{sl}^o and γ_{lv}^o decrease because of the free energy contribution of the reaction, but then they rapidly approach their static interfacial tension values γ_{sl} and γ_{lv} (Fig. 1). During the initial reaction stage, thus, an instantaneous lowering of the contact angle or spreading may be observed[17] which is immediately

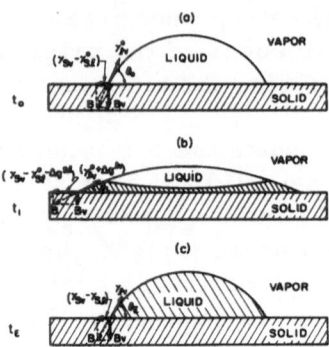

Fig. 3. Schematic representation of the dynamic stages of a sessile drop when the initial liquid is not saturated with some or all of the components of the solid.

followed by the drop pulling back to an equilibrium contact angle θ_e which is retained until the system reaches equilibrium (Figs. 3b and 3c). With high-viscosity liquids and fast diffusion rates, however, the initial spreading may not be realized because the static interfacial tensions are attained faster than the liquid can spread. Again, as in the previous case, the amount of solid dissolved by the liquid was considered to be small enough to be neglected. In actual fact, however, as the reaction proceeds, the solid-liquid interface will drop below the solid-vapor surface[20], complicating the analysis of mechanical equilibrium.

A type (iii) reaction is expected to be similar in behavior either to type (i) or (ii) reaction depending on whether an increase or decrease of volume of the solid occurs at the interface, but the kinetic analysis of the reaction and determination of the nature of the physical configuration become more complicated. The formation of a compound at the interface (type (iv) reaction) is also expected to cause the mechanical behavior of the system to be similar to that for one with type (i) reaction. The analysis in this case could be even more complicated, particularly if the compound should isolate the liquid from direct contact with the reacting solid.

LIQUID PHASE SINTERING UNDER CHEMICAL
NON-EQUILIBRIUM CONDITIONS

It has been reported[21] that liquid phase sintering kinetics can be divided into three densification stages: rearrangement, solution-precipitation, and coalescence. Upon formation of a liquid phase in a porous non-sintered compact, relative motion of particles occurs by rearrangement. If sufficient liquid is present, complete densification may occur by this process alone with the kinetics of densification being those of viscous flow[21, 22]. With insufficient liquid, residual porosity remains. When the pore shape reaches a steady state configuration, the rearrangement stage is completed. Further densification must occur by another mechanism. The second stage, solution-precipitation, is characterized by a zero dihedral angle which causes liquid to penetrate along solid-solid contacts. Due to capillary pressure effects caused by porosity, a solubility gradient of the components of the solid phase in the liquid is established causing diffusion of material through the liquid. The final stage, coalescence, is characterized by the formation of a finite dihedral angle and solid-solid contacts. Densification is then due to diffusion of material through the solid phase because of the existence of a vacancy concentration gradient established by the curvature of the solid-liquid and the liquid-vapor interfaces. It can be shown, however, that in systems at bulk chemical equilibrium only two of the three proposed densification stages can be realized; i.e. when the equilibrium dihedral angle, Φ_ϵ, is zero only the rearrangement and the solution-precipitation stages exist, and when $\Phi_\epsilon > 0$, only the rearrangement and coalescence stages exist.

Kinetics of liquid phase sintering are determined by the relative magnitudes of the interfacial tensions and pore shapes. Compacts containing large volume fractions of liquid have no sv surfaces and form spherical porosity within the liquid phase directly upon completion of the rearrangement stage. On the other hand, compacts with insufficient liquid normally have sv surfaces and form toroidal shaped porosity after the rearrangement stage; as solution-precipitation proceeds, liquid is squeezed into the void space, and eventually spherical porosity forms.

Therefore, if one defines a stage in sintering as an interval during which the pore shape remains constant, as well as an interval during which the sintering mechanism remains constant, there are clearly two stages of solution-precipitation for compacts containing small liquid volumes. The first stage is characterized by toroidal shaped porosity, and the second stage is characterized by the presence of spherical porosity.

Kingery[21] has presented a kinetic analysis for the second stage of solution-precipitation. His method has been extended by Hoge and Pask[23] to describe the kinetics of both stages of solution-precipitation. Results for densification as a function of time and the initial volume are given in Table I.

Table I. Values of the time (t) exponent, x, in $(\Delta V/V_o)$ $\sim t^x$ for various densification stages of liquid phase sintering. V_o denotes the original volume.

Densification Stage	Time Exponent, x
Rearrangement (viscous flow[21, 22])	1.50
Solution precipitation:	
Initial stage	0.24
Final stage	0.36
Coalescence (solid phase sintering in the presence of a liquid phase):	
$\gamma_s/\gamma_{lv} = 0.05$	0.437
$\gamma_s/\gamma_{lv} = 1.0$	0.464
$\gamma_s/\gamma_{lv} = 2.0$	0.471
$\gamma_s/\gamma_{lv} = 3.0$	0.475

If $\gamma_{ss} < 2\gamma_{sl}$, an equilibrium dihedral angle greater than zero degrees and solid-solid contacts are formed. Under these conditions solution-precipitation can cause surface rearrangement, but does not lead to densification. Sintering kinetics are then characterized by solid state diffusion mechanisms, either bulk or grain boundary. However, the liquid as well as the magnitude of the dihedral angle affect the sintering kinetics. Hoge and Pask[23] have presented a kinetic analysis for the non-zero dihedral angle case, corresponding to small liquid volumes and bulk diffusion. The values of the time exponent, x, are given in Table I for several ratios of $\gamma_{s\ell}/\gamma_{\ell v}$. An analysis for grain boundary diffusion has been given by Gessinger, et al[24].

All kinetic analyses discussed above assume static interfacial tensions. In real compacts, interfacial tensions initially change continuously, as do corresponding dihedral angles, as the system moves to chemical equilibrium. Under such conditions, the dihedral angle equation, Eq. (2), is only valid in terms of the dynamic interfacial tensions.

Let us now, as in the previous section, consider the specific effects of several types of chemical reactions on the dihedral angle, assuming that chemical equilibrium exists between the vapor and the condensed phases but not between the solid and the liquid. The first type of reaction corresponds to the case where only the solid is not saturated with some or all of the components of the liquid. Two dynamic sequences associated with this type of reaction are shown schematically in Fig. 4. Let us assume firstly that the diffusion rates of the reacting components in the solid are slow compared to the penetration rate of the liquid along the solid-solid interface. Under these conditions, the solid-liquid interfacial tension is reduced by the free energy of the interfacial reaction (Section II), but γ_{ss}^{o} is unaffected since the reaction product has not moved ahead of the triple point. Thus, if the contribution of the interfacial reaction is large enough in magnitude, the reduction in the solid-liquid interfacial tension can be sufficiently large to cause a dynamic zero dihedral angle, as shown in Fig. 4a. At this instant, it is thermodynamically favorable for the solid-solid interface to be replaced by two solid-liquid interfaces. As this happens, the liquid is continuously exposed to an unreacted solid-solid interface and continues to penetrate it.

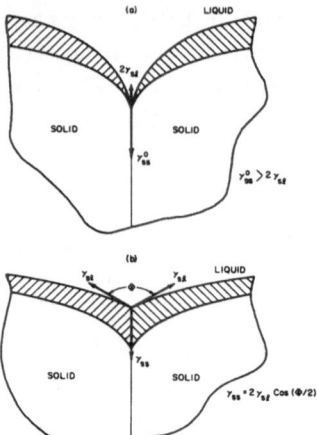

Fig. 4. Schematic representation of the two dynamic stages of a solid-liquid-solid system when the solid is not saturated with some or all of the components of the liquid and when the growth rate of the reaction product is (a) slower and (b) faster than the penetration rate of the liquid along the solid-solid interface[3].

On the other hand, when the diffusion rates of the reacting components in the solid are fast relative to the penetration rate of the liquid along the solid-solid interface, both $\gamma_{s\ell}^{o}$ and γ_{ss}^{o} are affected simultaneously by the interfacial chemical reaction. Thus, the combined reduction of $\gamma_{s\ell}^{o}$ and γ_{ss}^{o} by the interfacial reactions can cause the dihedral angle to be relatively unaffected by the reaction, resulting in a finite dihedral angle as the system moves to chemical equilibrium, as shown in Fig. 4b.

During these reaction stages, if the sintering compact contains very small amounts of liquid relative to the solid, complete dissolution of the liquid in the bulk solid can result as the interfacial reaction continues, causing the liquid to be entirely eliminated. At this point, a finite dihedral angle, determined by the resulting γ_{sv} and γ_{ss}, would appear. On the other hand, if the amount of material going into solid solution is large, volume changes of the solid also have to be considered in the analysis.

The second type of reaction corresponds to the case where only the liquid is not saturated with the solid and dissolution of the solid occurs until the liquid is saturated. The dissolution reaction at the solid-liquid interface causes $\gamma_{s\ell}$ to be reduced instantaneously relative to γ_{ss}, which in turn causes the dihedral angle to be lowered for a transient period. In this stage, the penetration of the liquid along the solid-solid interface is further aided by the preferential dissolution of the solid at the grain boundaries if they are poorly formed and thus are high energy sites. Liquid then will tend to penetrate the grain boundaries as long as the liquid is unsaturated with the solid. Upon saturation, the equilibrium dihedral angle will form, recreating solid-solid contacts, and possibly trapping liquid in isolated pockets.

When both phases are unsaturated with respect to the other, the dynamic reaction stages are similar in behavior to either one of the above reaction types depending on whether an increase or decrease of the volume of the solid occurs at the interface. If a compound forms due to interfacial reactions, the mechanisms are similar to those for the first type of reaction.

It was stated earlier that the magnitude of the dihedral angle determines the sintering mechanism and influences the sintering kinetics. In the absence of bulk chemical reactions, after the initial viscous flow rearrangement, the densification kinetics will be either those of solution-precipitation (zero dihedral angle) or those of solid phase sintering in the presence of a liquid phase (non zero dihedral angle), Table I. In the presence of chemical reactions, however, a transient penetration of the liquid along the solid-solid boundary may occur in all of the chemical non-equilibrium cases discussed above. All three densification stages can then be realized as the system approaches chemical equilibrium. As shown in Table I, the sintering kinetics will differ markedly during this process. During the coalescence stage, however, the effect of interfacial reactions on the $\gamma_{s\ell}/\gamma_{\ell v}$, and thus the sintering kinetics, is not significant since for practical purposes, the time exponent, x, is not appreciably affected by the $\gamma_{s\ell}/\gamma_{\ell v}$ ratio, Table I.

WETTING OF MAGNESIUM OXIDE

Wetting of magnesium oxide by $CaO-MgO-SiO_2-R_2O_3$ (Al_2O_3, Cr_2O_3, or Fe_2O_3) liquids has been studied by Aksay, et al[17]. Experiments to be discussed and evaluated were performed at 1550° in air on cleaved (001) surfaces of MgO single crystals. The liquid of monticellite, $CaO \cdot MgO \cdot SiO_2$ (CMS), composition is in equilibrium with MgO after a slight precipitation of MgO in the CMS liquid[25]. The CMS-MgO equilibrium, however, is disturbed with the addition of Al_2O_3, Cr_2O_3, or Fe_2O_3. The solubilities[26] of Al_2O_3, Cr_2O_3, and Fe_2O_3 (FeO) in MgO are <0.5, 9.0, and 51.0 wt%, respectively. The interfacial reaction to consider between the liquid and the solid is

$$MgO_{(s)} + R_2O_3 \text{ (in liquid)} \longrightarrow MgR_2O_4 \text{ (in MgO)}, \qquad (8)$$

which results in the formation of an MgO solution with MgR_2O_4. At 1823K, an average value for the standard free energy of this reaction[27], ΔG°, is -8,000 cal/mol. The free energy of the reaction (8), then, is

$$\Delta G = RT \ln \frac{a_{MgR_2O_4}}{a_{R_2O_3}} - 8,000 , \qquad (9)$$

where R is the gas constant, T is the absolute temperature, and a is the activity of the designated component. Assuming that Raoult's law is applicable, $a_{MgR_2O_4}$ is equal to the concentration of MgR_2O_4 in the MgR_2O_4 solid solution at equilibrium with the MgO solid solution. A typical value[26] for $a_{MgR_2O_4}$ is ≈ 0.9. Similarly, $a_{R_2O_3}$ depends on the concentration of R_2O_3 in the liquid phase. Assuming a value of 0.5 for $a_{R_2O_3}$, $\Delta G = -6,050$ cal/mol, and the corresponding transient reduction in the solid-liquid interfacial tension could be as high as 1,000 dynes/cm^2. As the concentration of R_2O_3 in the liquid increases and the liquid becomes saturated with respect to MgR_2O_4, a second possible interfacial reaction is

$$MgO \text{ (sat. with } MgR_2O_4) + R_2O_3 \text{ (in liquid)} \longrightarrow$$

$$MgR_2O_4 \text{ (sat. with MgO)} \qquad\qquad (10)$$

which results in the formation of a spinel compound at the interface. The contribution of the free energy of this reaction to the reduction of any of the interfacial energies, however, is zero since all the phases involved in the reaction are at equilibrium with each other. The interfacial reaction (10) does not contribute to additional changes in $\gamma^{o}_{s\ell}$ other than the changes due to the precurser reaction (8).

As pointed out in Section III, in addition to the contribution of the interfacial reaction on $\gamma^{o}_{s\ell}$, one also has to consider the kinetics of the growth of the reaction product relative to the flow rate of the liquid drop in order to determine if γ^{o}_{sv} will simultaneously be affected by the same reaction. Diffusion kinetics of Al^{3+}, Cr^{3+}, and Fe^{3+} in bulk MgO have been reported. At $1550°C$, the volume diffusivities of Al^{3+} and Fe^{3+} in MgO are 1.6×10^{-9} and 9.55×10^{-9} cm^2/sec, respectively[28, 29], whereas the diffusivity of Cr^{3+} is 6.6×10^{-12} cm^2/sec[30]. Assuming that the surface diffusivities, D_S, of these cations in MgO are proportional to the volume diffusivities, $D_S^{Cr^{3+}} << D_S^{Al^{3+}}$ and $D_S^{Fe^{3+}}$. The growth rate of the MgO ($MgCr_2O_4$) solution at the interface, therefore, is expected to be considerably slower than the growth rate of the MgO ($MgAl_2O_4$) or MgO ($MgFe_2O_3$) solid solution[3, 17].

The results of the sessile drop experiments as shown in Fig. 5 support the above discussion. Liquids with Cr_2O_3 additions in excess of 3 wt% showed spreading for periods of up to 6 h. With Al_2O_3 and Fe_2O_3 additions, the contact angle was lowered but no spreading took place. The microstructure studies[17] showed the growth of the MgO ($MgAl_2O_4$) and MgO ($MgFe_2O_3$) solid solutions into the sv interfaces as schematically shown in Fig. 2b'. The contact angle values

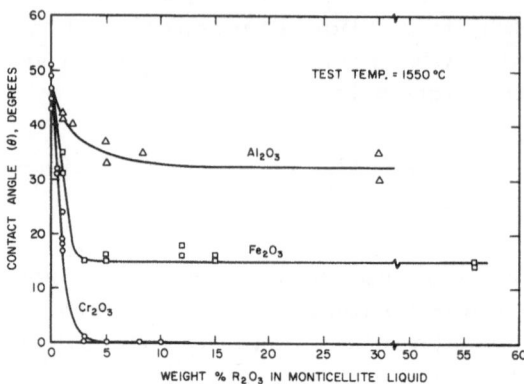

Fig. 5. The effect of R_2O_3 additions to monticellite liquid on contact angle, after 3 h at $1550°C$[17].

of Fig. 5 for Al_2O_3 and Fe_2O_3 additions correspond, essentially, to the equilibrium values, θ_e, Fig. 2d. The liquids with Cr_2O_3 additions spread since the growth rate of the solid solution product is slower than the spreading rate of the liquid drop and thus the driving force for wetting, as shown in Fig. 2b, is high due to an unaffected γ_{sv}^o. When the experiments with Cr_2O_3 additions, however, were repeated with much smaller drops and larger substrates, the spreading drop did not reach the edges of the substrate. The drops with 3.0 and 5.0 wt% Cr_2O_3 additions showed some recession from the periphery of furthest spreading and formed contact angles of 7° and 6°. This effect would be expected when the Cr_2O_3 content in the liquid is reduced resulting in a substantial reduction of the ΔG of reaction (8), Eq. (9). Similarly, the drop with 15 wt% Cr_2O_3 initially spread and segregated into small segments and droplets as it tried to pull back when the ΔG of reaction (8) was reduced.

PERMEATION OF SILICATES INTO MAGNESIA

The permeation of the $CMS-R_2O_3$ liquids of the previous section into sintered MgO compacts of 92.5% of the

theoretical density and 3.1% open porosity has been studied
by Wong[8]. The experiments were performed at 1550°C for
periods of 5 min to 2 h. The permeation distance of the
liquids into the compacts as a function of the annealing
time at temperature is shown in Fig. 6. Since the penetration rate of a liquid along a capillary is directly proportional to the square root of $(\gamma_{sv} - \gamma_{s\ell})^3$, the direct correlation between the contact angles of the previous section
and the penetration kinetic data shown in Fig. 6 then is in
agreement with the capillary penetration theory[31].

A more important aspect of these studies, however, is
that the microstructures developed after the liquid permeation into the MgO compact were in direct support of the
discussions of Section IV. Figure 7 shows the microstructure of an MgO compact permeated by a liquid containing 15 wt% Cr_2O_3 after 5 min at 1550° C. A thin film
of liquid is present along most of the grain boundaries although the reported[32] equilibrium dihedral angle for the
system is 45°. Since the solid-solid contacts existed[8]

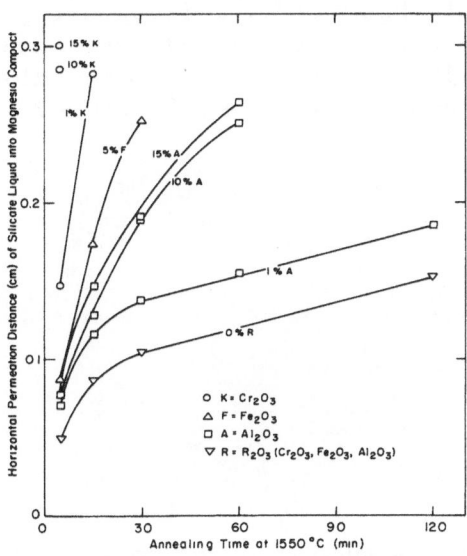

Fig. 6. The permeation distance of the $CaO-MgO-SiO_2-R_2O_3$
liquids into MgO compacts as a function of time at 1550° C. [8]

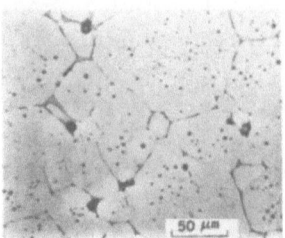

Fig. 7. The microstructure of an MgO compact after permeated by a liquid of 15 wt% Cr_2O_3 + 85 wt% $CaO \cdot MgO \cdot SiO_2$
composition for 5 min at 1550° C[8].

prior to the permeation of the liquid, the complete penetration of the liquid along the solid-solid grains is explained
by the transient reduction of the $\gamma_{s\ell}$ during chemical reaction, Fig. 4a.

No permeation of the liquids with high Fe_2O_3 additions
(45.0 and 55.0 wt% Fe O) was observed although the liquid
with 5.0 wt% Fe_2O_3 addition permeated the MgO compacts
as expected on the basis of the corresponding contact angle
value (Figs. 5 and 6). The microstructure shown in Fig. 8
shows the MgO-liquid interface of a permeation couple with
a liquid composition of 45 wt% Fe_2O_3 after 2 h at 1550° C.

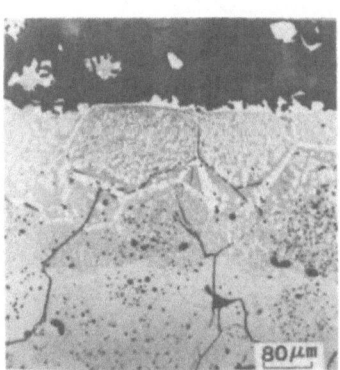

Fig. 8. The microstructure of the solid-liquid interface of
an MgO-liquid permeation specimen with an original liquid
composition of 45 wt% Fe_2O_3 + 55 wt% $CaO \cdot MgO \cdot SiO_2$ after
2 h at 1550°C[8].

The lack of permeation into MgO is clearly explained by the closing off of the channels at the interface due to the extensive volume increase with the formation of MgO ($MgFe_2O_4$) solid solution. $MgFe_2O_4$ precipitation (white particles) took place during cooling of the specimen. Any liquid that may have permeated into the compact before the channels were closed off probably disappeared as it reacted with the solid to form solid solution.

SUMMARY

Under bulk chemical equilibrium conditions, Young's and dihedral angle equations in terms of the static interfacial tensions outline the condition of mechanical equilibrium in a solid-liquid-vapor system. Under bulk chemical non-equilibrium conditions, mass transfer across an interface results in a transient decrease in the corresponding interfacial free energy and thus the interfacial tension by an amount equal to the free energy of the effective chemical reaction per area at that interface. When the chemical reaction is between the solid and the liquid, in sessile drop experiments, a transient lowering of the contact angle or spreading is observed if the growth rate of the reaction product is slower than the spreading rate of the liquid. Similarly, in a solid-liquid-solid system, in the presence of chemical reactions, a transient penetration of the liquid along the solid-solid boundary may occur. This transient stage then corresponds to the solution-precipitation stage of liquid phase sintering in systems that are characterized with equilibrium non-zero dihedral angles.

REFERENCES

1. T. Young, Phil. Trans. Roy. Soc. London 95, 65 (1805).
2. C.S. Smith, Trans. AIME 175, 15 (1948).
3. I.A. Aksay, C.E. Hoge, and J.A. Pask, "Thermodynamics of Wetting" submitted to J. Appl. Phys.
4. J.W. Gibbs, Trans. Conn. Acad. 3, 343 (1878); also included in Scientific Papers of J.W. Gibbs (Dover Publ., Inc., New York, 1961) Vol. 1.

5. R. Defay, I. Prigogine, A. Bellemans, and D. H. Everett, Surface Tension and Adsorption (John Wiley and Sons, Inc., New York, 1966).

6. D. A. Netzel, G. Hock, and T. I. Marx, J. Colloid Sci. 19, 774 (1964).

7. R. E. Johnson, Jr., J. Phys. Chem. 63, 1655 (1959).

8. B. Wong, "Permeation of Silicate Liquids into Sintered Magnesia Compacts," (M. S. Thesis) LBL-877, University of California, Berkeley, August 1972.

9. A. A. Leont'eva, Koll. Zhurn. 11, 176 (1949).

10. P. Kozakevitch, G. Urbain, and M. Sage, Rev. Met. 52, 161 (1955).

11. S. I. Popel, O. A. Esin, G. F. Konovalov, and N. S. Smirnov, Dokl. Akad. Nauk SSSR 112, 104 (1957), or Proc. Acad. Sci. USSR - Phys. Chem. 112, 27 (1957).

12. M. Humenik, Jr. and W. D. Kingery, J. Am. Ceram. Soc. 37, 18 (1954).

13. W. M. Armstrong, A. C. D. Chaklader, and M. L. A. DeCleene, J. Am. Ceram. Soc. 45, 407 (1962); W. M. Armstrong, A. C. D. Chaklader, and D. J. Rose, Trans. AIME 227, 1109 (1963); and A. C. D. Chaklader, W. M. Armstrong, and S. K. Misra, J. Am. Ceram. Soc. 51, 630 (1968).

14. A. A. Zhukhovitskii, V. A. Grigorian, and E. Mikhalik, Dokl. Akad. Nauk SSR 155, 392 (1964), or Proc. Acad. Sci. USSR - Phys. Chem. 155, 255 (1964).

15. J. E. McDonald and J. G. Eberhart, Trans. AIME 233, 512 (1965).

16. V. I. Kostikov and B. S. Mitin, Sb., Mosk. Inst. Stali Splavov, No. 49, 114 (1968).

17. I. A. Aksay, A. P. Raju, and J. A. Pask, "Wetting of Magnesium Oxide by Silicate Liquids," submitted to J. Am. Ceram. Soc.

18. J. T. Davies and E. K. Rideal, Interfacial Phenomena (Academic Press, New York, 1963), 2nd ed., p. 360.

19. The free energy change at the solid-vapor interface is designated as $\Delta g^{s\ell}$ instead of Δg^{sv} since, assuming that composition of the liquid at the interface remains essentially constant, compositional variations are only in the solid due to components diffusing from the liquid and thus the nature of the reactions at the solid-vapor and the solid-liquid interfaces do not differ appreciably.

20. R.D. Carnahan, T.L. Johnston, and C.H. Li, J. Am. Ceram. Soc. 41, 343 (1958); and J.A. Champion, B.J. Keene, and J.M. Sillwood, J. Mater. Sci. 4, 39 (1969).

21. W.D. Kingery, J. Appl. Phys. 30, 301 (1959).

22. J. Frenkel, Zhur. Tech. Fiz. 9, 305 (1945), or J. Phys. (USSR) 9, 385 (1945); and F.V. Lenel, The Physics of Powder Metallurgy, W.E. Kingston, ed. (McGraw-Hill Book Company, New York, 1951), p. 238.

23. C.E. Hoge and J.A. Pask, "Kinetics of Liquid Phase Sintering," unpublished.

24. G.H. Gessinger, H.F. Fischmeister, and H.L. Lukas, Acta Met. 21, 715 (1973).

25. E.F. Osborn and A. Muan, Phase Diagrams for Ceramists, E.M. Levin, C.R. Robbins, and H.F. McMurdie, eds. (The Am. Ceram. Soc., Inc., Columbus, Ohio, 1964), p. 210, Fig. 598.

26. A.M. Alper, R.N. McNally, P.H. Ribbe, and R.C. Doman, J. Am. Ceram. Soc. 45, 263 (1962); A.M. Alper, R.N. McNally, R.C. Doman, and F.G. Keihn, J. Am. Ceram. Soc. 47, 30 (1964); and J.C. Willshee and J. White, Trans. Brit. Ceram. Soc. 66, 541 (1967).

27. A. Navrotsky and O.J. Kleppa, J. Inorg. Nucl. Chem. 30, 479 (1968).

28. W.P. Whitney, II, and V.S. Stubican, J. Phys. Chem. Solids 32, 305 (1971).

29. S.L. Blank and J.A. Pask, J. Am. Ceram. Soc. 52, (1969).

30. H. Tagai, S. Iwai, T. Iseki, and M.S. Tokio, Radex-Rundschau 4, 577 (1965).

31. E.W. Washburn, Phys. Rev. 17, 273 and 374 (1921).

32. B. Jackson, W.F. Ford, and J. White, Trans. Brit. Ceram. Soc. 62 577 (1963); and B. Jackson and W.F. Ford, Trans. Brit. Ceram. Soc. 65, 19 (1966).

EVIDENCE OF INVERTED UNDERCUTTING OF A SOLID METAL SUBSTRATE BY A LIQUID CERAMIC SESSILE SPECIMEN AT ELEVATED TEMPERATURES

L. D. Lineback, LeRoy Hoenycutt III,
C. R. Manning & K. L. Moazed
Department of Materials Engineering
North Carolina State University
Raleigh, North Carolina

When a drop of liquid is placed on a solid substrate, a metastable configuration, illustrated in Fig. 1, is often achieved instantaneously. This configuration may be maintained for a very long period of time, if the temperature is substantially below the melting point of the solid. This configuration is only an approximation of the final equilibrium state, because the surface free energy vectors do not balance. However, as θ decreases toward zero the approximation becomes closer. As the magnitude of the vector G_{lv} decreases relative to G_{ls} and G_{sv}, once again, the approximation becomes closer. As we approach the melting point of the solid, mass transport in it becomes

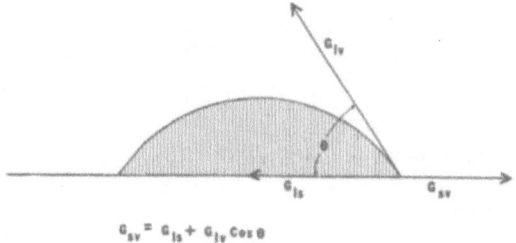

$$G_{sv} = G_{ls} + G_{lv} \cos \theta$$

Fig. 1. Metastable configuration for a liquid placed on a solid substrate (Young's expression).

323

measureable, and the rate of achieving equilibrium becomes
much greater.

The first possibility for achieving equilibrium[1], illustrated in Fig. 2, will be realized if the liquid can dissolve
the components of the solid, that is, if alloying takes place.
This configuration will also be realized when the liquid cannot dissolve components of the solid, if the rate of bulk
diffusion in the solid is relatively large compared to the
rate of vaporization of the solid.   However, if the rate of
vaporization of the solid is high relative to the rate of mass
transport through the solid and if alloying does not occur,
or when the rate of alloying is low, the configuration illustrated in Fig. 3 will be realized.

In the situations illustrated in Figs. 2 and 3, not only
the horizontal components of the vectors but also the vertical
components are zero - a requirement for equilibrium.   In
contrast to this self-consistent approach, other means must
be sought to rationalize the metastable configuration of Fig.
1 as a true equilibrium state.   This is usually accomplished
by introducing ill-defined parameters such as, gravitational
attraction, forces of adhesion and cohesion, etc. to account
for the fact that the vertical components of the vectors in
Fig. 1 do not balance - a condition that is not compatible
with a state of equilibrium.

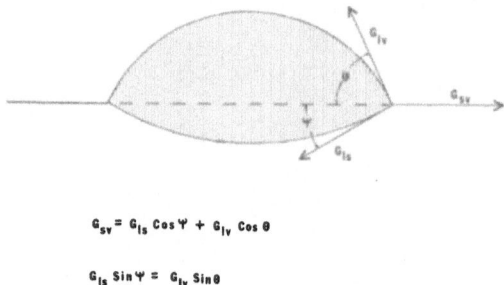

$$G_{sv} = G_{ls} \cos \Psi + G_{lv} \cos \theta$$

$$G_{ls} \sin \Psi = G_{lv} \sin \theta$$

Fig. 2.   Equilibrium configuration for a liquid drop on a
solid substrate when penetration or undercutting is achieved
by the liquid phase.

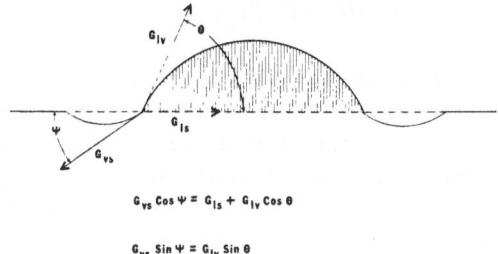

$$G_{vs} \cos \Psi = G_{ls} + G_{lv} \cos \theta$$

$$G_{vs} \sin \Psi = G_{lv} \sin \theta$$

Fig. 3. An alternate equilibrium configuration for a liquid drop on a solid substrate.

LITERATURE

Hodkin et al.[2] determined the surface energy of tungsten and compared it with previously reported values. His value of 2230 $\pm$ 180 dynes/cm at 1500°C agrees with the theoretical values of 2490[3] and 2300[4]. For our purposes a value of 2400 dynes/cm at 2000-2400°C was used.

A number of surface tension values have been reported for molten Al_2O_3 as shown in Table I. The most recent are those of Rasmussen and Nelson[5]. Their pendant drop value does not agree with that of Kingery[6], but does agree with the sessile drop values of Bartlett and Hall[7] and Wartenburg et al.[8] It has been well established that alumina does not completely wet tungsten and this may cast some doubt on the values of Rasmussen and Nelson obtained by the capillary method.

Table I. Reported values of G_{lv} for Molten Al_2O_3.

Investigator	G_{lv}	Method
Bartlett & Hall[7]	550	Sessile on inert substrate and pendant
Kingery[6]	690	Sessile drop on inert substrate
Sokolov[9]	700	Theoretical
Wartenburg, et al.[8]	577	Sessile drop on inert substrate
Rasmussen & Nelson[5]	638	Capillary
" "	574	Pendant

EXPERIMENTAL

Powder X-ray diffraction showed the alumina to be of
the α phase both before and after melting. X-ray diffraction
records of the tungsten strip indicated a rolling texture.
Although no pole figures were determined, analysis of the
diffraction lines indicated that the texture of the tungsten
strip changed in an unpredictable manner with time at temperature. Microprobe analysis of the Al_2O_3-W system
indicated no alloying of either tungsten in Al_2O_3 or Al_2O_3
in tungsten (Fig. 4).

Liquid drops of Al_2O_3 were found to partially wet
tungsten in the temperature range of 2050°C - 2350°C. It
was observed that the contact angle changed as a function
of time. A groove with a constant radius of curvature
cross-section was formed around the outside periphery
of the sessile drop in the solid tungsten substrate. This
groove increased in depth and width as the contact angle
and the time at temperature increased.

The change in contact angle θ at a temperature of
2225°C $\pm$ 150°C as a function of time is shown in Fig. 5
for both polycrystalline and single crystal starting material.

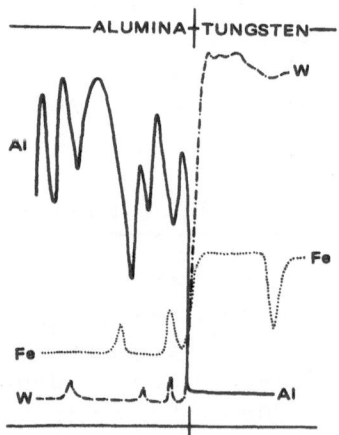

Fig. 4. Typical microprobe line scan in the Al_2O_3-W
system.

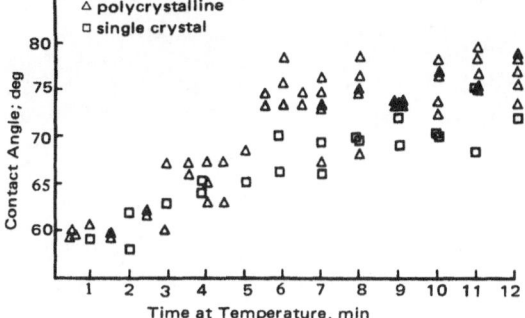

Fig. 5. Contact angle θ as a function of time at temperature for both polycrystalline and single crystal Al_2O_3.

The contact angle was initially about 59° for both polycrystalline and single crystal material. The polycrystalline material approached a value of 70-80° and the single crystal material approached 70-75°.

Microscopic investigation indicated that grooving of the tungsten substrate proceeded slowly for 4 minutes, then rapidly, and reached an equilibrium value at 6 minutes. Typical cross-sections at times of 2, 4, 6 and 12 minutes are shown in Fig. 6. Cross-sections of the cooled sessile drop and substrate showed the groove to be relatively irregular but satisfactory for measuring the solid-vapor contact angle by the same method used for the liquid-solid contact angle θ. The ψ angles for grooves formed with single crystals at times greater than or equal to 6 minutes was measured. The root mean square and mean values of these values was 12.5°. Using 72° for the liquid-vapor contact angle of single crystal material, the surface tension G_{lv}, for molten Al_2O_3 (single crystal) in the temperature range of 2200 $\pm$ 150°C is calculated to be 545 $\pm$ 64 dynes/cm, in good agreement with values cited in Table I.

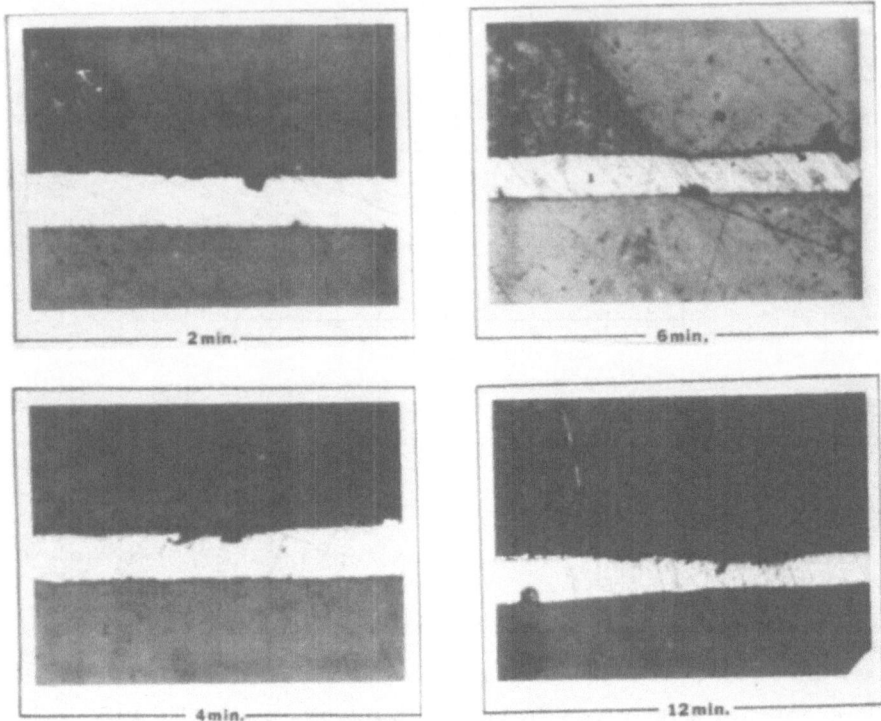

Fig. 6. Cross-section of Al_2O_3 drops on W substrate.

DISCUSSION OF RESULTS

The observation that the contact angle θ changed as a function of time for molten Al_2O_3 on solid tungsten indicated that some time-dependent phenomenon was occurring in the specimen or its environment. X-ray diffraction and electron microprobe analysis indicated that this change was not due to solution formation in liquid or substrate or to a reaction between liquid and solid resulting in a third phase. While some reduction of the Al_2O_3 did result, not enough oxygen was removed to induce a phase change upon solidification. In general the chemical stability of the system was felt to be sufficient not to cause these changes in the contact angle.

Porosity apparently played a small role. While both polycrystalline and single crystal starting materials had initial θ of approximately 59°, the final θ for the polycrystalline material was slightly higher than for the single crystal. The implication is that the melt from the polycrystalline materials contained voids although this was not confirmed from cooled specimen cross-sections due to the results of the polishing technique.

It is believed that the tungsten partial pressure reached equilibrium early as condensation of tungsten on the interior of the furnace body was observed after short exposure times. It is doubtful that the Al_2O_3 equilibrium partial pressure was ever reached due to limited surface area, its approach to the equilibrium value is not believed to be responsible for the change in the contact angle as that pressure is very small.

It is believed that the change in the contact angle is the direct result of the groove formation in the tungsten substrate on the periphery of the Al_2O_3 sessile drop. It was shown in Fig. 3 that the formation of such a groove should increase the angle.

CONCLUSIONS

The phenomenon of inverted undercutting may occur to produce a thermodynamic equilibrium between surface tension, surface energy, and interfacial energy. In certain instances inverted undercutting may be more favorable than undercutting because of the kinetics involved. It is concluded that molten aluminum oxide on solid tungsten is such a case.

REFERENCES

1.   Moazed, K. L. , "The Application of the Theory of Heterogeneous Nucleation to Epitaxy", in The Use of Thin Films in Physical Investigations, J. C. Anderson ed. , Academic Press (1966).

2. Hodkin, E. N., Nicholas, M. G. and Poole, D. M.,
    "The Surface Energies of Solid Molybdenum, Niobium,
    Tantalum, and Tungsten," J. Less Common Metals
    20 (1) 93-103 (1970).
3. Avraamov, Y. S. and Grozdev, A. G., "Calculation of
    the Surface Energy of Metals with a Body-Centered
    Cubic Lattice," Fiz. Metal Metalloved, 23 (3), 405 - 8
    (1967).
4. Zudremkin, S. N., Izv. Akad. Nauk SSSR Otd. Tekhn.
    Nauk Met. i Toplivo 1, 55 (1961).
5. Rasmussen, J. J. and Nelson, R. P., "Surface Tension
    and Density of Molten Alumina," J. Am. Ceram. Soc.
    54 (8) 398 - 401 (1971).
6. Kingery, W. D., "Surface Tension of Some Liquid
    Oxides and Their Temperature Coefficients," J. Am.
    Ceram. Soc. 42 (1) 6 - 10 (1959).
7. Bartlett, R. W. and Hall, J. K., "Wetting of Several
    Solids by Al_2O_3 and BeO Liquids," Am. Ceram. Soc.
    Bull. 44 (5) 444-48 (1965).
8. Wartenburg, H. V., Wehner, G. and Saran, E., "The
    Surface Tension of Molten Al_2O_3 and La_2O_3," Nachr.
    Ges. Wiss. Göttingen, Jahryesber, Math-physik. Kl.,
    Fachgruppen II [N. F] 2, 65 - 71 (1936).
9. Sokolov, O. K., "Surface Tension of Melts," Izv. Akad.
    Nauk SSSR Met. Gorn. Delo 4 59-64 (1963).

DISCUSSION

R. Atkin (IBM): Why is there preferential evaporation of
tungsten near the liquid alumina droplet? Could surface
or bulk diffusion of oxygen from the liquid be responsible?
Author: We believe that this is solely a result of the
approach to thermodynamic equalibrium. Whether oxygen
or any other element plays a role is irrelevant.
D. E. Smith (Owens-Illinois): Do you know that the grooving
of the substrate is due to volatilization by measurement of
the volume of the liquid drop, thereby elimination of dissolution?
Author: No. The work should be done, however.

INTERFACES IN CERAMIC NUCLEAR FUELS

K.D. Reeve

Materials Division, Australian Atomic
Energy Commission Research Establishment
Sydney, Australia

Internal interfaces such as grain boundaries, pore
surfaces and phase boundaries play a major role in
determining the mechanical and to some extent thermal
properties of most ceramic bodies. Other important
interfaces are those between regions of different density,
chemical composition or stress state. One major problem
of the ceramic engineer is to prevent the ceramic component from failing at these interfaces under service conditions. The nuclear ceramist has to face most problems
encountered by his non-nuclear colleague, but the need
to predict and allow for radiation-induced changes as well
as those due to temperature and stress effects during
service life often makes his task more difficult. This
paper explores some of these problems, using the interfaces in selected all-ceramic nuclear fuels as examples.

The fuel material for current light water reactors is
a ceramic - UO_2 - but the fuel as a whole does not fit into
the pattern outlined above. The single phase UO_2 pellets
are fully contained within a metal can, operate under a
large thermal gradient and, because they are contained
by the can, are allowed to crack in service. The pellet-
can interface and the UO_2 cracking pattern are technologically very important in determining the temperature
profile of the pellet and thence the stress in the cladding

and the extent of fission gas release. Further, the temperature profile and radiation environment may produce various combinations of pore shrinkage, pore growth, pore migration, grain growth, crack healing and solid fission product deposition. The initial grain size of the pellet appears to be of secondary importance in determining in-service behaviour but the initial pore fraction, size and shape help to determine the thermal conductivity, the temperature profile, and the balance between in-pile sintering and swelling. The situation is similar for $(U, Pu)O_2$ fuels to be used for fast reactors, although there are additional problems arising from temperature gradient induced plutonium redistribution in an initially homogeneous single phase solid solution. However, in water and fast reactor fuels, most of the resultant problems are quite specific to nuclear fuel technology.

On the other hand, all-ceramic dispersion fuels, such as those for high temperature gas-cooled reactors (HTGCRs), contain interfaces which are of much broader interest; the body of this paper will concentrate upon such fuels.

At least three non-fissile ceramic materials have been proposed as matrices for HTGCR fuels, viz BeO, SiC and graphite. During the 1960s BeO-based fuels were studied extensively in a number of countries for use in land-, sea- and air-based reactors. The author was much concerned with one of these studies - the Australian pebble-bed reactor project[1, 2] - which highlighted the general problems of handling multiple interfaces in a ceramic body subjected to neutron-induced changes, and the results will be discussed in some detail. A SiC-clad SiC-based dispersion fuel has also been developed[3], but the currently-accepted HTGCR fuel[4, 5, 6] consists of multiple-coated microspheres of fissile ceramic dispersed in a graphite matrix. The development of these coated particles has been a major success story, and the design, function and in-service behaviour of the various coating layers and the interfaces between them will be the second main topic discussed.

BeO-BASED DISPERSION FUELS

The Australian work on BeO fuels was initially based on a conceptual 200 MW(e) CO_2-cooled pebble-bed HTGCR[1,2] for which the fuel element was a 25 mm diameter sphere containing the fuel dispersed as $(U, Th)O_2$ particles in the BeO moderator. Fission product retention was to be provided by a fuel-free coating of dense BeO around the fuelled core. Fig. 1 shows this fuel element in section, together

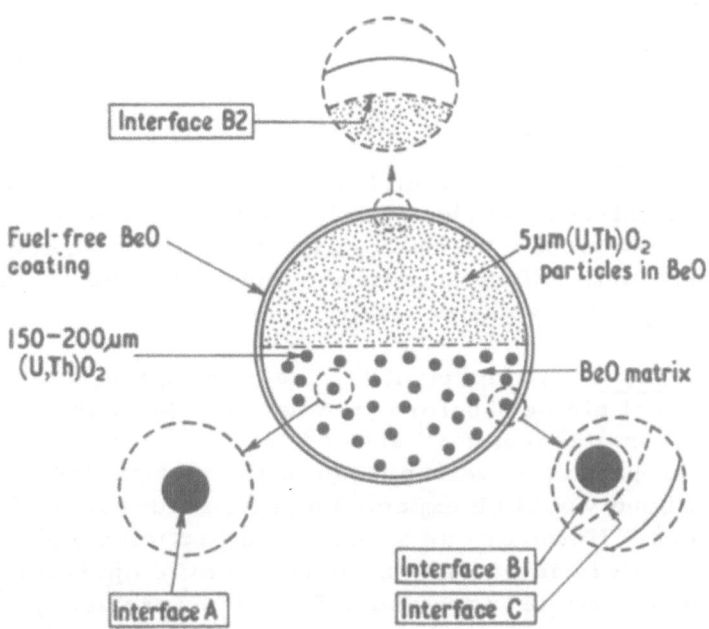

Fig. 1. Schematic cross-section of spherical BeO-based fuel element. Interface A: fuel particle - BeO (unirradiated); Interface Type B: BeO (fission recoil damage) - BeO (n + β damage); Interface C: BeO (n + β damage).

with the interfaces which will be discussed. Note that the
size of the fuel particles was not decided - both "coarse"
(150-200 μm) and "fine" (< 5 μm) dispersions were considered. These fuel elements would be required to operate
at 600-800°C at a moderate power density to a burnup of
10-20% of U+Th in the fuel particles; the resultant fast
neutron dose would be 4-5 x 10^{20} nvt.

Later, the focus of Australian work was changed to a
small, air cooled pebble-bed reactor with a somewhat
larger BeO-based fuel element. The temperature was in
the range 500-1200°C, the power density was lower, the
burnup was only 5-10% of U+Th and the fast neutron dose
was ∼ 2 x 10^{20} nvt. The fuel element was specified to
operate in moist (ambient) air and thus required a protective coating to prevent BeO volatilisation as $Be(OH)_2$.
This fuel element is shown, with the relevant interfaces,
in Fig. 2.

Interfaces A, B1, B2 and C

The fuel element of Fig. 1 was required to be fission
product retentive throughout its life and as it was to
operate under tensile thermal stresses of up to 20,000
psi, its strength before irradiation and to the end of its
life had to be as high as possible.

Interface A is important in determining the strength
of such a fuel element before irradiation. Mechanical
property measurements[7, 8] showed that in the fuel dispersion the particles were bonded to the matrix. No
chemical bond would be expected and its nature is still
unexplained. However, as a result the particles and
Interface A were in radial tension at room temperature
and for fuel particle sizes above 5 μm fracture always
initiated in the fuel particles, the weaker phase of the
dispersion. Fine dispersions were found to be at least
as strong as unfuelled BeO.

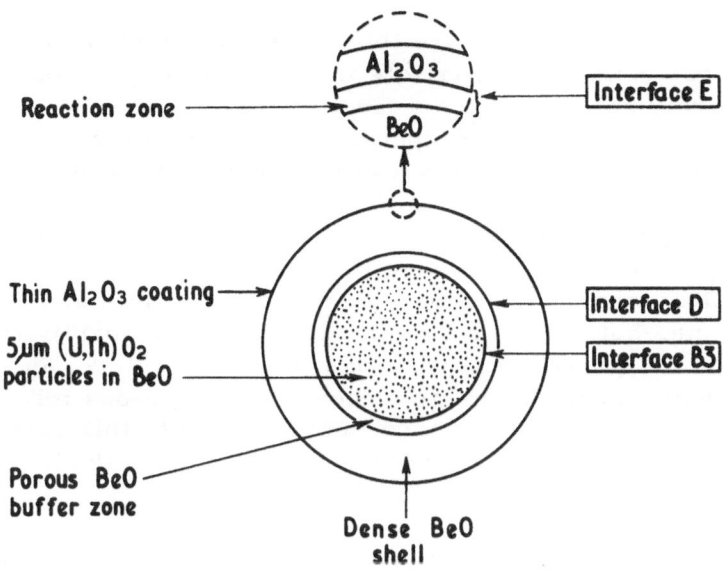

Reaction zone

Al_2O_3

Interface E

BeO

Thin Al_2O_3 coating

5μm $(U,Th)O_2$
particles in BeO

Interface D

Interface B3

Porous BeO
buffer zone

Dense BeO
shell

Fig. 2.   Schematic cross-section of modified spherical
BeO-based fuel element.   Interface B3: BeO matrix of
fine dispersion (fission recoil damage) - porous BeO
(n + β damage); Interface D: Porous BeO (n + β damage)
BeO (n damage); Interface E: BeO-Al_2O_3 reaction zone.

It was concluded that this system violates Weber's
design principle of dispersion fuels[9] where the assumption
is that the strength of the matrix controls the strength of
the composite. For a brittle ceramic composite this is
true only if the dispersed particle is stronger than the
matrix, is not bonded to it, or has a size below the critical
crack length of the matrix.

When a coarse dispersion fuel is irradiated, Interface
B1 takes over the formerly significant role of Interface A.
Type B interfaces are unique to nuclear fuels. They arise
because fission fragments with high energies recoil from
the fuel particle surface; the range of fission recoils is

small ($\sim$ 15 μm) and therefore they cause a thin shell of
matrix surrounding each particle in a coarse dispersion
to be severely damaged. To a first approximation, we
can think of this as the introduction of a layer of new material with density significantly lower than that of the matrix[10].
In an irradiated dispersion, internal stresses at Interface
B1 will now control mechanical strength. A tangential tensile stress would be expected in the matrix BeO and compressive stresses would be expected in the particle and the
damage zone, but there are insufficient data for quantitative
analysis of any particular situation. However, Hickman
et al.[11] found that the strength of coarse dispersions fell
rapidly with irradiation and they could correlate this loss
in strength with the fission product recoil dose to the
damage zone.

In the same work, Hickman et al. found that uncoated
fine dispersions of (U, Th)O$_2$ in BeO (which were initially
stronger) lost little strength during irradiation. However,
the density decrease was now considerably greater, since
the particles were so closely spaced that all the BeO was
damaged by fission recoils. Thus, in a BeO-coated fine
dispersion fuel there will be differential expansion at Interface B2, and this may eventually cause the BeO coating to
fail in tangential tension.

As well as fission recoils, fast neutrons and energetic
β-particles originate in the fuel particle and both cause
volume changes in the BeO. The range of fast neutrons is
large, and the neutron fluence is close to uniform over the
whole volume of a fuel element. However, β-particles have
a range of only $\sim$1.5 mm in dense BeO; this means that
while the additional volume change ($\sim$ 25%)[10] caused by
β-particles over that caused by fast neutrons is nearly
uniform throughout the dispersion, there is a problem at
Interface C. This again is specific to the nuclear application. In an unirradiated dispersion, Interface C has no
precise identity because the BeO structure is continuous
from matrix to coating (which is applied by isostatic pressing and co-sintering). However, because the thickness of
the coating (1.25 mm) is comparable with the range of β-

particles, there will be a progressive decrease in β-
fluence from Interface C to the outer surface. To a first
approximation, the increase in volume of the inner half
of the coating will be 25% above that of the outer half,
stressing the surface in tangential tension.

The behaviour of Interfaces B1, B2 and C which could
be predicted from the above considerations was confirmed
qualitatively in an irradiation test[12] on twelve Fig. 1-type
fuel elements; the experiment was commenced before the
dominant role of these interfaces was understood. Outer
surfaces of all irradiated spheres were cracked or crazed
at the design burnup and fast neutron dose. There were
insufficient data for confident calculation of the tangential
tensile stresses developed in the coating, but these have
been estimated as ∼ 60,000 psi. For this calculation (in
which the coating was considered as two integral shells
with different expansions), and all others reported in this
paper for BeO-based fuels, a small computer program
based on Love's treatment[13] of the case of a shell bonded
by concentric spherical surfaces was used.

Interfaces B3 and D

As a result of the work discussed above, and from
changes in the conceptual reactor design, a new fuel design[14]
was evolved as shown in Fig. 2. As mentioned earlier,
this was required to operate at only moderate thermal
stresses and to moderate burnups, but required compatibility with a moist air coolant. This fuel contained:

(i)   A fine dispersion core to maximise strength
      before and after irradiation.

(ii) A porous BeO buffer zone, 0.9-1.5 mm thick.

(iii) A relatively thick (5-8 mm) coating of dense
      BeO to ensure fission gas retention.

(iv) An outer coating (0.25 mm) of another oxide,
      non-reactive to air or moisture, to restrict

access of moisture in the air coolant to the BeO.
The oxide chosen for initial coating development
was Al_2O_3.

The functions of the porous BeO buffer zone were:

(i)   To absorb (by creep) the larger volume expansion
of the fuelled BeO core at <u>Interface B3</u>.

(ii) To absorb all fission recoil and β-damage, ensuring
that <u>Interface D</u> has no "type C" component and that
the shell is not stressed from differential irradiation-induced expansion effects.

As before, the fuel element (Fig. 2) was made by
isostatic pressing and co-sintering[14]. The buffer zone
and Interfaces B3 and D as fabricated are shown in Fig.
3.

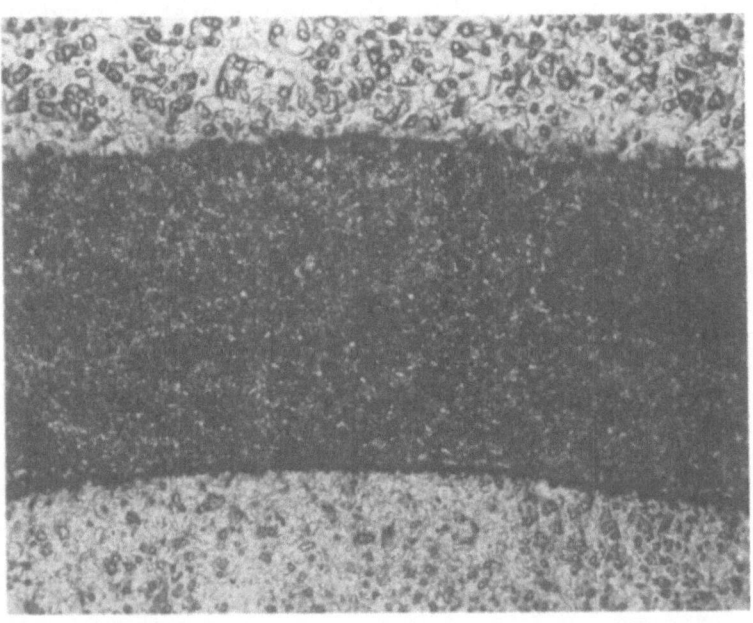

Fig. 3. Buffer Zone (centre) and Interfaces B3 (lower
centre) and D (upper centre) from the BeO-based Fuel
Element of Fig. 2 (X 45)

An irradiation test[15] of twelve fuel elements of the new type (but without alumina coatings) did not completely prove the design. Eight spheres irradiated to the design burnup and fast neutron dose at 500° and 750°C had cracked or partly detached coatings; this was attributed to some β-damage to the shell caused by an insufficiently thick (0.9 mm) buffer zone. The c-lattice expansion at the inner shell surface was measured as 18% above that at the outer surface. The resultant tangential tensile stress at the outer surface was estimated as ∼ 30,000 psi. Thus the buffer zone appeared to be functioning correctly at Interface B3 but not at Interface D. Four spheres irradiated at 1000°C showed no shell cracking, probably because at 1000°C actual and differential expansions are lower and stresses are more likely to be relieved by creep.

It was tentatively concluded that given a thicker buffer zone (e.g. ∼ 2mm) the concept embodied in Fig. 2 should be viable at all three temperatures provided that thermal stresses during operation are kept within reasonable limits.

Interface E (unirradiated)

This interface typifies the problems of reaction-bonding two ceramics with different thermal expansion behaviour and the subsequent use of the composite at temperatures such that further reaction at the bond might occur. These problems are of general interest in ceramic technology. The special problems of the reaction of this composite to fast neutron irradiation will be dealt with separately.

The reasons for the choice of Al_2O_3, the method of applying it to BeO as a "moisture-proof" coating, and its efficacy in this role have been documented already[16,17]. Briefly:

(i) The thermal expansion mismatch is such that the coating will be in moderate compression (now recalculated as 84,000 psi) at room temperature when cooled from fabrication above 1000°C. This is the correct type of stress for the thin, brittle coating.

(ii) The 250 μm coating is applied by spraying an aqueous slurry of Al_2O_3 plus 2 weight % MgO onto an as-pressed BeO or fuel element sphere and co-sintering with the sphere at 1450-1500°C, the sintering shrinkages having been carefully matched.

(iii) In accelerated corrosion tests in moist flowing air at 1300°C the rate of loss of BeO was reduced by factors of between 10 and 130 by the Al_2O_3 coating.

A small amount of chrysoberyl (CB) formation during sintering was expected to assist in bond formation. Metallographic examination suggested that a good bond had formed (Fig. 4) and that the reaction zone was 25-50 μm thick. This led to consideration of the possibility that further reaction might occur during the required 2 years' service at up to

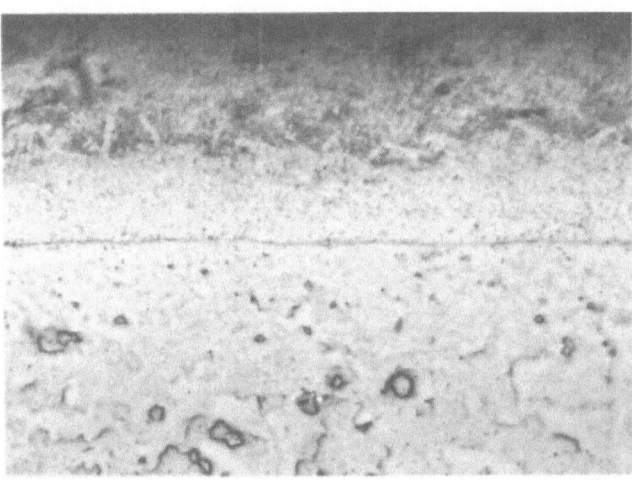

Fig. 4. Typical interface (centre) between Al_2O_3-2%MgO coating (above) and BeO (below) in as-sintered condition. (X 250)

1200°C. If the reaction zone (assumed to be CB) reached the surface of the sphere, loss of BeO as Be(OH)$_2$ would recommence.

To assess this possibility, accelerated annealing tests[18] were carried out on coated spheres at 1300°C, 1400°C and 1500°C. The main results and conclusions from this work were:

(i) The reaction bond increased in thickness according to a parabolic growth law. For example, at 1500°C:

$$X^2 = 530 + 102t$$

where X is the bond thickness in μm and t the annealing time in hours. The activation energy over the temperature range studied was 125 ± 5 kcal/mole.

(ii) Extrapolation of these results to 1200°C led to the prediction that a 250 μm coating should not be completely consumed by reaction for 80 years.

(iii) After about 120 hours at 1500°C, when the coating was less than half consumed, the coating-to-BeO bond was so weakened that coatings flaked off easily. The same occurred at 1400° and 1300°C but after correspondingly longer times. After 240 hours at 1500°C the coating had expanded and buckled and was detached from the sphere.

(iv) Taking (iii) into account, the useful life of the coating was predicted as 2 years at 1200°C.

(v) The annealed coating and the reaction bond consisted of three and two clearly-defined zones, respectively (Fig. 5). The composition of each zone was determined by electron microprobe analysis and X-ray diffraction as:

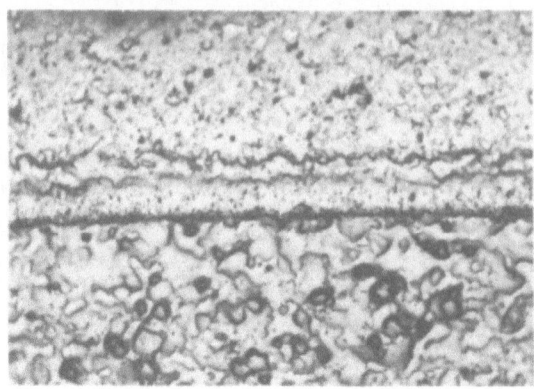

Fig. 5. Al_2O_3-2%MgO coating on BeO sphere annealed at 1500°C for 24 hours showing two zones of the reaction bond between BeO (below) and unchanged Al_2O_3 (above). (X 90)

Zone   I : Al_2O_3 + segregated $MgO.Al_2O_3$

Zone   II : The ternary compound
             $3BeO.8Al_2O_3.5MgO$ ("TC")

Zone III : $BeO.Al_2O_3$ (CB)

(vi)   The proposed reaction mechanisum involved Al^{3+}, Mg^{2+} and Be^{2+} diffusion in a basically continuous oxygen sub-lattice, with Al^{3+} diffusing inwards and Mg^{2+} and Be^{2+} diffusing outwards.

(vii) Eventual failure of the bond, known to occur at the BeO-CB interface, was thought to result from the slightly more open oxygen sub-lattice of CB compared with those of BeO and Al_2O_3. (The structure of TC is not known).

In recent work by the author (unpublished), designed to assess the role of MgO in reaction zone growth, planar BeO-Al_2O_3 and BeO-(Al_2O_3+2wt.%MgO) diffusion couples cut from hot pressed composites were annealed at 1500°C, but in each case the bond failed after only 3 hours. In further experiments, annealing was carried out under a pressure of 2000 psi at 1600°C. A platinum foil marker at the original BeO surface was included in one BeO-$(Al_2O_3+2\%MgO)$ couple.

The main observations from this work were:

(i) Multiple-zone reaction bonds formed in each case, although they differed in detail from those previously observed.

(ii) The platinum marker experiment was not definitive but suggested that the reaction zones grow into the Al_2O_3 very much faster than into the BeO.

From the point of view of bond integrity the second observation is highly significant and it is now easy to suggest why the bond eventually fails. Formation of BeO. Al_2O_3 in dense Al_2O_3 by reaction with excess diffusing BeO would involve a volume increase of approximately 34%. In a sintered coated sphere, some of this volume is found by densification in the coating (for example, in Fig. 4). Once available porosity in the coating has been used, plastic deformation of the unreacted Al_2O_3 may occur, but eventually the coating buckles and contact is progressively lost. Apparently, plastic deformation or creep also operates to some extent at lower temperatures, since no radial cracks indicating tangential tensile failure were ever observed. It is noteworthy, however, that bond failure occurred at lower reaction zone thicknesses at the lower temperatures.

Despite the unfavourable nature of the interfacial reaction in this system, Al_2O_3-2%MgO coatings on BeO spheres have a much longer endurance than would be predicted from the results on planar diffusion couples, and the predicted endurance of 2 years at 1200°C appeared to satisfy the proposed reactor application. This is in part

due to the favourable contribution of spherical geometry
to the maintenance of contact between coatings and substrate sphere.

Interface E (irradiated)

The endurance of the coating-to-BeO bond was also
tested by neutron irradiation of coated spheres at various
temperatures, and by corrosion-testing of some neutron-
irradiated spheres. These requirements are of course
specific to the nuclear application, and so is the failure
mechanism. The coating is initially in compression, as
it should be, but during irradiation critical interfaces go
into excessive tension and failure results.

Because of the differing rates of fast neutron-induced
expansion of BeO, Al_2O_3, CB and probably TC, the dose
endurance of a coating was not expected to be high. However, at the outset of the development it was predicted[16]
that the required dose of 2 x 10^{20} nvt should be attainable.

In the following analysis any effect from TC is ignored,
because its expansion rate is unknown. The reaction zone,
50 μm wide, is assumed to be CB.

Fig. 6 shows the expansion rates of BeO, CB, Al_2O_3
and of 20 coated spheres irradiated at pile temperature[19].
Fig. 7 shows the calculated stresses in the Al_2O_3 and CB
layers. Depending upon whether the initial compressive
stress of 84,000 psi in the Al_2O_3 is taken into account or
not, an estimated tensile failure stress in the 93% dense
alumina of 30,000 psi will be reached at 2.8 or 2.2 x 10^{20}
nvt. Fig. 6 shows that at 2.2 x 10^{20} nvt and above, the
expansion of the sphere follows that expected of BeO, which
suggests that the Al_2O_3 has indeed cracked. Metallographic
evidence confirmed that at 2.2 x 10^{20} nvt and above (but not
at 1.9 x 10^{20}) coatings were badly cracked, the most characteristic crack being a radial crack in the coating not
extending far into the BeO (Fig. 8). Thus the results at
pile temperature are qualitatively consistent with expect-
ation and the quantitative agreement is reasonable.

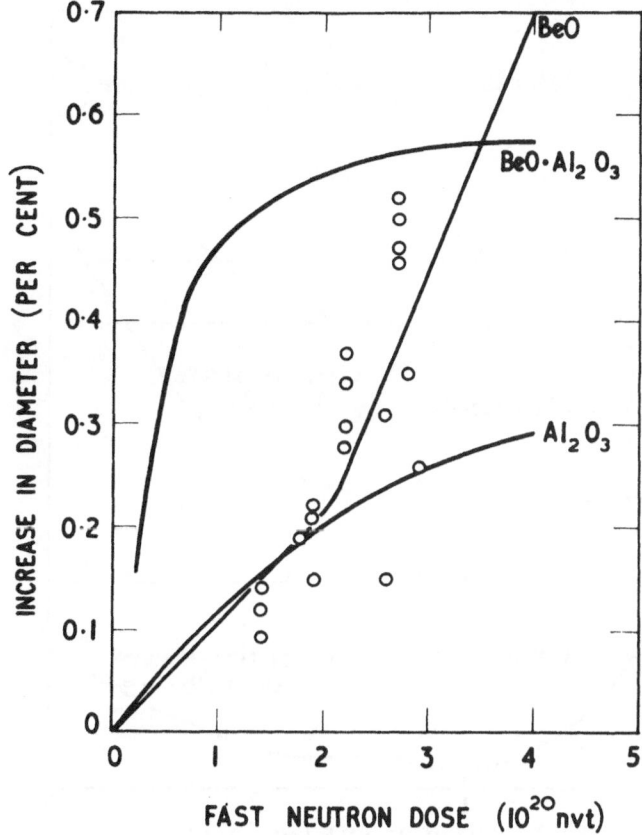

Fig. 6. Increase in diameter versus irradiation dose at pile temperature for alumina-coated BeO spheres. Full lines are predicted expansions.

Fig. 9 shows the expansion rates of BeO, Al_2O_3 and of six coated spheres irradiated[19] at 700°C. The expansion rate of CB has not been measured at 700°C but, on the basis of the easy annealing of pile temperature damage, Jostsons and Hickman[20] predicted that the expansion rate at higher temperatures would be less than that for BeO.

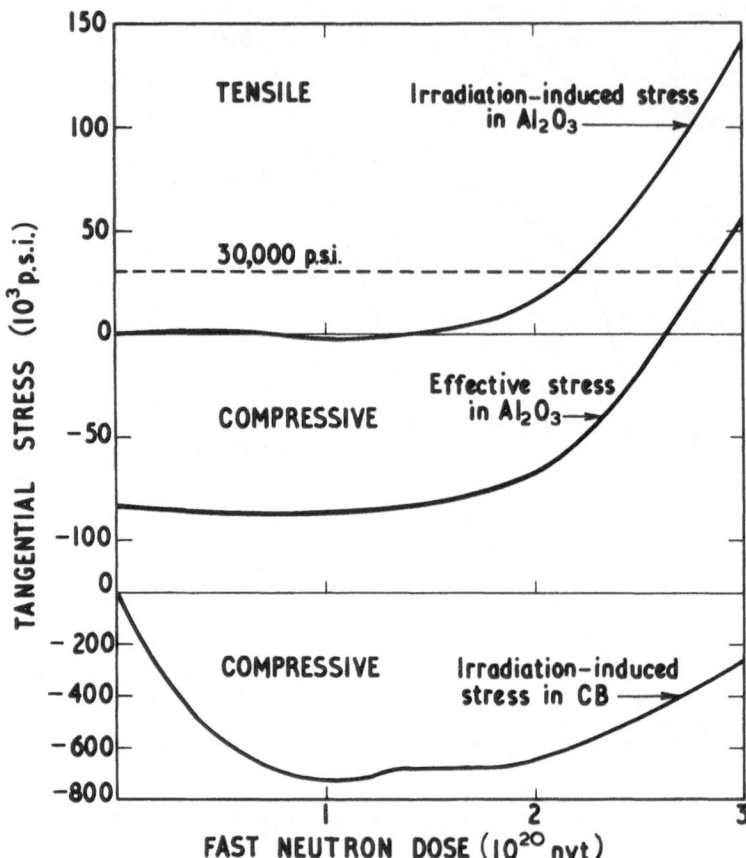

Fig. 7. Tangential stresses in Al_2O_3 and CB on alumina-coated BeO sphere versus fast neutron dose at 75-100°C.

In the following analysis it was assumed to be half that for BeO. Fig. 10 shows the calculated stresses in the Al_2O_3 and CB layers, assuming that the initial thermal expansion mismatch stresses are no longer operative at 700°C. From Fig. 10, tensile cracking would be expected in the CB below 10^{20} nvt. Only spheres irradiated to 4.0 and 4.8 x 10^{20} nvt were examined directly after irradiation

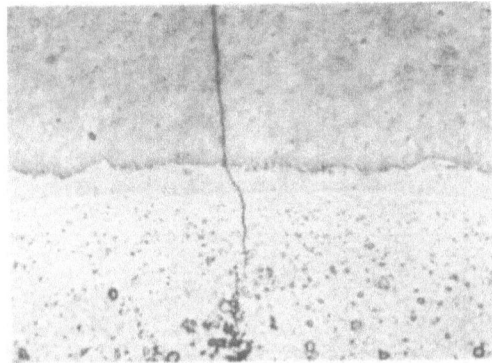

Fig. 8. Alumina-coated BeO irradiated at pile temperature to 2.2×10^{20} nvt. Phases (from top) Al_2O_3, CB, BeO. (X 85)

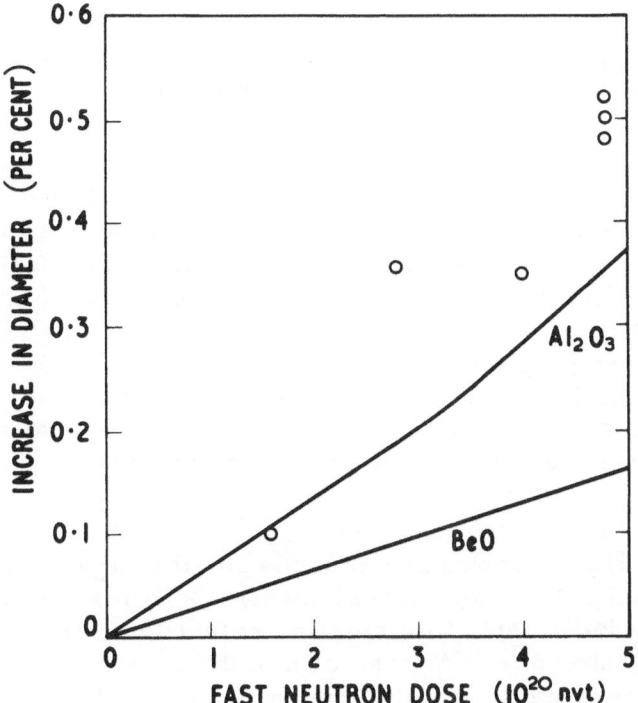

Fig. 9. Increase in diameter versus irradiation dose at 700°C for alumina-coated BeO spheres. Full lines are predicted expansions.

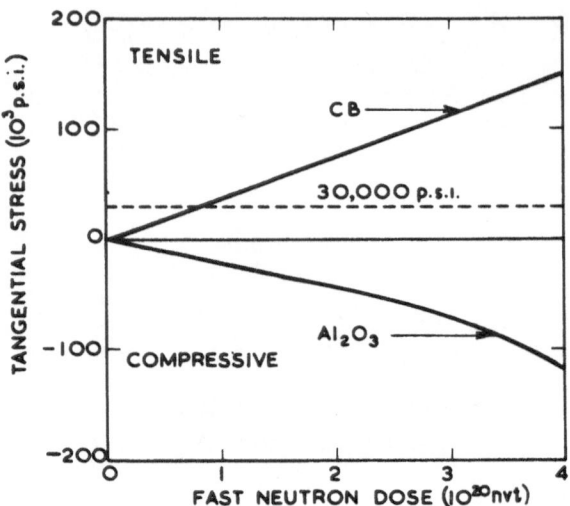

Fig. 10. Tangential stresses in Al_2O_3 and CB on coated
BeO sphere versus fast neutron dose at 700°C.

and the metallography (Fig. 11) indeed showed radial cracking in the chrysoberyl. The two spheres irradiated to 1.6
and 2.8×10^{20} nvt were corrosion-tested after irradiation.
The first initially showed good protective behaviour but
after 28 weeks' testing the coating was completely detached
from the sphere. The second sphere showed only fair
protective behaviour and on examination after 8 weeks was
found to have a system of fine cracks at the outer coating
surface. The sphere was then sectioned and polished (Fig.
12).

Fig. 12 clearly shows the reasons for the high weight
loss of this sphere in the corrosion test. The loss is of
BeO from the bulk BeO, to a smaller extent from the CB,
and from the interface. Access of moisture and escape
of $Be(OH)_2$ presumably occur through those cracks which
extent to the surface. There were a few cracks which did
not extend to the outside surface of the Al_2O_3, similar to
those in Fig. 11, and these showed no cavity in the BeO.
It is postulated that the damaging cracks were formed

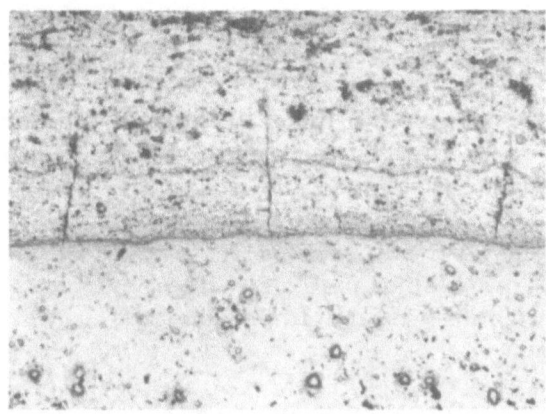

Fig. 11. Alumina-coated BeO irradiated at 700°C to 4.8
x 10^{20} nvt. Phases (from top) Al_2O_3, CB, BeO. (X 95)

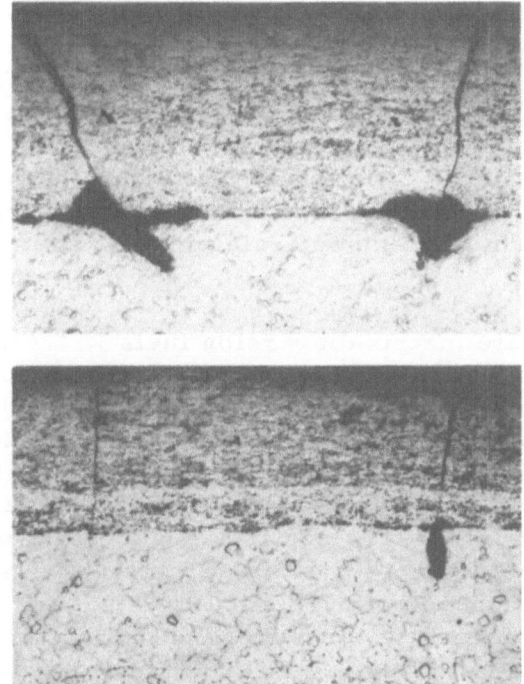

Fig. 12. Alumina-coated BeO irradiated at 700°C to
2.8 x 10^{20} nvt and then exposed to moist air at 1200°C.
Phases (from top of each field) Al_2O_3, CB, BeO. (X 63)

by the propagation of Fig. 11-type cracks to the surface
during thermal cycling.

Fig. 12 also demonstrates in dramatic fashion the
excellent protective behaviour of crack-free regions of
the coating, and by inference the protective capability of
crack-free coatings. However, Al_2O_3 coatings on BeO
would be of very limited use in-pile, and had the technological requirement continued, development of an alternative coating would have been necessary.

COATED PARTICLES FOR GRAPHITE-BASED NUCLEAR FUELS

The development of all-ceramic graphite-based
nuclear fuels for HTGCRs commenced in the 1950s and
proceeded in parallel with BeO-based fuel development
during the 1960s. During this period, small prototype
reactors such as DRAGON, AVR and PEACH BOTTOM,
each employing closely-related fuel concepts, were built
and operated successfully. A further stage has now been
reached where large commercial units have been ordered
in the U.S.A., a medium-sized prototype is being built
in West Germany, and commercial designs are available
in the U.K.

The graphite matrix dispersion fuels[4,5,6] for these
reactors differ geometrically, but the common feature is
that the fuel (which may be a fissile, fertile, or fissile-
fertile oxide or dicarbide) is dispersed in the matrix as
multiple-coated microspheres. The purpose of the coatings is to retain fission products, both gaseous and solid,
within each particle. The coatings consist of two or more
layers of pyrolytically-deposited carbon (PyC), with or
without an inter-layer of pyrolytically-deposited silicon
carbide. The pyrolytic carbon successfully retains gaseous and some solid fission products, but a silicon carbide
layer is necessary for complete retention of the troublesome fission products, Ba, Sr and particularly Cs.

The development of coated particles for this task proceeded rapidly and very successfully during the 1960s as as more experience was gained in coating technology, as irradiation results were built up, and as mathematical design methods for multi-layer coatings were developed. Some of the earlier types containing no silicon carbide layer are still in use for special purposes, but the development has culminated in the "TRISO" particle (Fig. 13); this, with slight variations, is employed in all the reactor designs mentioned above. The various coating layers must retain integrity to high fuel burnups and high fast neutron doses. These requirements vary with the reactor and fuel design and with the particle type (e. g. fissile or fertile) but may be up to 70% burnup of all heavy metal atoms and a fast neutron dose of 6×10^{21} nvt. Each coating layer also contains residual fabrication stresses and

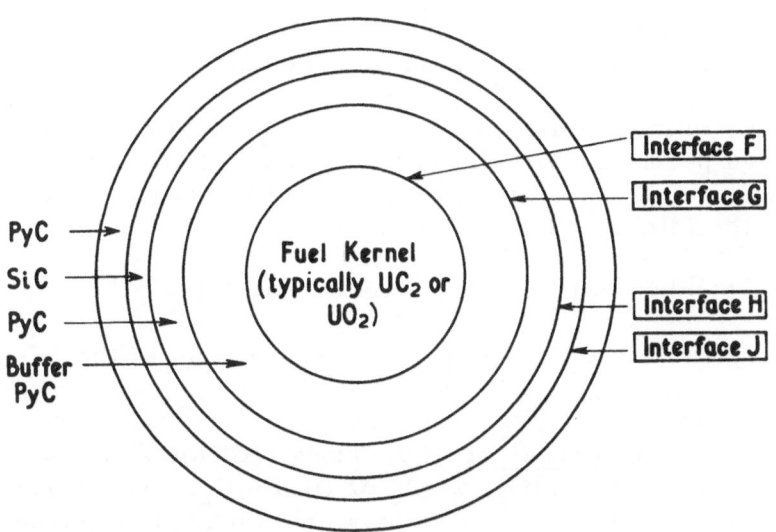

Fig. 13. Schematic cross-section of "TRISO" coated particle for HTGCR reactors. Interface Fl: UC_2-PyC; Interface F2: UO_2-PyC; Interface G: Porous PyC - dense PyC; Interfaces H and J: Dense PyC - SiC.

is subject to mechanical damage during incorporation in the dispersion, in-service thermal and irradiation-induced stresses, and chemical changes by interaction with its immediate environment.

The coated particle concept is of course highly specific to the particular nuclear application but, given the basic features of the design and the known reaction of each component of the coated particles to the appropriate radiation environment, the resultant fabrication technology and methods of design and analysis are highly sophisticated, but logical, extensions of more widely used methods. For example, deposition of carbon and silicon carbide pyrolytically is used in non-nuclear technologies, but for coated particles precision control of thickness, textures, densities and homogeneity must be absolutely assured. Also, the general principle that brittle ceramics should be placed in compression rather than tension applies no less here, but sophisticated design methods are required to ensure that the most brittle component, the silicon carbide, always remains in compression in-service.

In the following discussion, the mechanical design of the coated particles will first be discussed, followed by consideration of the chemical stability of the coating layers and the interfaces between them.

Mechanical Design of Coated Particles

A typical TRISO particle (Fig. 13) consists of the following:

(i) The fuel kernel, of UC_2, ThC_2, $(UTh)C_2$, UO_2, ThO_2, $(U, Th)O_2$, PuO_2 or $(U, Pu)O_2$. The kernel may be fully dense or may contain built-in porosity to provide space for fission gas accumulation and fuel swelling. Its diameter is specified at some value within the range 100-800 μm, depending on the fuel design philosophy.

(ii)   The buffer layer, of lower density PyC. This
must be thick enough to absorb all fission recoil
damage and, if the kernel is dense, to provide
for fission gas accumulation and kernel swelling.
Its thickness is in the range 30-60 μm.

(iii) The inner high density PyC layer (30-50 μm
thick). This is the primary pressure vessel
within which gaseous fission products are retained. It also protects the SiC layer from reaction with the fuel and helps to place the SiC
in compression.

(iv)   The SiC layer (25-30 μm thick). This is necessary
to retain Ba, Sr and Cs fission products which
diffuse more or less rapidly through PyC.

(v)   The outer high density PyC layer (30-60 μm
thick). This protects the SiC layer by preventing
loss of silicon at service temperatures and by
placing the Si C in compression.

Various mathematical models for the mechanical design and analysis of coated particles have been developed[21-26]
but the following description of coating behaviour will refer
to the model of Kaae[25, 26], which is typical of those in current use.

The buffer layer is assumed to have no strength at
either Interface F or Interface G. Interface F is somewhat
similar in function to Interface B3 in BeO-based fuels, but
now fission recoils cause densification of the buffer layer.
This layer is also compressed by eventual outward movement of Interface F due to fuel swelling and inward movement of Interface G due to irradiation-induced dimensional
changes in the PyC. The buffer layer is not called upon to
absorb all the β-particle damage as it was in the BeO-based
fuel element; the scale is now such that the range of β-
particles far exceeds the overall coating thickness and
β-damage to the coating will be not far from uniform.

In calculating the fission gas pressure at Interface G, the fission gas release fraction and the volume available for its accumulation must be known. The latter requires allowance for the volume occupied by solid fission products as a function of burnup, for fission recoil-induced buffer layer shrinkage and for inward movement of Interface G. The resultant fission gas pressure (to which must be added any pressure from other gases entrapped during or after deposition of the PyC) deforms the inner PyC, but because of the low elastic modulus of PyC compared with SiC, most of the load is borne by the SiC layer which as a result tends to be stressed in tangential tension.

The dimensional changes in PyC as a result of fast neutron irradiation are complex, depending on the fast neutron dose and on the density and the extent of anisotropy of the PyC. However, a typical situation[27] is as shown in Fig. 14. There is initially an isotropic shrinkage of the PyC, but at higher doses this situation is replaced by tangential shrinkage with radial expansion. The extent of

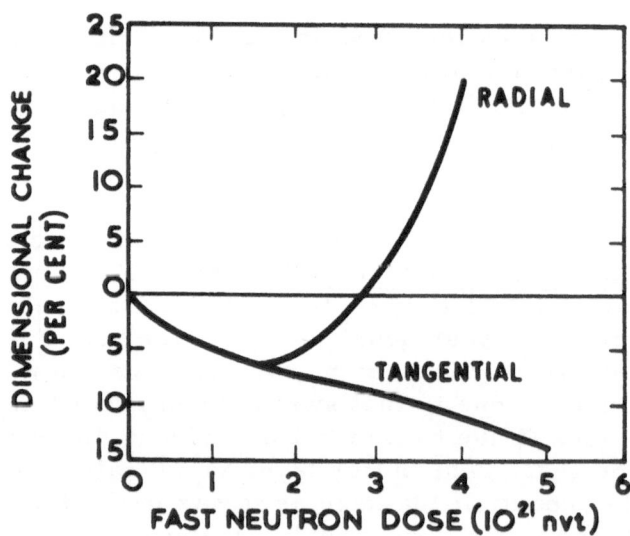

Fig. 14. Typical dimensional changes in PyC (density 1.80/cm^3) versus fast neutron dose at 1200°C (after Rose[27])

these changes increases as the density of the PyC increases and thus density is a valuable design variable.

The thermal expansion mismatch between PyC and SiC at Interfaces H and J is small but is allowed for in the models. The irradiation-induced dimensional changes in SiC are very small and are assumed to be zero. The tangential shrinkage of the inner PyC at all doses puts the C-SiC bond at Interface H in radial tension but this stress seldom exceeds the measured strength of the bond, which is ∼ 8000 psi. The SiC thus tends to be placed in tangential compression by the PyC at Interface H. At Interface J, the C-SiC bond is placed in radial compression, and the SiC is stressed in tangential compression by the PyC.

The stresses in the PyC, but not in the SiC, relax progressively due to creep at high temperatures, strongly enhanced by the fast neutron flux. Thermal stresses in the particle due to thermal gradients during operation are assumed to be zero. The net result of increasing fast neutron fluence and burnup as far as the SiC is concerned is that the tangential stress throughout is initially highly compressive, but becomes less so as the fission gas pressure increases during burnup. However, by correct particle design the stress in the SiC can be maintained compressive.

The resultant calculated stress pattern in a particle depends on the burnup, the fast neutron dose, the thicknesses and densities of the various layers, the data and assumptions on fission gas release from the kernel, the assumed fate of solid fission products, and on the creep data used for the PyC. A typical result (after Kaae[25]) is shown in Fig. 15.

Fig. 15 shows that the most dangerous stresses occur in the inner PyC layer. The fracture strength at temperature of the PyC is often assumed to be 40,000 psi but there is evidence[28] from the application of Kaae's model to irradiation results on PyC-coated particles that its strength may be nearer to 25,000 psi. The stress in this layer can be reduced, at least to moderate burnups, by reducing the

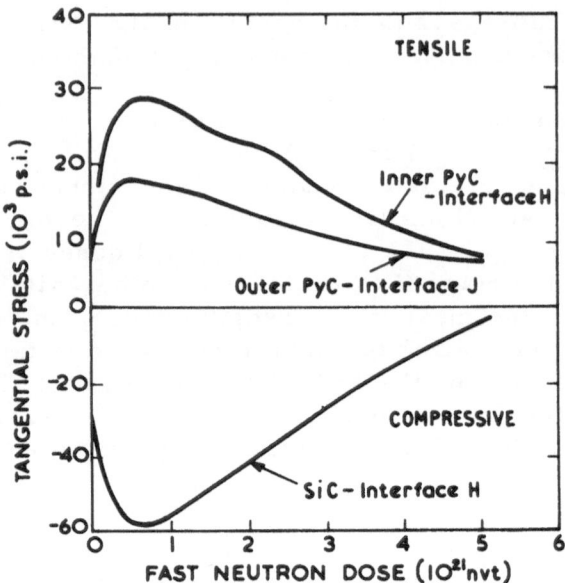

Fig. 15. Tangential stresses in coating layers on TRISO particles with strong inner PyC-SiC interface versus fast neutron dose (after Kaae[25]).

strength of Interface H, e. g. by depositing a weak PyC layer there. This reduces the compressive stress in the SiC, and the tensile stress in the inner PyC early in life; however, the inner PyC now takes the fission gas pressure load and therefore there is an increasing tensile component as burnup proceeds. This may eventually reach the failure stress, and therefore interface separation is not necessarily beneficial (Fig. 16).

Theoretically, a particle can be designed so that tensile stresses in the PyC are moderate, and so that stresses always remain compressive in the SiC. However, this is not entirely necessary and may be expensive in terms of the fabrication controls required. Direct strength measurements have been made[29] on hemispherical SiC shells extracted from coated particles, and these results suggest

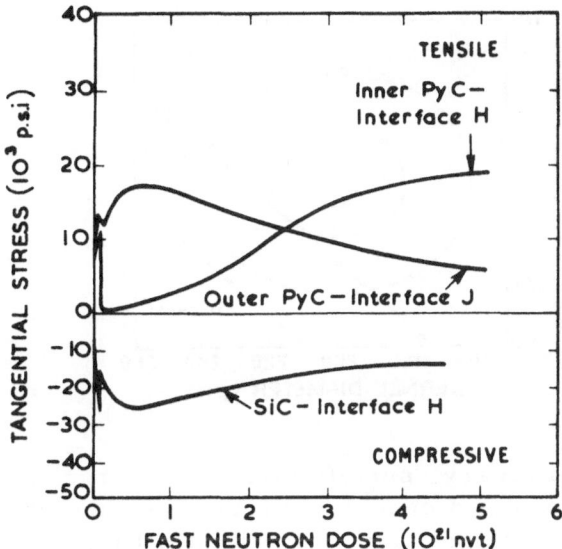

Fig. 16. Tangential stresses in coating layers on TRISO particles with weak inner PyC-SiC interface versus fast neutron dose (after Kaae[26]).

that a 1 in 10^4 failure strength of 30,000 psi can be achieved by good control during the coating operation. Gulden et al.[30] have pointed out that in quantity production there are statistical variations in kernel diameter and coating layer thicknesses, and have advocated a statistical approach to particle design. In practice, if the kernel diameter is at the top end of its range and the buffer thickness is at the low end, tensile stresses can be set up in the SiC. They were able to correlate the failure rate of irradiated coated particles having closely controlled initial properties with the calculated tensile stress in the SiC, assuming a failure stress of 30,000 psi for the SiC. An example of a practical specification[30] for a coated particle is shown in Fig. 17.

Chemical Stability of Interfaces

Provided there is no thermal gradient across the coated particle, and neglecting the complicating effects of fission

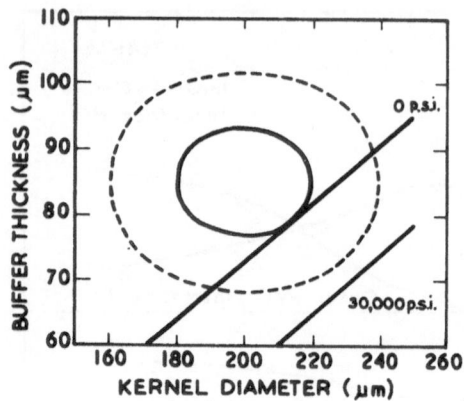

Fig. 17. Preliminary "specification" for kernel and buffer layer dimensions of a coated particle, showing 50% (full circle) and 90% (dotted circle) confidence limits, and zero and 30,000 psi (tensile) iso-stress lines for the SiC layer (after Gulden et al. [30])

products, all types, whether with carbide or oxide kernels, are reasonably stable chemically to temperatures well above the present maximum service temperature of 1350°C.

In the following analysis the fuel kernels are assumed to be of the fissile type and to be of either UO_2 or UC_2. The interfaces can then be classified into three types, viz:

(i)   The PyC-SiC interfaces (H and J) where there is little tendency for reaction. Slight diffusion of Si into the PyC can occur during deposition of the SiC.

(ii)   The UC_2-PyC interface (F2), across which uranium and carbon diffusion can occur.

(iii)   The UO_2-PyC interface (F2), which is thermo-dynamically unstable and where reaction is arrested only by the build-up of reaction product CO.

The diffusion of uranium from UC_2 kernels into the adjacent PyC is a potential problem which increases with increasing fabrication or service temperature. Significant uranium contents in the adjacent PyC have been measured[31] by electron microprobe up to 50 μm from the kernel in both irradiated (< 1400°C) and unirradiated particles. Uranium diffusion measured on coated particles[32] increased rapidly over the range 1600-1800°C but from diffusion couple work Evans[33] concluded that uranium migration was confined to grain boundaries and microcracks. It appears[34] that with a dense coating of optimum structure, uranium diffusion at currently-used service temperatures is of little significance.

Carbon diffusion across Interface F1 is of significance only when there is a thermal gradient across the particle; this subject is treated separately later.

At the UO_2-PyC interface (F2) the following reactions[27] can occur during fabrication and/or service:

$$UO_2 + xC = UO_{2-x} + xCO \qquad\qquad (1)$$

$$UO_2 + 4C = UC_2 + 2CO \qquad\qquad (2)$$

During coating deposition, the buffer layer is usually deposited at 1200-1400°C where the rates of reactions (1) and (2) are low. However the dense PyC is applied at about 1800-1900°C where excessive reaction could occur. To prevent this, a thin (10-15 μm) "seal layer"[27] may be applied to 1500-1800°C before the first dense PyC layer. Once the first impervious PyC layer has been applied this acts as a pressure vessel; the CO pressure builds up and then ceases after a very small amount of UO_2 reduction by reaction (1). The formation of UC_2 would be expected only in the case of cracked or poorly sealed coatings. The resultant CO pressure must be added to the fission gas pressure in the mechanical model. This CO pressure is very sensitive to initial stoichiometry and to prevent high CO pressures the use of hyperstoichiometric "UO_{2+x}" must be avoided in favour of stoichiometric UO_2. In $(U, Pu)O_2$ fuels, the initial oxygen content should be[35] in the substoichiometric $(U, Pu)O_{2-x}$ range, since $(U, Pu)O_2$ is more easily reduced than UO_2 and will have a higher equilibrium CO pressure, and since the $O/(U+Pu)$ ratio increases during burnup much more than does O/U ratio of UO_2.

Kernel Migration - The "Amoeba Effect"

The "amoeba effect" - unidirectional kernel migration through the coatings in a thermal gradient exposing fuel and, potentially, fission products to the porous graphite matrix - was first observed[36] on PyC-coated $(U, Th)C_2$ particles heated out-of-pile in a thermal gradient at temperatures between 1600° and 1900°C. The effect has been studied[39] in more detail recently, both out-of-pile and in-pile, and the mechanism has been studied in planar geometry by diffusion couple techniques[40]. An example of the effect is shown in Fig. 18. An apparently similar phenomenon has been observed in PyC-coated oxide particles[37, 38], both with and without an SiC interlayer, irradiated at temperatures between 1400° and 1900°C in a thermal gradient. In

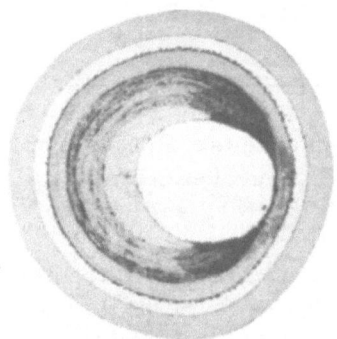

Fig. 18. Kernel migration in unirradiated TRISO-coated UC_2 particle annealed for 32 hours at 1775°C with $\Delta T \simeq$ 1.5°C (from Scott and Stansfield[39]). The hot side is to the right. On the cold side of the kernel there is a crescent-shaped zone of rejected graphite. (X 90)

typical coated UO_2 particles from irradiation tests, times to failure were found[37] to decrease with increasing rating of the particle, temperature and temperature gradient and were correlated empirically with these parameters. The amoeba effect is not troublesome in either carbide or oxide fuel particles at present HTGCR fuel operating temperatures. However, it could cause problems at the higher power densities and higher temperatures being considered for future HTGCR designs.

Although there are some conflicting results, it appears to be generally agreed that both dicarbide[39] and oxide kernels[27] move up the temperature gradient. However, the mechanisms differ in the two cases.

In dicarbide particles, the basic process is carbon diffusion down the thermal gradient[39,40]. Carbon from the PyC is dissolved in the UC_2 on the hot side, diffuses across the particle, and is rejected as graphite on the cold side (Fig. 18). Slight uranium diffusion into the PyC on the hot side probably helps to maintain good contact at this interface[40], which moves up the gradient "dragging" the fuel kernel behind. The observation[40] that the grain and crack structure of UC_2 in $PyC/UC_2/PyC$ diffusion couples remains unaltered as it moves supports this mechanism.

Kernel migration in oxide particles has been observed only in particles irradiated in a thermal gradient. No detailed discussion and analysis of the mechanism has yet been published in the open literature, but it seems to be accepted[27] that mass transfer of carbon from the hot side to the cold side of the kernel again occurs, but this time through the gas phase inside the miniature pressure vessel formed by the inner PyC layer. This gas will contain both CO and CO_2, and mass transfer is driven by gradients in the equilibrium CO/CO_2 ratio and in the corresponding pressures of CO and CO_2 from the hot side to the cold side. At the hot side, PyC is consumed by the reaction:

$$C + CO_2 = 2CO \qquad\qquad (3)$$

while PyC is deposited at the cold side by the reaction:

$$2CO = C + CO_2 \qquad\qquad (4)$$

This cycle is commonly considered to be the primary cause of failure of carbon filament light bulbs[41]. Note that carbon transport may or may not cause kernel migration - this may not occur if carbon deposition on the cold side merely densifies the buffer layer.

Once the PyC on the hot side is consumed, the SiC is subject to chemical attack by the CO. It has been reported[42] that measured CO pressures in irradiated TRISO particles are in the range of CO pressures over the $SiC-SiO_2-C$ equilibrium, suggesting that this system may be chemically buffering the pressure of CO in the particle. However, the local oxidation of SiC which does occur on the hot side probably causes stresses in the coating in a significant proportion of particles sufficient to fracture the SiC and allow the outer PyC coating to be eventually breached.

Methods of reducing the amoeba effect in oxide particles for future applications are under study. One approach being used[43] is to reduce the equilibrium CO pressure by "chemical buffer" additions to the oxide fuel kernel. Among the additions under study are Ce_2O_3, La_2O_3, Y_2O_3 and ZrC. It is also claimed[43] that UN and U(N, C) have attractions as kernel materials, and coated UN particles are being irradiation tested. An earlier proposal[44] that PyC-coated UC particles were sufficiently stable for reactor use at temperatures up to 2000°C has not been taken up. This proposal was based on out-of-pile annealing studies; as expected, the UC and PyC reacted at the interface to form a layer of UC but this was relatively thin even at 2000°C and was claimed not to damage the particle.

CONCLUSION

The above discussion has highlighted some of the general principles and problems of internal interfaces in ceramic

bodies together with some of the special problems faced by
the designer of ceramic nuclear fuels. The following general points can be made:

(i) The strength of a two phase ceramic dispersion
is controlled by the strength of a weaker dispersed
phase bonded to the matrix, provided that the
particle size of the dispersed phase is above the
critical crack length of the matrix.

(ii) Differential expansion between two phases of a
ceramic composite, from whatever cause, may
be controlled by an intermediate buffer zone of
low density to reduce stresses in the two main
phases.

(iii) A thin ceramic coating on a ceramic body should
be in compression as-fabricated and tensile
stresses should not be allowed to develop in-
service.

(iv) A small amount of reaction between coating and
substrate during fabrication may be allowable, or
even beneficial. However, further reaction during service may be damaging and this possibility
should be taken into account.

(v) In a sealed system, potentially damaging reactions
between coating layers may be arrested after a
small amount of reaction, by reaction product
build-up.

(vi) In a thermal gradient, damaging mass transfer
may occur in systems which would be stable under
isothermal conditions. A thorough knowledge of
the chemistry and thermodynamics of these systems
is necessary if such processes are to be prevented
or reduced to insignificance. Introduction of chemically buffering additions is one route to greater
control.

The special problems of the nuclear fuel designer are
of two kinds, viz:

   (i) The requirement for virtually complete retention
       of fission products within all-ceramic nuclear
       fuels demands the application of multiple fuel-
       free ceramic coatings with low permeability to
       the fission products concerned. These coatings
       must be of very high quality, and coating thick-
       nesses must be maintained within close tolerances.

  (ii) The need for all the interfaces, including those
       between the various coating layers, to withstand
       the changes caused by irradiation by neutrons,
       fission recoils and β-particles places added
       restrictions on the design of the system, bearing
       in mind that each material may respond in a
       different way to the same irradiation, and that
       the same material may respond in different ways
       to the three types of irradiation, whose zones of
       influence may not always overlap.

In this paper, two ceramic fuel systems which exemplify
these special problems and at the same time embrace some
more general ceramic problems have been discussed. In
each system both mechanical stresses and chemical reactions
between coating layers should be considered as potentially
damaging. Chemical reactions will become increasingly
important as fuel operating temperatures are increased.

ACKNOWLEDGEMENTS

Thanks are due to those who assisted in various ways
during the preparation of this paper, particularly Dr. D. G.
Walker, Dr. M. J. Bannister, Mr. R. J. Hilditch and Mr.
J. Whatham (all of the AAEC Research Establishment,
Lucas Heights), Dr. P. R. Kasten and Dr. T. B. Lindemer
(both of ORNL) and Dr. O. M. Stansfield (Gulf General
Atomic). Mr. E. J. Ramm, Mr. C. E. Webb and Mr. J. G.
Napier assisted in unpublished work on the BeO-Al$_2$O$_3$ re-
action.

REFERENCES

1. W. H. Roberts, J. Nucl. Mater., 14, 29-40 (1963).
2. D. R. Ebeling and J. E. Hayes, The Instn. of Engnrs.,
    Australia, Mech. and Chem. Eng. Trans., 1967 (May),
    119-137.
3. R. W. M. D'Eye, pp. 617-628 in Proc. Symposium on
    Advanced and High-Temperature Gas-Cooled Reactors,
    Julich, 1968. IAEA, Vienna, 1969.

4. H. B. Stewart, R. C. Dahlberg, W. V. Goeddel, D. B.
    Trauger, P. R. Kasten and A. L. Lotts, pp. 433-447
    in Vol. 4 of Proc. Fourth International Confce. on the
    Peaceful Uses of Atomic Energy, Geneva, 1971. UN,
    New York and IAEA, Vienna, 1972.
5. R. W. M. D'Eye and T. J. Heal, pp. 449-458 in Vol. 4
    of Proc. Fourth International Confce. on the Peaceful
    Uses of Atomic Energy, Geneva, 1971. UN, New York
    and IAEA, Vienna, 1972.
6. L. Aumüller, K. H. Hackstein, M. Hrovat, E. Balthesen,
    B. Liebmann, K. Ehlers and K. Röllig, pp. 415-432 in
    Vol. 4 of Proc. Fourth International Confce. on the Peace-
    ful Uses of Atomic Energy, Geneva 1971. UN, New York
    and IAEA, Vienna, 1972.
7. K. Veevers and W. B. Rotsey, Australian A. E. C. Re-
    port TM338 (1966).
8. K. Veevers and W. B. Rotsey, J. Mater. Sci., 1, 346-
    353 (1966).
9. C. E. Weber, p. 295 in Progress in Nuclear Energy,
    Series V, Vol. 2, Pergamon Press, New York, 1954.
10. D. G. Walker and B. S. Hickman, J. Nucl. Mater., 24,
    60-68 (1967).
11. B. S. Hickman, W. B. Rotsey, R. J. Hilditch and K.
    Veevers, J. Amer. Ceram. Soc., 51 (2), 63-69 (1968).
12. R. J. Hilditch, unpublished work.
13. A. E. H. Love, A Treatise on The Mathematical Theory
    of Elasticity, Dover Publications, New York, 1944. pp.
    104 and 142.
14. K. D. Reeve and E. J. Ramm, Australian A. E. C. Report
    TM521 (1969).

15. G. L. Hanna and K.D. Reeve, Australian A.E.C. Report E239 (1973).

16. K.D. Reeve, Australian A.E.C. Report TM334 (1966).

17. K.D. Reeve, E.J. Ramm and C.E. Webb, Australian A.E.C. Report E216 (1971).

18. K.D. Reeve, E.J. Ramm and W.J. Buykx, J. Australian Ceram. Soc., 6 (2), 39-50 (1970).

19. K.D. Reeve, J. Nucl. Mater., 44, 285-294 (1972).

20. A. Jostsons and B.S. Hickman, J. Nucl. Mater., 25, 278-283 (1968).

21. J.W. Prados and J.L. Scott, Nuclear Applications, 2, 402-414 (1966).

22. J.W. Prados and J.L. Scott, Nuclear Applications, 3, 488-494 (1967).

23. J.M. Thomson, pp. 35-48 in Vol. 2, Part C, Proc. First International Confce. on Structural Mechanics in Reactor Technology, Berlin, 1971. Commission of European Communities, Brussels, 1972.

24. G.K. Williamson and P. Horner, pp. 383-389 in Proc. Third. Confce. on Industrial Carbons and Graphite, London, 1970. Society of Chemical Industry, London, 1971.

25. J.L. Kaae, J. Nucl. Mater., 29, 249-266 (1969).

26. J.L. Kaae, J. Nucl. Mater., 32, 322-329 (1969).

27. K.S.B. Rose, J. Inst. Nucl. Eng. (U.K.), 1971 (July/August), 95-100.

28. J.L. Kaae, D.W. Stevens and C.S. Luby, Nuclear Technology, 10, 44-53 (1971).

29. A.G. Evans, C. Padgett and R.W. Davidge, J. Amer. Ceram. Soc., 56 (1), 36-41 (1973).

30. T.D. Gulden, C.L. Smith, D.P. Harmon and W.W. Hudritsch, Nuclear Technology, 16, 100-109 (1972).

31. D. Quataert and H.W. Schleicher, J. Nucl. Mater., 19, 221-233 (1966).

32. R.W. Dayton, J.H. Oxley and C.W. Townley, J. Nucl. Mater., 11, 1-31, (1964).

33. R.B. Evans, ORNL-3619 (1964).

34. R.A.U. Huddle, p. 637 in Proc. Symposium on Advanced and High-Temperature Gas-Cooled Reactors, Julich, 1968. IAEA, Vienna, 1969.

35. M. Dalle Donne and G. Schumacher, J. Nucl. Mater., 40, 27-40 (1971.

36. W. V. Goeddel, GA Reports 3588 (1963) and 2880 (1963).
37. L. W. Graham, pp. 500-503 in Proc. Gas-Cooled Reactor Information Meeting, Oak Ridge, 1970 (CONF-700401). USAEC, Oak Ridge, 1970.
38. H. Nickel, KFA Report JUL 687 (1970).
39. C. B. Scott and O. M. Stansfield, Gulf-GA-A12081 (1972).
40. T. D. Gulden, J. Amer. Ceram. Soc., 55 (1), 14-18 (1972).
41. H. Schafer, p. 40 in Chemical Transport Reactions, Academic Press, New York, 1964.
42. T. B. Lindemer and H. J. de Nordwall, 74th Annual Meeting Abstract, in Amer. Ceram. Soc. Bull., 51 (4), 390 (1972).
43. O. E. C. D. (Nuclear Energy Agency); p. 39 in "Dragon" High Temperature Reactor Project Thirteenth Annual Report, 1971-1972.
44. A. Auriol, C. David, A. Fillâtre, G. Kurka, E. Le Boulbin and J. Rappeneau, pp. 533-537 in New Nuclear Materials including Non-Metallic Fuels (Conference Proceedings, Prague 1963). IAEA, Vienna, 1963.

DISCUSSION

A. Choudry (Univ. Rhode Island): (1) You reported radiation damage with nvt $\sim 10^{21}$; with a core flux of $\sim 10^{13}$ one needs ~ 10 years. How were the measurements carried out, i. e., was a single sample or batch of samples followed? (2) Is it not more desirable to make a 'homogeneous' pellet by dispersing some U-compound in a ceramic matrix and thus avoid the problem of fission-gas pressure buildup and the stresses that ensue at various interfaces?

Author: (1) Most, if not all, measurements on coated particles have been on separate batches each irradiated to a particular dose. (2) There are several problems with your suggested approach, viz. (i) The difficulty of making the matrix completely dense and fission product retentive, (ii) Release of fission products from the fuel at or near the surface by recoil and/or diffusion, (iii) The probability of cracking of the matrix if it is operated at a high power density. This approach is therefore not favoured. How-

ever, as pointed out in the text of my paper, a SiC-clad
SiC-matrix dispersion fuel has been proposed.

J. Wong (GE): Are the various coatings deposited after
sintering of the fuel ceramics or co-sintered?

Author: In the BeO-based fuel element the coatings are
applied to an unsintered core, and the whole fuel element
is co-sintered. In the case of coated particles, the pyrolytic carbon and SiC coatings are applied to sintered cores
by vapour deposition.

P. S. Kotval (Union Carbide): Could you comment on some
of the crystallographic characteristics of the SiC coatings
in concentric multilayer particles. Secondly, how, if at
all, do the stresses and irradiation treatments affect the
crystallographic features of the SiC.

Author: The pyrolytic SiC is dense, fine grained, isotropic
β-SiC. Neutron irradiation cuases a small expansion of the
SiC lattice due to the formation of point defects, but at fuel
operating temperatures this expansion saturates at very low
doses. The crystal structure is not affected. Unless the
stresses in the SiC layer of the particle exceed the fracture
strength of the SiC they will have little effect on its structure. The creep rate of pyrolytic SiC is so low that it is
usually neglected in the computation of particle behaviour.

GRAIN BOUNDARY PRECIPITATION IN ZIRCONIA

D. J. Green, D. R. Maki and P. S. Nicholson

Department of Metallurgy and Materials Science
McMaster University
Hamilton, Ontario, Canada

Recent work[1,2] has shown calcia partially stabilised
zirconia (PSZ) to have good thermal shock resistance.
This indicates a degree of toughness in the material. However, the heat treatments needed to produce the optimum
microstructure for strength and toughness have not been
investigated. Garvie[3] intuitively expected this two-phase
ceramic to consist of a grain boundary precipitate of pure
ZrO_2 (10-20 microns) and a much finer dispersion of monoclinic ZrO_2 (0.1 - 0.2 microns) within the cubic grains.
The improved properties of calcia PSZ were thought to be
due to the finer precipitate of pure ZrO_2.

The work here reported* concerns the process of microstructural development in PSZ and how it relates to heat
treatment. The role of grain boundaries in this development
and the consequent material properties are discussed.

EXPERIMENTAL

Test bars were made from micronized-grade ZrO_2
(99.7%) and reagent grade calcia. All specimens contained
3.4 wt. % calcia. Bars 4 x 1 x 1/2 in. were pressed iso-

---

*Supported by the Atomic Energy of Canada, Ltd.

statically at 20,000 psi and fired for approximately 5 hours at 1850°C. After firing, the system was cooled to room temperature by 3 methods. These methods are schematically illustrated in Fig. 1. Batch 1 was cooled slowly to 1300°C (cooling rate 100°C/hr.), annealed for 24 hrs. and then slow-cooled to room temperature. Batch 2 was also slow-cooled to 1300°C (100°C/hr.) but was then air-quenched to room temperature with no anneal. Batch 3 was quenched from 1850°C to 1300°C and then annealed and slow-cooled to room temperature.

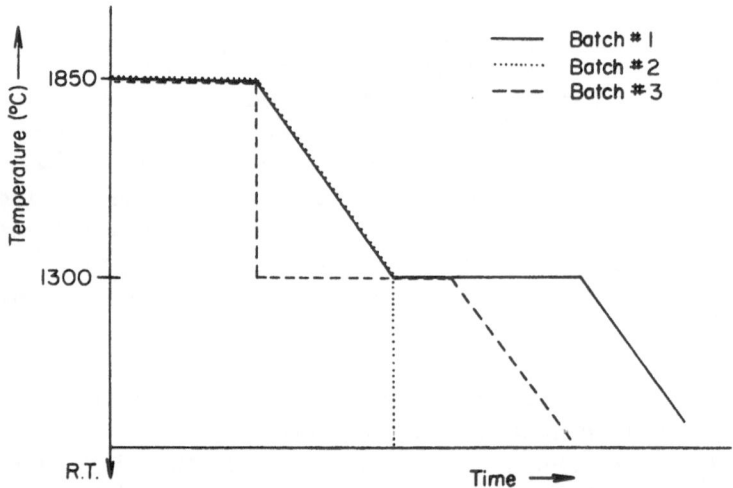

Fig. 1. Schematic cooling cycles for CaO-PSZ batches.

The apparent densities of the four batches of calcia PSZ were determined by a water displacement technique and the "theoretical" densities by pycnometry on material which had been ground and passed through a -325 sieve.

The fracture strengths were determined in four-point
bending on an Instron testing machine using Lucalox knife
edges and self-aligning alumina columns to transmit the
load. The fracture surface energy[4] and work of fracture
parameters[5] were determined for batch 2, also in four-
point bending. The samples were diamond machined and
polished from as-fabricated bars. A sharp, central notch
was introduced into the sample using an altrasonic impact
grinder and the depth measured with a travelling microscope. The crack length was taken to be the notch depth
plus the size of the observed machine flaws.

The different batches of calcia PSZ were ground to a
fine powder (-325 mesh) and set onto microscopic slides
using nail polish. These samples were placed in a Norelco
X-ray diffractometer and were irradiated by CuK_a radiation using a graphite crystal monochromator. The samples
were scanned in the range $2\theta = 25$ to $35°$, at a scanning
rate of $1/2° \ 2\theta/min$. The monoclinic and cubic zirconia
peaks were identified and the integrated intensities were
estimated, using a planimeter to measure the area under
the diffraction peaks. The total weight percentage monoclinic ZrO_2 was then estimated using the calibration of
Garvie and Nicholson[6].

Fracture surfaces of the different batches of calcia
PSZ were studied in a scanning electron microscope and
an X-ray analyser was used in conjunction with the microscope to obtain qualitative information on the composition
of the phases observed at the fracture surface.

MICROSTRUCTURAL DEVELOPMENT IN PSZ

The properties of the four PSZ batches are summarized
in Table I. The fracture surfaces of batches 1 and 2 typically show structure to be bimodal in both cases, with
smaller grains located on the grain boundaries of the larger grains (Figs. 2 and 3). Fracture of batch 1 shows a
mixed mode with the large grains failing transgranularly

TABLE I.   Properties of Calcia PSZ Batches

Batch	#1	#2	#2 Annealed	#2 Powder	#3	#3 Powder
Density, g/cm^3	5.49	5.76	5.75	5.79	5.27	5.58
Wt. % CaO	3.4	3.4	3.4	3.4	3.4	3.4
Wt. % Monoclinic	40	27	27	27	28	28
Ave. Grain Size, μ						
Large Grains	80	80	80		40	
Grain Boundary Precipitate	15	15	14			

Fig. 2. Composite of fracture surface: first batch of material.

and the small grains intergranularly. Batch 2 material exhibited intergranular fracture only. Batch 3 showed no grain boundary phase development and the presence of a liquid phase at the boundaries. A typical fractograph of

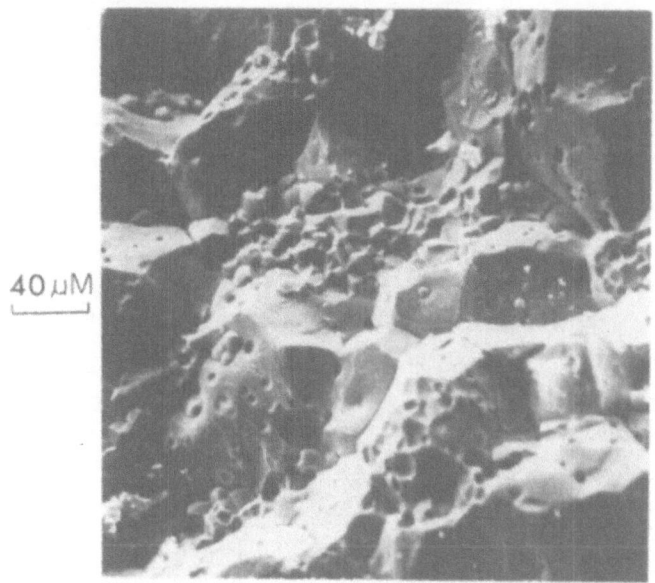

Fig. 3. Fractograph of batch 2 showing bimodal grain structure and intergranular fracture.

of batch 3 is shown in Fig. 4a and a heavily etched polished section in Fig. 4b. The grain sizes of batches 1 and 2 were estimated from fractographs by a linear intercept technique and are also included in Table I. Grain sizes were estimated from fractographs because attempts to produce satisfactory polished sections failed. The degree of cohesion between the fine grain boundary phase and the larger grains was such that polishing removed the fine grains so rendering meaningful analysis impossible. The grain size of batch 3 was estimated ceramographically. The approximate nature of the fractographic technique used for batches 1 and 2 may explain the difference of average grain size between these materials and batch 3. The transgranularly fractured large grains in batches 1 and 3 show a fine internal structure. This structure is presently being examined in detail in a separate investigation[7] and indications are that no such structure exists within the large grains of batch 2.

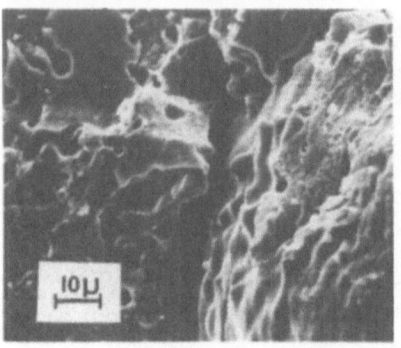

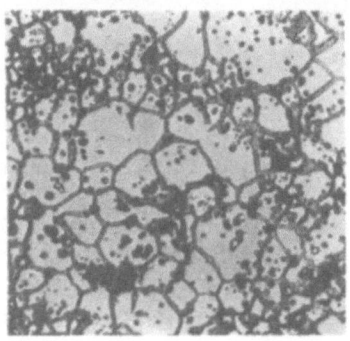

Fig. 4. Microstructure of batch 3 PSZ: (a) fractograph;
(b) polished section (heavily etched with HF).

        Utilizing the phase diagram as proposed by Garvie[8]
(Fig. 5), it is possible to follow the steps involved in the
production of the observed microstructures. On cooling
from the sintering temperature (X on diagram) the cubic
phase becomes supersaturated with pure zirconia, which
therefore precipitates out. This precipitation must involve
the counter diffusion of calcium and zirconium ions. The
low volume diffusion coefficients[9, 10] in calcia PSZ would
seem to militate against the production of a large grain
boundary phase. However, microstructural observations
and those on the role of the silica impurity in the ZrO_2
source powder in the sintering of CaO-PSZ to satisfactory
densities[11], indicate that the process is one of liquid phase
sintering. It has been observed, for instance, as shown in

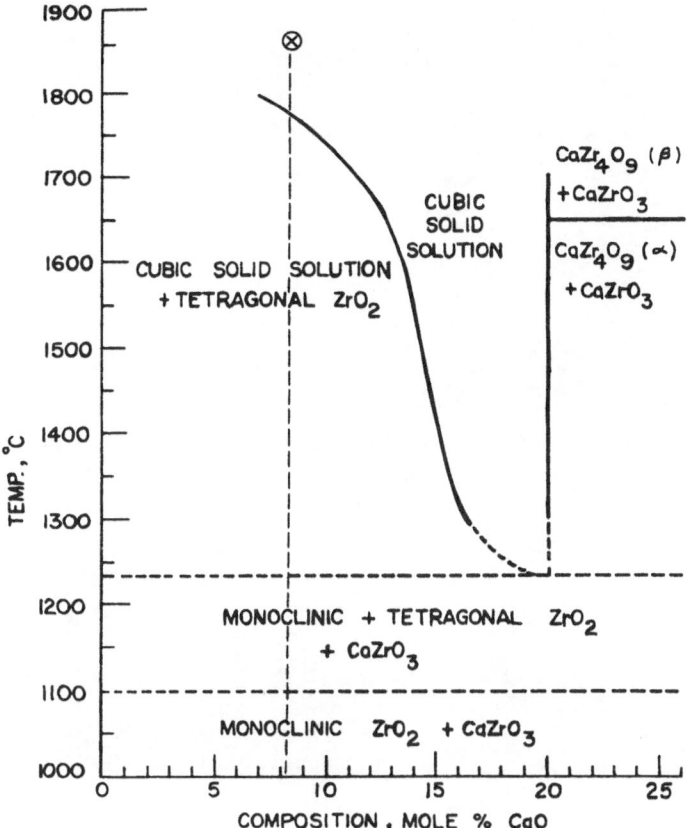

Fig. 5.   Partial phase diagram for the CaO-ZrO$_2$ system.

Fig. 6, that CaO-PSZ made with silica-free ZrO$_2$ will not
sinter at 1900°C whereas the addition of 1 wt. % colloidal
SiO$_2$ initiates sintering at ~ 1300°C with completion at 1650°C.
The presence of the liquid phase and its demonstrated role
in the sintering process could allow abnormal grain growth
of the grain boundary phase.   In batch 3 the liquid phase was
quenched by the heat treatment and can therefore be seen in
the microstructure (Fig. 4).

It is doubtful that the grain boundary phase consists of
undissolved grains of the original ZrO$_2$ since batch 3 shows
none of this phase.   The sintering at 1850°C must therefore

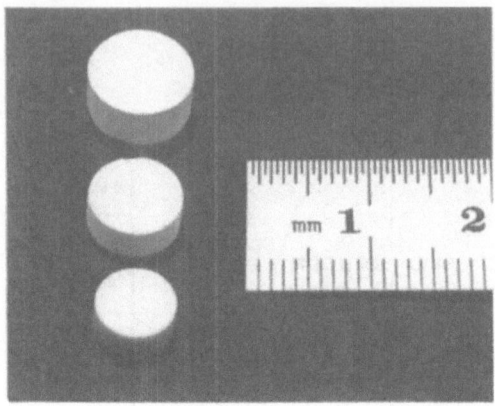

Fig. 6. Influence of silica on sinterability of CaO-ZrO_2:
M - 0% SiO_2; T - 1 wt. % SiO_2

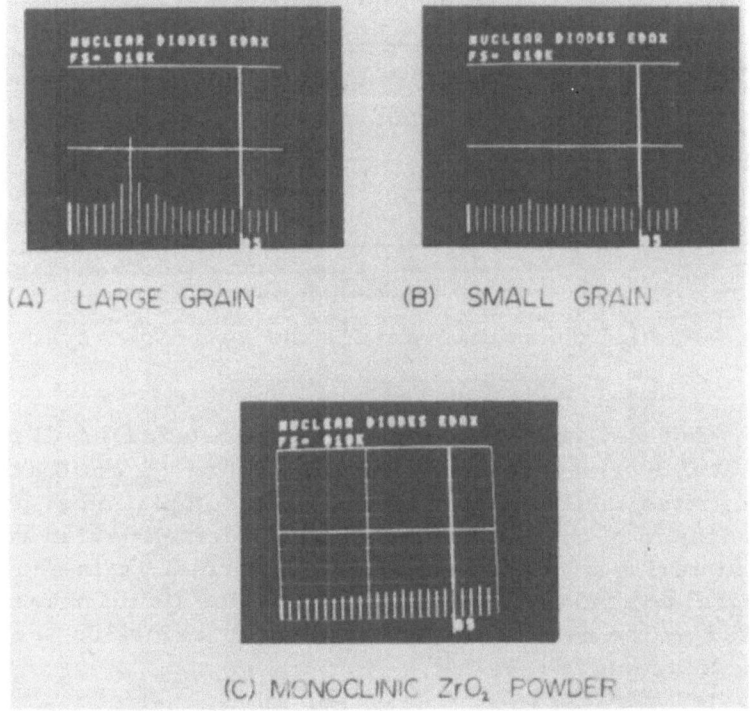

Fig. 7. Detection of calcium in calcia-PSZ using X-ray
analysis on scanning electron microscope (batch 2).

take place in the single-phase region, in agreement with the phase diagram. This suggests that the small-grained material present at the grain boundaries is monoclinic ZrO_2. X-ray dispersive analysis of Ca-K_α carried out on the SEM gave results as shown in Fig. 7. The amount of calcium present in the small grains is negligible compared with the large grains and since monoclinic ZrO_2 has almost zero solid solution limit for calcia, the grain boundary phase would appear to be pure monoclinic ZrO_2. Concurrent X-ray analysis was also carried out on pure monoclinic ZrO_2 powder and the results are also included in Fig. 7.

Analysis of many fractographs of batches 1 and 2 indicated that the level of grain boundary phase in both is substantially the same. This equality is likely since both batches were cooled to 1300°C at the same rate and have the same grain size. The difference in level of pure ZrO_2 detected in these two batches is probably associated with the finer intragranular precipitate. In like manner, the difference in ZrO_2 levels between batches 1 and 3 is probably associated with the absence of grain boundary phase in batch 3.

Summarizing the microstructural observations, it would appear that the slow cool from 1850 to 1300°C and the anneal at this lower temperature play an important role in the precipitation of the pure ZrO_2 from the saturated cubic solid solution. Batch 1 underwent a heat treatment involving both of these stages and both grain boundary and intragranular ZrO_2 precipitates were developed. The level of pure ZrO_2 in this material was the maximum observed and this conforms with the presence of both forms of precipitate. Batch 2 was slow cooled from 1850 to 1300°C but quenched thereafter. The grain boundary phase in this material developed to the same extent as in batch 1 but little intragranular fine precipitate development occurred. It appears therefore that the 1300°C anneal step is important in the development of the fine precipitate of ZrO_2. Finally, batch 3 was quenched from 1850 to 1300°C, which procedure eliminated the development of the grain boundary phase. In this material the pure ZrO_2 must be confined to the fine intragranular precipitate.

GRAIN BOUNDARIES AND MECHANICAL PROPERTIES

On cooling through the phase transformation region, the pure ZrO_2 component of each microstructure will transform to the monoclinic modification. This transformation involves an expansion of 3 volume %. The grain boundary precipitate will transform and the grain boundaries in the PSZ will therefore be weakened or decohered. It would be expected that this weakening will be more marked in batch 2, as it was shocked through the transformation; the slow cool of batch 1 should produce less damage. Batch 3 should have the least damage, as it is devoid of grain boundary phase. These expectations are born out by the relative extent of dye penetration into the PSZ materials, and by their fracture strengths. The rate of dye penetration and its extent were substantially more in batch 2 than in batch 1 and penetration into batch 3 was least of all. The fracture strengths of batches 1, 2 and 3 are 5000, 2500 and 22,000 psi respectively, which figures support the predicted extent of damage. The relative densities of batches 1 and 2 are high, indicating that the damage results in partially decohered grain boundaries rather than microcracks or cracks of finite volume. Further evidence in favour of the decohesion model is given in Fig. 8, which shows the fracture edge of a bar from batch 2, which has been set in hard epoxy resin after partial crack propagation. The loose monoclinic grains and the formation of cracks along the grain boundaries near the fracture surface are evident. The fractographs of the batches are also consistent with the picture described.

The most pronounced influence of the damaged grain boundaries is to hinder crack propagation in the material. The interaction of cracks with weak interfaces in brittle materials can lead to an increase in the resistance of the material to crack propagation[12, 13]. Fracture toughness tests were undertaken on batch 2, as this batch would tend to maximize any crack retardation effects. Crack propagation in this material was found to be always stable, even when the bend samples were unnotched. A partially broken

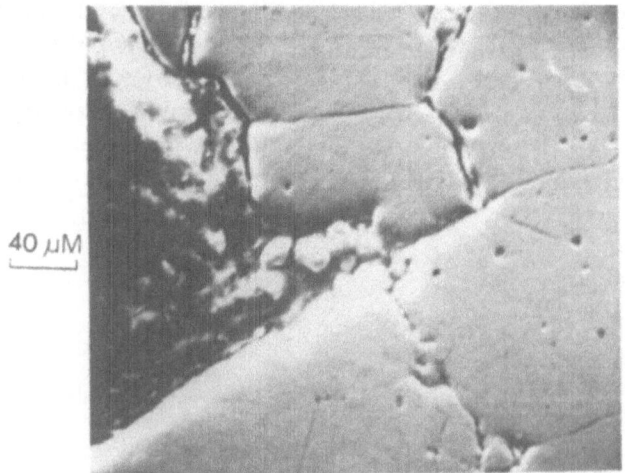

Fig. 8. Scanning electron micrograph of fracture edge showing typical damage and loose small grains (batch 2).

sample is shown in Fig. 9 and although the crack can be seen to have propagated across a significant portion of the ligament, the sample could be reloaded to a reasonable fraction of its fracture strength before crack propagation again commenced.

The many decohered and weak grain boundaries in the material will involve the presence of a microcrack zone at

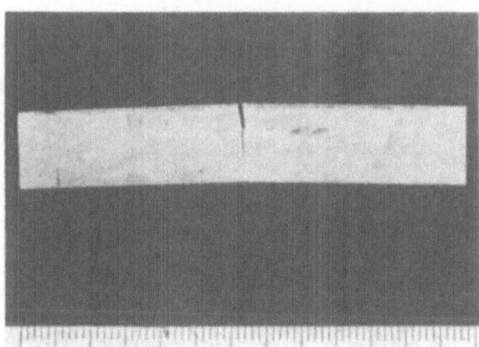

Fig. 9. Sample of calcia PSZ after partial crack propagation in four-point bend test.

the tip of any propagating crack. This zone will contain microcracks associated with the "opening" of partially decohered boundaries, and microcracks formed by the application of stress. A micrograph of such a zone at the tip of a crack is shown in Fig. 10. The extent of crack branching associated with crack propagation through the

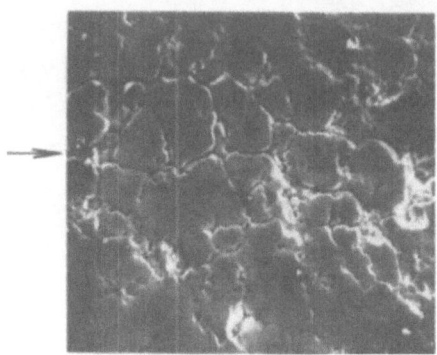

Fig. 10. Scanning electron micrograph of the microcrack zone at the tip of the main crack.

the material is well illustrated in Fig. 11. The size of the zone or the density of microcracks within it will increase in response to the increasing stress concentration at the crack tip as the main crack length increases. This response results in stable growth of the crack for a limited time. Eventually the equilibrium condition for crack propagation will be met and unstable growth will ensue. The work of fracture (γ_F) and fracture energy (γ_i) values for this batch are shown as a function of strain rate ($\dot{\epsilon}$) in Fig. 12. The drop in value of both parameters with increasing $\dot{\epsilon}$ could be associated with the time involved for microcrack zone development in response to an applied stress.

SUMMARY

By the choice of the appropriate heat treatment cycle it is possible to produce strong or tough materials. The nature of the toughening process, however, is such as to

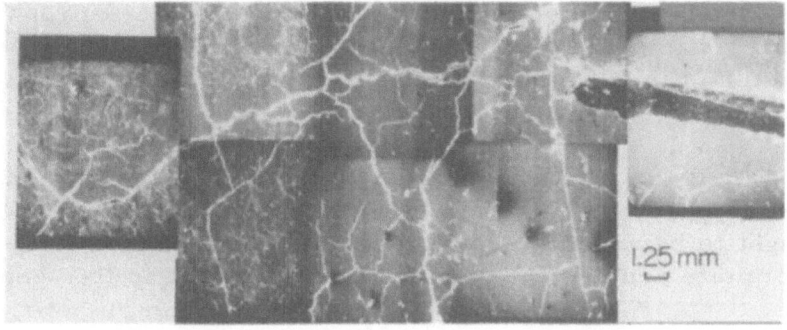

Fig. 11. Crack branching of a partially-propagated crack for a CaO-PSZ sample, vacuum-set in epoxy resin.

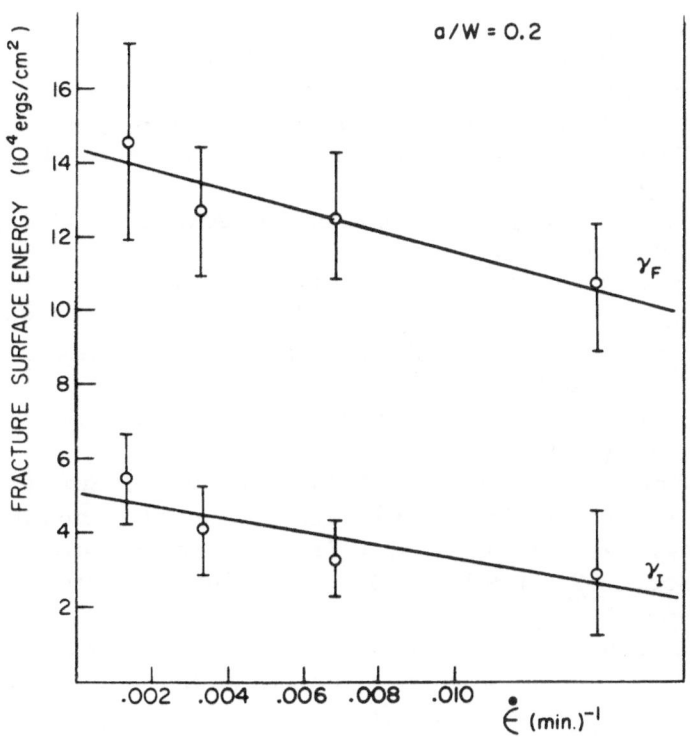

Fig. 12. Fracture surface energy and work of fracture as a function of outer-fibre strain rate.

lower the fracture strength as resistance to crack propagation is improved. Nevertheless, it is felt that heat treatments might lead to a reasonably strong and reasonably tough material.

Finally it should be pointed out that all these materials show superior thermal shock resistance. This property is thought to be associated with the fine intragranular ZrO_2 precipitate and has therefore not been treated in this paper.

REFERENCES

1. R. C. Garvie and P. S. Nicholson, J. Am. Cer. Soc., 55 (3) 152-57 (1972).
2. R. G. Cooke and D. E. Lloyd, Special Ceramics, 5, 125-34 (1972). Ed. P. Popper, Brit. Cer. Res. Assoc., Proc. 5th Symposium on Special Ceramics.
3. R. C. Garvie, U. S. Patent (11) 3,620,781 (1971).
4. W. F. Brown, Jr., and J. E. Srawley, ASTM Special Technical Publication No. 410, pp. 9-16 (1967).
5. H. G. Tattersall and G. Tappin, J. Mater. Sci., 1 (3) 296-301 (1966).
6. R. C. Garvie and P. S. Nicholson, J. Am. Cer. Soc., 55 (6) 303-305 (1972).
7. G. K. Bansal and A. Heuer, to be published.
8. R. C. Garvie, J. Am. Cer. Soc., 51 (10) 553-56 (1968).
9. W. H. Rhodes and R. E. Carter, J. Am. Cer. Soc., 49 (5) 244-49 (1966).
10. L. A. Simpson and R. E. Carter, J. Am. Cer. Soc., 49 (3) 139-144 (1966).
11. J. F. Shackelford, P. S. Nicholson and W. W. Smeltzer, Paper #15-B-73, American Ceramic Society Annual Meeting, Cincinnati, May, 1973. Abstract: Bull. Am. Cer. Soc., 52 (4) 341 (1973). To be submitted to J. Am. Cer. Soc.
12. J. Glucklich, Eng. Fract. Mech., 3 (3) 333-344 (1971).
13. J. Cook and J. E. Gordon, Proc. Roy. Soc., London, Series A, 282, 508-520 (1964).

DISCUSSION

K. D. Reeve (Australian AEC): In the case of precipitation of the second phase within grains, presumably mismatch stresses develop and this is the reason for the relatively low strength of this material?

Author: The grain boundary phase precipitates on a much larger scale than the intragranular phase. This leads to low strength due to decohesion of this grain boundary phase. In the high strength material, where there is only the intragranular precipitate, there is no evidence of this decohesion such as low strength and low elastic moduli etc. This type of phenomenon is observed in several ceramic systems where the release of internal stress into fracture surface area leads to critical size effects. That is, above a certain inclusion size, decohesion or microcracking will occur while below this size the developed stresses are elastically stored.

C. R. Manning (N. C. State Univ.): When looking at the thermal shock resistance of ZrO_2 if you plot E_{eff} vs. fracture strength you will find the E_{eff} dropping faster than σ_F. When E_{eff} reaches a minimum the thermal shock resistance should reach a maximum.

Author: The data which we have on thermal shock would seem to suggest that for the high strength batch 3 material the effective modulus drops slower than the fracture stress. However, we have not made an extensive study of the relationship between these two parameters. For the other two batches the thermal shock picture is complex since grain boundary decohesion occurs during fabrication. We believe that your final statement is reasonable for the batch 3 material since a microcracking mechanism of thermal shock resistance has been proposed for this material[1].

J. A. Pask (UC Berkeley): What is the source of the liquid phase in the specimens fired at the high temperatures? If it is due to an impurity, is there an appropriate phase diagram to substantiate the appearance of the liquid phase? What has happened to the liquid phase in the specimens that

do not show it on examination at room temperature?
Author: The source of the liquid phase is thought to be the
1% silica impurity in the source zirconia. The observation
of this liquid phase is consistent with the low eutectics
found in the ZrO_2-CaO-SiO_2 diagram (1300-1400°C). (Ref.
K. Matsumoto, T. Sawamoto and S. Karde, Asahi Garasu
Kenkyu Hokoku, 4 (2) 8 (1954). The specimens which do
not show the liquid phase were slow-cooled from 1850°C to
1300°C. We believe that the phase probably had time to
crystallise during this slow cool and was therefore not
identifiable in the room temperature microstructure.

GRAIN BOUNDARY MOBILITY IN ANION DOPED MgO

Cawas M. Kapadia and Martin H. Leipold

Department of Metallurgical Engineering and
Materials Science, University of Kentucky,
Lexington, Kentucky

Impurities are known to control the microstructure of
ceramic materials. Impurities tend to reside at grain
boundaries in ceramics, even in relatively high purity materials[1,2]. Extensive work on the role of cation impurities
on the grain growth kinetics has been reported[3]. Anions
in general have received little attention even though there
are suggestions as to their importance in the literature.
For example, fluoride additions significantly enhance the
fabricability of MgO by hot pressing[4]. Ceramic surfaces,
qualitatively similar to grain boundaries, show a strong
affinity for gases (Cl_2, F_2, H_2O)[5,6]. One reason for the
lack of attention to anion impurities is the analytical problems involved in their detection; routine survey analyses
are insensitive to their presence. However, studies have
shown that they are present and often exist as a major impurity when cation impurities are reduced to 0.01% or less[7].
With the exception of water vapor[8], few data are available
concerning the influence of anions on grain growth in MgO.

The purpose of the present work is to define the effect
of anions on grain boundary migration in a typical ceramic
oxide, MgO. The influence of anions on the grain growth
kinetics in MgO must be considered in light of solubility in
the matrix and vapor pressure which will determine the
nature in which they exist in the oxide, namely, as a gas,

385

a precipitate, or a liquid phase. Since most anions, as elements or as magnesium compounds, exhibit high vapor pressures at temperatures of interest, an entrapped gas phase exists within the microstructure which retards densification and gives some residual porosity[9]. As pores also influence grain boundary migration rates, the role of anions themselves on grain growth must be studied in conjunction with their effect on intrinsic pore mobility and pore removal. This effect can be clearly defined if such pore-controlled grain growth in the undoped MgO is compared with that in anion-doped material and toward that end, such investigations are included here.

THEORY

Grain boundary migration in real ceramics must allow for the various modifications to theory due to material factors, i.e., porosity and second phase particles. When the growth rate depends strictly on boundary curvature, the theoretical equation for isothermal grain growth[10] is

$$D^2 - D_o^2 = Kt \tag{1}$$

where K is the temperature-dependent rate constant, D is the average grain size at anneal time t and D_o is the initial grain size. In the presence of grain growth inhibiting effects, the growth law when $D_o << D$ is empirically represented by

$$D^n = Kt \tag{2}$$

where the observed grain growth exponent, n, is commonly greater than the theoretical value of 2. Since Eq. 2 can also be written as $D = Kt^{1/n}$, grain growth shows a $t^{1/n}$ time dependence.

Grain boundary velocity in porous MgO depends on pore concentration and is controlled by (i) pore removal for interconnected porosity ($\geq 5\%$) where the observed value[11] of n equals 2, and (ii) pore mobility or boundary mobility for

isolated pores lying along the grain boundaries (porosity
$\leq 5\%$) where $n \geq 2$. Thus, in the presence of porosity, n
may or may not equal 2, and where n = 2, the controlling
mechanism could differ. Since most compacts in this
study were hot pressed well into the final stage of densification (< 5% porosity), the latter case is of primary
significance here. Isolated pores at grain boundaries
migrate with the boundary with velocity given by[12]

$$V = \frac{M_b F_b}{1 + N M_b / M_p} \tag{3}$$

where M_b is intrinsic grain boundary mobility, M_p the
average pore mobility, F_b the total force driving the average grain boundary and N the average number of pores per
boundary. Eq. 3 reduces to two important forms depending on relative magnitudes of M_b and M_p[12]. These are:
(i) $V = M_b F_b$ when $M_p >> N M_b$ and corresponds to grain
growth controlled by boundary mobility alone. Such behavior is present at low temperatures and small grain
sizes. (ii) $V = M_p F_b / N$ when $M_p << N M_b$ for pore-controlled growth, applicable at high temperatures and larger
grain sizes.

For case (i) the impurities influence the grain boundary
velocity by controlling both M_b and F_b depending on their
location (grain boundary vs. grain interior) and their form
in the matrix, namely, solid solution, discontinuous second
phase (precipitate) and continuous boundary phase which
could be a liquid at annealing temperature. The mechanism that controls grain growth in each case is different but
in general gives n = 3[13,14].

For case (ii) the boundary velocity can be expressed in
terms of M_p and N, both of which are related to the average
grain size. M_p is proportional to $1/r^m$ where r, the pore
radius, is some function of the grain size and m is an integer depending on mechanism of material transport responsible for pore migration[15]. The variation of N with D depends on whether the pores are located at grain corners or

on individual grain boundaries[12]. It can be shown[16] that
depending on the mode of material transport, two sets of
grain growth exponents are possible, namely, n = m for
pores at boundary intersections and n = m-1 for pores on
individual grain boundaries. Since pores actually lie both
on the grain boundaries and at boundary intersections, the
above two cases of pore location may be somewhat ideal-
ized and one could observe an intermediate behavior.
Table I gives values of n predicted for both cases of pore
location, for each pore transport mechanism. Although
one normally expects n > 2 for pore-controlled grain
growth, there are certain pore transport mechanisms
wherein n ≤ 2 is predicted.

The above discussion was based on 1/D dependence of
the driving force which is true when D << r/f, where f is
the volume fraction of intergranular pores. However, for

Table I. Values of m for Different Pore Transport Mechanisms and Corresponding Grain Growth Exponents, n.

Mechanism	m	n		Conditions when applicable		
		$n=m^{+}$	$n=m-1^{*}$	Pore radius, r	Temp.	Vapor Press.
Surface diffusion	4	4	3	small	low	low
Volume diffusion	3	3	2	medium	medium	low
Vapor trans- port (p=constant)	3	3	2	Large ($\geq 1\mu$)	high	high
Vapor trans- port ($p=2\gamma_{sv}/r$)	2	2	1	Large ($\geq 1\mu$)	high	high

+ = pores at boundary intersection
* = pores on individual grain boundaries

large D, the driving force for grain growth is proportional to $(1/D - f/r)$[17]. As annealing progresses, the inhibiting effect of pores becomes more pronounced until eventually the grains cease to grow when

$$D \approx \frac{r}{f} \tag{4}$$

Limiting grain size $(GS)_L$ is said to have been reached and corresponds to $n = \infty$. For random dispersion of pores[18] with inhibition by those at the grain boundaries,

$$(GS)_L \cong 0.15 \frac{r}{f} . \tag{5}$$

The effectiveness of pore control on grain growth in MgO depends on the annealing temperature; at high temperatures ($\sim 1500°$ C) inhibition by pores is observed at very small annealing times. Limiting grain size is reached after a few hours since a rapid increase in D satisfies criterion (4) for limited growth. At lower temperatures ($1300°$ C) where grain growth is slower, pore control and growth-limiting effects require very long anneals. At such high anneal times, pore coalescence and simultaneous densification may increase $(GS)_L$ to a value such that limiting grain size may not occur. Also, at a given annealing temperature, changes in microstructure with time can vary the growth kinetics, with different n values over the whole annealing period. After limiting grain size is observed one expects further grain growth if annealing is continued, owing to coalescence of intergranular pores and decrease in porosity, giving a higher value of $(GS)_L$.

In conclusion, grain growth kinetics are influenced by porosity and impurities. A combination of mechanisms of comparable importance, involving pore-grain boundary interactions, may operate simultaneously, in which case the time-dependence of grain growth is insufficient to determine the atomic processes. Only by observing the microstructure and its change with time can one eliminate alternate hypotheses and determine the rate-controlling process or processes. In the present study, the microstructural features of both pure and anion-doped MgO were investigated and used as a basis for interpreting grain growth behavior.

EXPERIMENTAL PROCEDURE

Grain growth anneals were done with two types of hot pressed MgO compacts, i.e., 1) <u>Undoped</u>, wherein two grades of powder were used: High Purity JPL and Fisher M-300 electronic grade. The preparation, chemical analysis and hot pressing conditions are given in an earlier paper[19]. The Fisher impurity levels were typical of commercial magnesia. 2) <u>Doped</u>, wherein known amounts of S^{-2}, Cl^-, OH^-, OD^-, F^- were added to the Fisher MgO. Details are given in Reference (9). OD^- was chosen to differentiate between the additive and OH^- from the atmosphere. However, OD^- gave negative results, apparently because exchange with an unknown source occurred.

Anneals were done at $1300°$ C and $1500°$ C in air in an Al_2O_3 tube furnace with SiC elements. Temperature was measured by a Pt/Pt-10% Rh thermocouple within $10°$ C. The specimens were removed periodically, examined microscopically, photographed and returned to the furnace for continued annealing. To get an average behavior, new surfaces were sectioned, polished[20] and etched and the average grain size was determined with ± 10% by lineal analysis[21]. Table II gives characteristics of hot-pressed specimens used. Bulk densities were taken by kerosene displacement.

RESULTS AND DISCUSSIONS

Undoped MgO

To bring out the influence of anions alone on the pore and grain boundary mobility, it was necessary to separate their effect from that of the inherent cation impurities ($\sim$ 1000 ppm Ca^{2+}, Si^{4+}, Al^{3+}, Fe^{2+}) in Fisher MgO. Grain

Table II. Composition and Fabrication Parameters of Specimens for Use in Grain Growth Studies.

No.	Dopant	Hot Pressing Conditions			As-Pressed	
		Temp. °C	Press. kpsi	At. *	Density % ≠	Analysis At. %
UK132	None	1000	15	A	99.6%	
UK130	None	1000	15	A	98.4%	
H120	3% OH	1100	15	V	98.8%	
H74	3% OH	1050	15	A	95.5%	
F114	3% F	1100	15	V	99.6%	0.68
F110	3% F	1100	15	V	98.7%	0.90
C48	3% Cl	1170	17	V	95.6%	0.22
C115	3% Cl	1100	15	V	99.6%	0.06
S33	1% S	960	20	V	95.8%	0.24
S96	3% S	1100	15	V	97.6%	0.31
S134	6% S	1100	15	A	99.3%	0.04

*A = Argon; V = Mechanical pump vacuum
≠ = Theoretical density of MgO = 3.576 g/cm^3.

growth in Fisher MgO is first compared with that in JPL MgO which has < 200 ppm cation impurity content.

Grain Growth in Dense (99.5%) MgO of Different Purities

Results at 1300° C and 1500° C for dense MgO specimens of both purities are given in Table III.

1300° C. In spite of the equivalent porosity levels, high purity JPL MgO (OP373) gave n = 2 whilst Fisher MgO (HP332) gave n = 3. Limiting grain size was not observed even after long anneal. Microstructure for Fisher material showed considerable liquid phase at triple points and grain boundaries owing to cation segregation. Probably growth is controlled by diffusion through the liquid phase.

1500° C. With the high purity material (OP373), limiting grain size was reached after only 1 hour. This is in

Table III. Grain Growth in Dense Undoped MgO.

No.	Purity (cation content)	Hot-pressed density	Grain Growth data[+]	
			1300°C	1500°C
OP 373[*] (JPL MgO)	~200 ppm	3.55	n=2 $K=1.5 \times 10^{-8} \frac{cm^2}{min}$	(1) $t < 10^3$ min Limiting grain growth (2) $t > 10^3$ min Abnormal grain growth
HP332[≠] (Fisher MgO)	~1000 ppm	3.56	n=3 $K=36 \times 10^{-12} \frac{cm^3}{min}$	(1) t<200 min n=3 (2) 200<t< 1500 min Limiting grain growth (3) t>1500 min n=3 $K=189 \times 10^{-12}$ cm^3/min

[*] hot pressed at 1025°C and 13 kpsi maximum pressure in Al_2O_3 die assembly.

[≠] hot pressed at 1270°C and 4 kpsi maximum pressure in graphite die assembly.

[+] n = grain growth exponent; K = rate constant.

contrast to grain growth with n = 3 observed for less pure material (HP 332) wherein impurities might have influenced pore mobility and also enhanced their removal.

Anion-doped specimens had two porosity levels, < 1/2% and 1 - 4%. For comparison, preliminary studies were conducted on undoped Fisher MgO of the same porosity levels. Any difference in behavior after anion addition with respect to the two porosity levels can be interpreted as a direct influence of the anions.

Grain Growth in Undoped Fisher MgO

1300° C. For the denser Fisher MgO material (UK 132; 0.4% porosity), microstructures showed few very fine pores along the grain boundaries at small anneal times (< 600 minutes). At longer heat treatments (1000 - 3000 minutes) large and hence, less mobile pores were evident at grain boundaries, especially at the triple points. Accordingly, square kinetics were obeyed up to 600 minutes while for larger anneal times a cubic growth law was observed due to pore control (Fig. 1). For the less dense Fisher MgO grain sizes are comparable to the denser material although considerable scatter makes prediction of kinetics difficult.

1500° C. For denser material (UK 132) an initial n = 2 behavior up to 150 minutes was followed by constant grain size (70 µm) up to 1000 minutes. Limiting grain size was

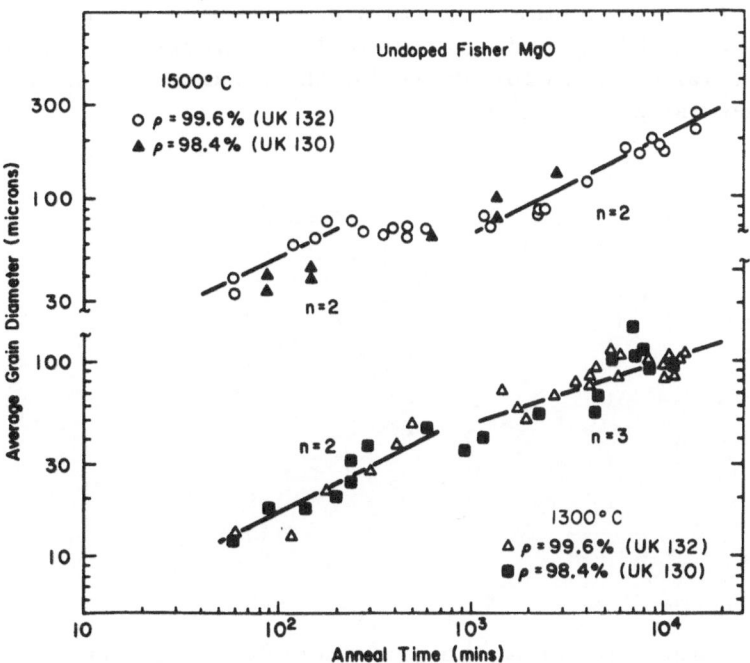

Fig. 1. Grain growth at 1300° C and 1500° C in undoped Fisher MgO with different densities.

reached more rapidly and at a smaller grain size than at
1300° C. Some intergranular pores were noted at 600 minutes, accompanied by considerable liquid phase wetting at
the grain boundaries and triple points at 6000 minutes
(Fig. 2). It is probable that the liquid phase enhanced
pore removal, thus reducing porosity to a very low level
which resulted in further grain growth after 1000 minutes.
Data at 1500° C for the less dense MgO (UK 130) is sufficient only to conclude that the grain sizes are comparable. Hence, the Fisher MgO with different porosity
yields comparable grain size at 1300° C and 1500° C.

For undoped material, in the presence of 1/2% porosity,
a square growth law was observed at 1300° C for high purity
JPL MgO while at 1500° C severe pore inhibition gave a
limiting grain size only after a one-hour anneal. For
Fisher MgO a cubic law was observed at both 1300° C and
1500°C, apparently as a result of impurity differences
between the materials. The growth rate with 1/2% porosity
was not very different from that with 2% porosity, making
the material suitable for observing the influence of anions
where porosity may vary.

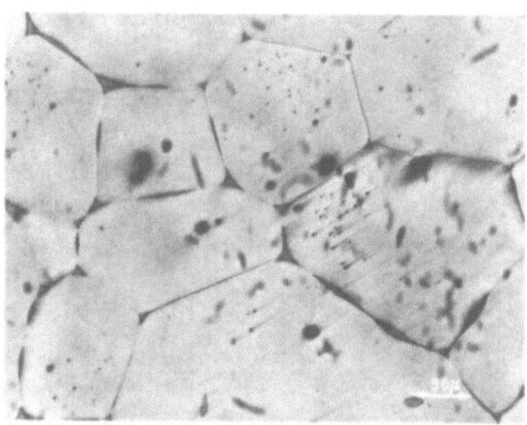

Fig. 2. Liquid phase at the grain boundaries and triple
points in dense Fisher MgO (UK 132) annealed at 1500° C
for 6000 min.

Anion-Doped Fisher MgO

Considerations of the anion sizes compared to that of oxygen suggests that the F^- and OH^- would be reasonably soluble in MgO, whereas Cl^- and S^{2-} would not. F^- and OH^- in solution should give cation vacancies, enhancing diffusion.

Cl^- and S^{2-}, if not in solution, would be expected to exist in second-phases or more likely as entrapped gas[9] both of which retard grain growth. The low levels of these impurities after hot pressing (Table II) are consistent with this. Microscopy did not reveal any appreciable second phase, and the presence of porosity suggests a gas phase as a likely site for these anions. Inhomogenities in the distribution of porosity and impurities led to considerable scatter in some data.

Fluorine-Doped Fisher MgO

1300° C. For the less dense material, the pores were found at grain boundaries and triple points. Pore dragging controlled grain growth, giving n = 3 which indicates that pore motion probably occurs by volume diffusion or vapor transport. For the denser material n = 2 up to a grain size of 150 μ. Microstructures revealed no porosity or second phase, suggesting that boundary mobility might be controlling. For grain sizes > 150 μ, n increases to 3.

1500° C. Data scatter for F110 (less dense) for anneal time less than 200 minutes makes prediction of the grain growth kinetics difficult although a value of n = 2 is most likely. Beyond this period, intergranular porosity inhibited growth up to 4000 minutes. After further heat treatment, the microstructure revealed very little porosity, residing mainly in the grains, with a second phase at grain boundaries and triple points. The value of n = 2 observed could correspond to grain boundary mobility controlling or diffusion across the liquid phase.

No porosity at all appeared in dense F114 for all the heat treatments; however, liquid phase was evident at

grain boundaries, especially near the triple points. Absence
of pores or second phase gave boundary-controlled growth
over the whole range of time. From the data at both temperatures (Fig. 3), one concludes that for F, the grain
growth rates are sensitive to porosity differences of 1%,
in contrast to the undoped material where such sensitivity
was not observed.

At 1500° C, F114 showed extremely large grain sizes
and absence of any grain size plateau. Such behavior could
be due to enhanced densification either by formation of a
liquid phase[22, 23] or defects due to solution of F^- in the
O^{2-} lattice[24]. This is supported by complete absence of
porosity even after long heating. Feasible reactions between MgF_2, residual carbonate and other cation impurities open up a range of possible low-melting phases. Also,

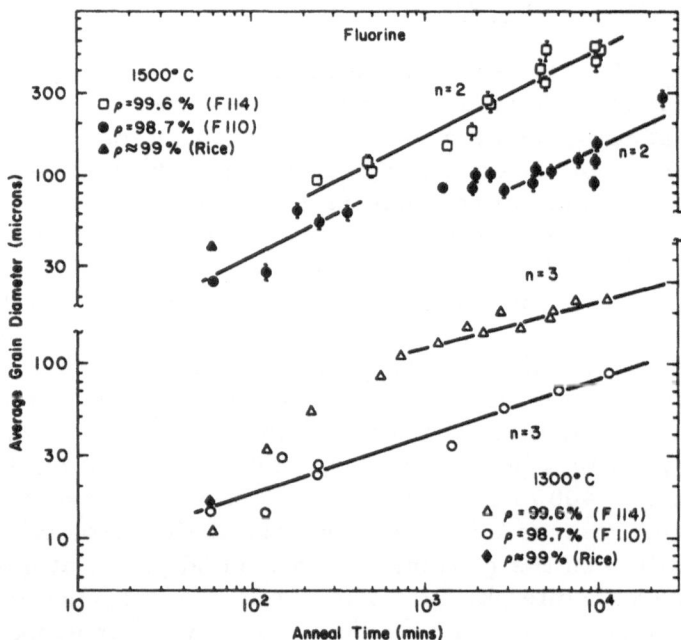

Fig. 3. Grain growth at 1300° C and 1500° C in fluorine-
doped Fisher MgO as a function of density.

since formation of the liquid at the grain boundary could only follow segregation of cations at grain boundaries, F^- in solid solution could be indirectly aiding liquid formation by creating lattice defects, thus increasing the cation diffusion rates. Others[25] have also observed grain growth in the presence of F and have proposed enhanced diffusion along and across a grain boundary layer of fluoride-rich phase.

Chlorine-Doped MgO

No evidence of appreciable liquid phase was present for the Cl specimens and the pores were present even after long anneals. The difference in behavior could be related to the size of the Cl^- ion relative to that of O^{2-}, which precludes its substitution for oxygen; it could then exist as entrapped gas. Also, the melting point of MgF_2 and $MgCl_2$ are $1396°C$ and $712°C$ respectively, while vapor pressures at $1000°C$ are $< 10^{-6}$ atm and 0.03 atm respectively. $MgCl_2$ or any other complex oxychloride will be unstable at sintering temperatures and will dissociate into vapor and form gas bubbles. This is evident in fabrication studies[9] where residual porosity was present in the Cl specimens.

$1300°C$. For less dense Cl-doped MgO (C48) no appreciable grain growth was observed up to 1000 minutes (Fig. 4), due to presence of growth-inhibiting interconnected porosity. Once porosity was reduced to a level where the pore phase was discontinuous, grain growth was controlled by pore mobility with n between 1 and 2. C115, a denser specimen, also gave n = 1 at $1300°C$ up to 300 minutes, followed by n = 2 up to a grain size of 200 μ. Intergranular pores (~ 1 μm size) makes boundary mobility controlled grain growth less likely. In both Cl-doped specimens, at higher temperatures and larger pore sizes, it is possible that pores move with the boundary by the evaporation-condensation mechanism previously suggested for UO_2 and and Al_2O_3[26]. If vapor transport is indeed rate-controlling, then $n \leq 2$ is possible. The possibility of enhancement of vapor transport of MgO by a chlorination reaction is sug-

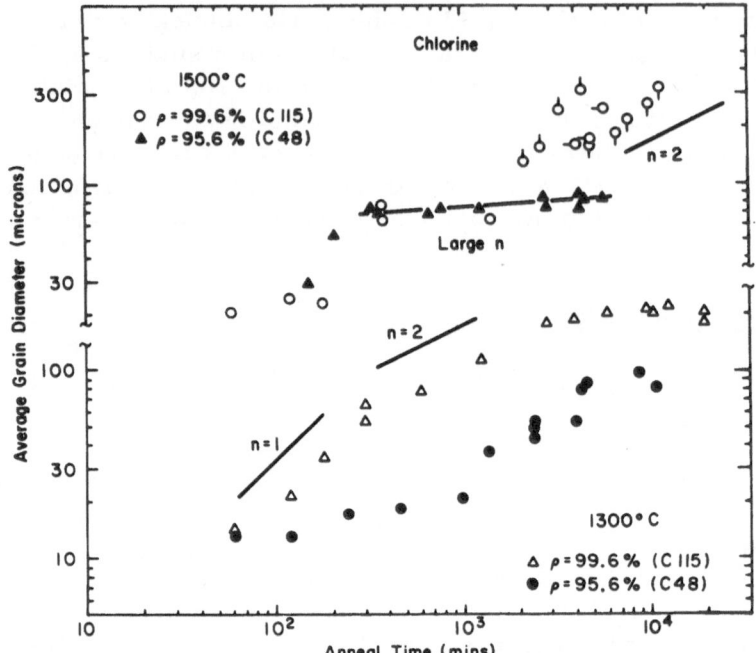

Fig. 4.   Grain growth at 1300° C and 1500° C in chlorine-doped Fisher MgO as a function of density.

gested.   Such transport is invoked in the formation of some minerals[27].   At heat treatments > 3000 minutes, the grain size for the denser material, C115, apears to stabilize at 200 μ.   It is not clear whether such limiting growth is due to the specimen size effect or a change in the pore behavior.

1500° C.   The porous specimen C48 (Fig. 5a) shows 3 μm pores at the grain boundaries, especially near triple points, and at grain interiors.   These large pores restrict the boundary mobility giving only slight increase in grain size over the whole anneal.   The less porous C115 showed appreciable grain growth with $t^{1/2}$ time dependence; however, growth rates seem to depend on the individual specimens (data points for 1500° C in Fig. 4 which correspond to the same specimen are identified by arrows pointing in the same direction).   As different pore densities were

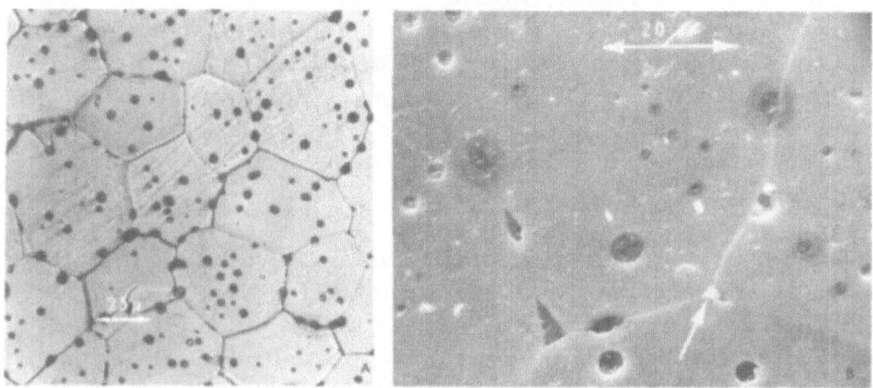

Fig. 5. (A) Porous chlorine-doped MgO showing pores at the grain boundaries and grain interiors. (B). Scanning electron micrograph of sulphur-doped MgO, annealed for 48 hours at 1500° C, showing precipitates at the grain boundaries and triple points. Arrow shows possible $CaSiO_3$ phase at the triple point.

observed in each of these specimens due to inhomogeneous pore distribution in the hot-pressed material, it appears that the larger grain sizes correspond to a lower pore content.

In conclusion, for Cl doped MgO, the presence of the pore phase controls the overall evolution of the microstructure. Grain growth rates are sensitive to pore content and vapor transport is the proposed mechanism behind migration of intergranular pores.

Sulphur-Doped MgO

Because of its large ionic radius, S^{2-} is not expected to go into solution; further, it cannot form simple compounds with Mg since these are not very stable at the sintering temperatures. Hence, it is expected that sulphur exists as vapor within the pores[9].

1300° C. A typical microstructure of S96 (97. 6% dense)
showed extremely fine pores both on the grain boundaries
and within grains. However, the mobility of these extremely fine pores was sufficiently large as not to hinder the
boundary mobility and n = 2. 2 was observed. For S134, a
denser material, n = 2 initially, followed by n = 3 beyond
1000 minutes (Fig. 6).

1500° C. Boundary mobility was rate controlling with
n = 2 at anneals less than 200 minutes for both S96 and
S134. However, at a grain size around 65 μ, inclusions,
suspected to be pores, severely inhibited boundary mobility
and very limited growth was observed for a period of 40
hours. On further heat treatment, the porosity being reduced to a low level, grain growth resumed with $t^{1/2}$ time
dependence; pores were mostly within grains, with pore-
free regions around the grain boundaries.

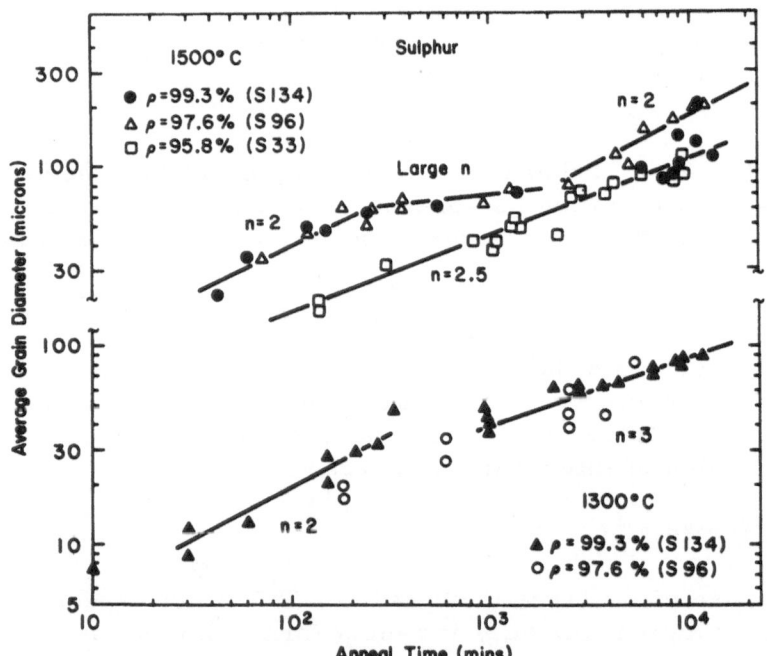

Fig. 6. Grain growth at 1300° C and 1500° C in sulphur-
doped Fisher MgO as a function of density.

There was no evidence of increase in the liquid phase other than normal amount in the undoped Fisher MgO; hence, densification was controlled by the intrinsic pore mobility. However, the SEM photos of the large-grain specimens (Fig. 5b) showed a second phase which gave large contact angles at the grain boundaries and triple points. Possibly the boundary migration here is controlled by the second phase precipitates together with the inter-granular pores. However, the $t^{1/2}$ growth dependence observed for the less dense S96 precludes drag of these second phase inclusions to be rate-controlling. Even though S96 and S134 had different densities, the actual grain sizes and the grain growth behavior were similar both at 1300° C and 1500° C.

However, for S33 (3.42 g/cc), the grain sizes at 1500° C are distinctly smaller. Here, since the porosity was inter-connected, the rate of boundary migration depends on the rate of pore removal rather than on the mobility of the in-dividual grain boundaries pores. After a sufficiently long anneal, a reduction in porosity gives a discontinous grain boundary pore phase, and the grain sizes become compar-able to that of S96 and S134.

As no densification enhancement was evident in the pres-ence of sulphur, residual porosity existed in the material giving primarily pore-controlled growth. Grain sizes were smaller than those in the other anion-doped specimens of comparable as-pressed densities. The grain growth rates for S were not nearly as sensitive to slight porosity differ-ences in starting material as observed for other anions.

OH⁻ Doped MgO

Earlier studies[8] indicate that both sintering and grain growth in magnesia are greatly enhanced by the presence of OH⁻ ion at around 1000° C. It is suggested[28] that solu-tion of hydroxide in MgO gives cation vacancies which in-crease all the diffusion processes. Also, OH⁻ can exist as a liquid, probably as $Ca(OH)_2$, which is stable at the sintering temperatures at internal pore pressures of 1000

psi or less[29]. On the other hand, if OH⁻ exists as H_2O
vapor it can hinder densification, giving limiting porosity
which is detrimental to grain growth[9]. As apparently
identical growth kinetics were observed at 1300° C and
1500° C, the discussion will be separated according to the
two observed time dependencies. Data for OH⁻-doped MgO
are given in Fig. 7.

$t^{1/2}$ Dependence (t < 1000 minutes). Grain growth for
H120 at 1300° C and 1500°C correspond to n = 2, a possible
mechanism being boundary mobility control. However, the
presence of grain boundary porosity means pore control is
more likely with volume diffusion as the probable mechanism for their motion. Another indication of pore control
is the data at 1300° C where two distinct growth rates are
observed corresponding to specimens with different porosity.

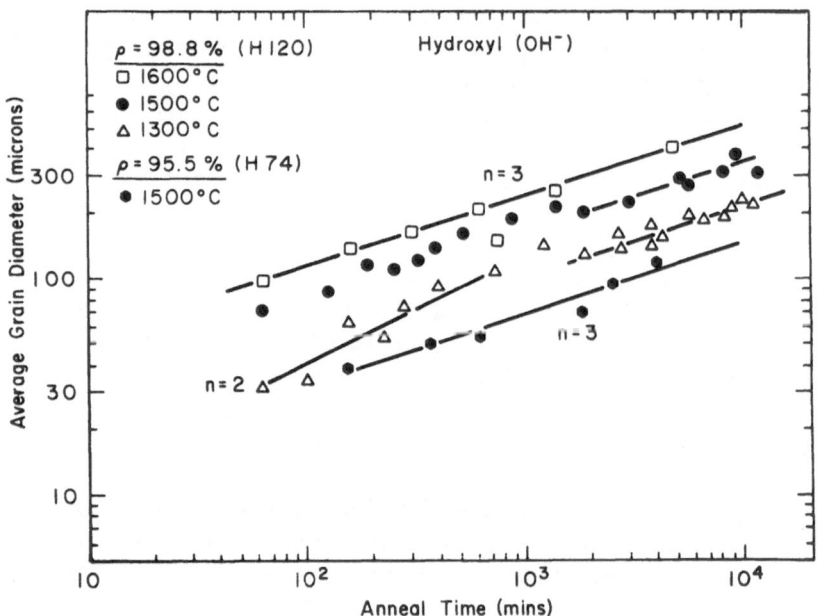

Fig. 7. Grain growth in OH⁻ doped Fisher MgO.

An approximate activation energy of 23 k cal/mole was obtained and was related to the possible rate controlling mechanism. It has been speculated[8a] that the surface absorption of OH^- enhances the surface mobility of O^{2-} in predominantly ionic crystals so that the cation migration would be rate controlling. Since the grain boundaries are qualitatively similar to ceramic surfaces and if one assumes sufficient OH^- at grain boundaries, a transport process for O^{2-} exists at grain boundaries, which makes Mg^{2+} diffusion rate controlling. However, this is not obvious from this study, as the observed activation energy is much lower than that reported for extrinsic Mg^{2+} diffusion in MgO.

$t^{1/3}$ Dependence (t > 2000 minutes). At 1300° C and 1500° C, the data obeyed the growth law $D^3 = Kt$ suggesting porosity- and/or impurity-controlled growth. A cubic law was observed at 1600° C even at small heat treatments. No porosity or second-phase inclusions were noted but liquid wetting was evident at grain boundaries and triple points. These microstructural observations coupled with $t^{1/3}$ dependence indicate that at large grain sizes, material transfer across the grain boundary liquid is rate controlling. Liquid control was observed only at long anneal times because 1) time is required to form the eutectic by diffusion of cations to the grain boundary and 2) the grain size must be large enough to give smaller grain boundary area so that appreciable solid-liquid interfaces are formed. Appreciable liquid phase was observed only after 60 minutes at 1600° C because of the large grain size ($\sim$ 100 µm). An activation energy of 35 k cal/mol was obtained for $t^{1/3}$ growth kinetics at 1300° C, 1500° C and 1600° C and could correspond to diffusion through the liquid phase. However, as these activation energies are extremely sensitive to the liquid composition, comparison with other liquid phase sintering is not meaningful.

Studies at 1500° C on a porous OH^--doped specimen gave n = 3. As intergranular porosity was evident, a pore mobility-controlled grain growth is possible.

Comparison Between Anions

For porous MgO (Fig. 8), density gradients in the hot
pressed compacts gave data scatter and only general trends
in grain growth were meaningful. There was very little
difference in the actual grain sizes at 1300° C and 1500° C
between the anion-doped and undoped Fisher MgO. Consequently, for the less dense materials (>1% porosity),
the kinetics were primarily pore-controlled and the effects
of anions, if any, minor. Such porosity was difficult to
avoid when these anions were present. Grain sizes for
the undoped material, UK130, at 1300° C were closer to
that for F material than for Cl or S, especially at low
anneals, probably because of similarity in densities. At
1500° C, in spite of differences in the density between the
different doped specimens, the limiting grain size was the
same. The grain growth rate increased more rapidly at

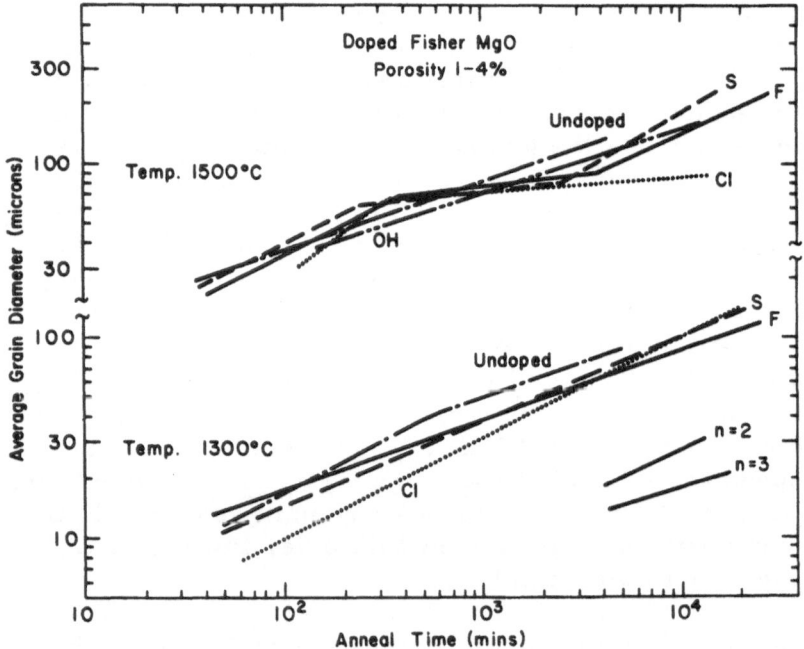

Fig. 8. Grain growth data at 1300° C and 1500° C for anion
doped and undoped Fisher MgO containing 1 - 4% porosity.

1500°C, but because of the greater limiting effects it slowed down and eventually became equivalent to that at 1300°C at heat treatments > 5000 minutes.

The influence of these anions with respect to pore removal and boundary migration was apparent only at very low porosity levels.   For the dense material, a very distinct difference in growth rates existed between OH⁻, F⁻ and Cl⁻ as a group (termed promoters) and S²⁻, which corresponded more to undoped Fisher MgO (Fig. 9).   As the densities of all these specimens were the same, the difference should be interpreted as an effect of the additive anions.

Larger grain sizes for the promoter anions compared to S²⁻ can be interpreted as (1) lack of growth inhibition by

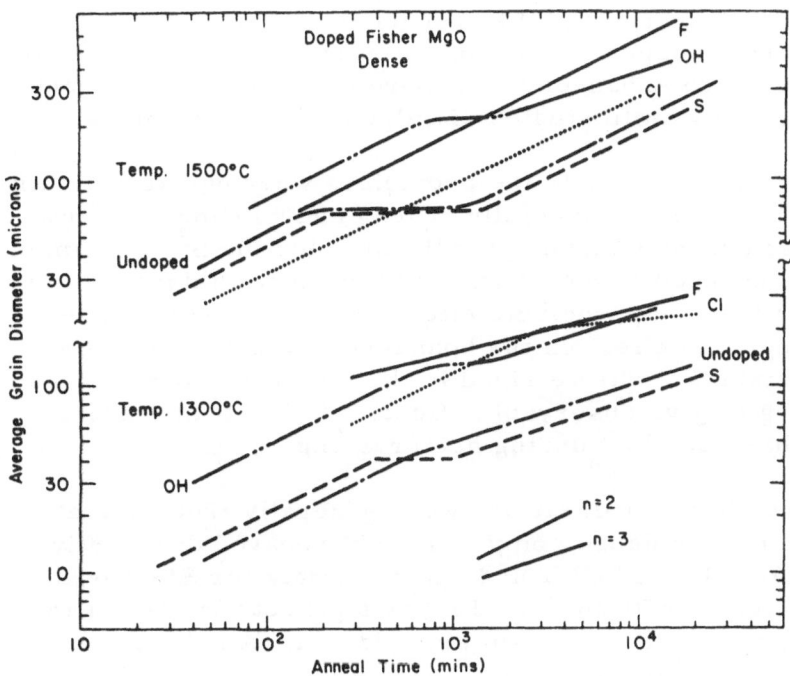

Fig. 9.   Grain growth data at 1300°C and 1500°C for anion doped and undoped Fisher MgO containing less than 1/2% porosity.

inclusions, especially intergranular pores, (2) the mass diffusion rates are higher. OH^- and F^- are known to enhance diffusion by creating lattice defects on substitution for O^{2-} and under certain conditions form grain boundary liquid phase. On the other hand, the mechanism for Cl^- is proposed to be that of pore transport along the grain boundary by evaporation-condensation process normally important only at high temperatures in MgO. At the sintering temperature, the gas phase (probably Cl_2) inside the pores is in equilibrium with the solid MgO in such a way that on the less concave pore surfaces the volatile compounds continuously form, move by vapor transport to more concave pore surfaces, and dissociate back into the oxide and Cl_2, the overall process resulting in pore motion. A similar mechanism was proposed by Eudier[30] for activated sintering of Cu by oxygen (confirmed later by experiments[31]) and iron in the presence of hydrogen halides. Bockstiegel[32] has reported such facilitation of gas phase transport by the action of sulphur on iron. However, no such enhancement of pore migration was evident in the grain growth studies of sulphur-doped magnesia.

At 1300°C (Fig. 9) the promoters were equivalent within scatter limits, especially at long anneal times. Identical time dependence and growth rates for the S^{2-} specimen and the undoped Fisher MgO could mean two things; either sulphur has no appreciable effect on the microstructure and hence, no effect on the boundary migration rate, or insufficient S^{2-} was retained in the material after hot-pressing to give observable effects. Table II indicates that S has been lost during hot-pressing.

The above observations were generally true even at 1500°C for the dense compacts. S^{2-} behaved differently from OH^-, F^- and Cl^-, and again the data for S^{2-} correspond closely with that for Fisher MgO both in growth rate and the presence of a grain size plateau. An obvious effect of F^- was the absence of such a plateau while OH^- showed limiting growth only at large grain size and very short time. The effect of these anions on grain growth is in the order $OH^- > F^- > Cl^- > S^{2-} \approx$ Undoped Fisher MgO.

Anion impurities were shown[9] to be deleterious to densification of MgO by hot-pressing in the order $S^{2-} > Cl^- > F^- > OH^-$. Hence, the effect of these anions on the grain growth can be related directly to their effect on densification.

Lastly, OH^-, F^- and Cl^- gave vastly different grain sizes for different porosity levels while S^{2-} behaved more like undoped material, where growth rates were insensitive to density differences.

SUMMARY AND CONCLUSION

(1) Porosity is invariably present in hot-pressed MgO doped with anion additives. The anions influence grain growth in MgO by controlling the pore mobility and the densification rates. For undoped materials, it was found that porosity as low as 1/2% controlled grain growth in very pure material. On the other hand, cation impurities in the less pure Fisher MgO gave grain sizes that were less sensitive to varying porosity.

(2) Anion additives are important when the porosity in MgO is sufficient low (< 1%); for porous material (>1%) boundary migration is predominantly pore-controlled.

(3) OH^-, F^- and Cl^- gave higher grain growth rates than undoped Fisher MgO, S^{2-} showed no effect. The effect on growth rates is $OH^- > F^- > Cl^- > S^{2-}$, the order of effectiveness can be related to position in the microstructure.

(4) Grain sizes in undoped and the S^{2-} doped MgO were not as sensitive to differences in specimen density as was MgO with OH^-, F^- and Cl^-.

(5) The growth kinetics are controlled by the overall evolution of the microstructure and can be expected to vary during annealing depending on changes in the pores and impurity distribution. Also, the observed kinetics may

not always be representative of a particular anion but can
vary with specimens depending on the pore and impurity
distribution. Hence, any interpretation of the rate controlling mechanism from observed kinetics must be done
in conjunction with the microstructural observations.

REFERENCES

1. M. H. Leipold, J. Amer. Ceram. Soc., 49 (9) 498-502
    (1966).
2. V. S. Stubican and D. Viechnicki, J. Appl. Phys., 37
    (7) 2751 (1966).
3. (a) G. C. Nicholson, J. Am. Ceram. Soc. 48 (10) 525-
        528 (1965).
    (b) G. C. Nicholson, ibid., 49 (1) 47-49 (1966).
    (c) I. Amato, R. L. Colombo, A. Petruccioli Balzari,
        J. Nucl. Mater. 19 (3) 252-260 (1966).
    (d) R. S. Gordon, D. D. Marchant, and G. W. Hollenberg,
        J. Am. Ceram. Soc., 53 (7) 399-406 (1970).
4. R. Rice, in Ultra Fine Grain Ceramics, J. J. Burke,
    ed., Syracuse Univ. Press, Syracuse, N. Y., 203
    (1970).
5. T. H. Nielson and M. H. Leipold, J. Am. Ceram. Soc.,
    49 (11) 626 (1967).
6. F. Freund, ibid., 50 (9) 493 (1967).
7. A. J. Socha and M. H. Leipold, J. Am. Ceram. Soc.,
    48 (9) 463 (1965).
8. (a) P. J. Anderson and P. L. Morgan, Trans. Farad.
        Soc., 60 (5) 930-937 (1964).
    (b) J. White, in High Temperature Oxides-Part I,
        A. M. Alper, ed., Academic Press, New York,
        p. 80 (1970).
9. M. H. Leipold and C. M. Kapadia, J. Am. Ceram. Soc.,
    56 (4) 200-203 (1973).
10. David Turnbull, Trans. A. I. M. E. 191 (8) 661-665 (1951).
11. (a) A. U. Daniels, Jr., R. C. Lowrie, Jr., R. L. Gibby
        and I. B. Cutler, ibid., 45 (6) 282-85 (1962).
    (b) T. K. Gupta, J. Mater. Sci., 6 (1) 25-32 (1971).
12. F. A. Nichols, J. Am. Ceram. Soc., 51 (8) 468-469
    (1968).

13. A. Mocellin and W. D. Kingery, J. Am. Ceram. Soc.,
    56 (6) 309-314 (1973).
14. R. J. Brook, J. Amer. Ceram. Soc., 52 (6) 339-340
    (1969).
15. P. G. Shewmon, Trans. A. I. M. E., 230 (5) 1134-1137
    (1964).
16. C. M. Kapadia, Ph. D. Thesis., Univ. of Kentucky.
17. P. G. Shewmon, Transformation in Metals, Chapt. 3,
    McGraw-Hill Book Co., New York, 1969.
18. N. A. Haroun and D. W. Budworth, J. Mater. Sci., 3
    (3) 326-328 (1968).
19. M. H. Leipold and T. H. Nielsen, Am. Ceram. Soc.
    Bull. 45 (3) 281-285 (1966).
20. M. H. Leipold and C. M. Kapadia, Tech. Rept. UKY
    25-70 - Met-12, June 1970.
21. J. E. Hilliard, "Grain Size Estimation," General
    Electric Research Rept. No. 61-RI-2898M, 1961.
22. G. D. Miles, R. A. J. Sambell, J. Rutherford and G. W.
    Stephenson, Trans. Brit. Ceram. Soc., 66 (7) 319-335
    (1967).
23. Edward Carnall, Jr., Mater. Res. Bull., 2 (12) 1075-
    1086 (1967).
24. Rinoud Hanna, Amer. Ceram. Soc. Bull., 49 (5) 548-
    549 (1970).
25. (a) R. Rice, J. Amer. Ceram. Soc. 54 (4) 205-207
        (1971).
    (b) R. K. Stringer, C. E. Warble, and L. S. Williams,
        in Kinetics of Reactions in Ionic Systems, T. J.
        Gray and V. D. Fréchette, eds., Plenum Press,
        New York 53-95 (1969).
26. F. A. Nichols, J. Appl. Phys. 37 (13) 4599-4602 (1966).
27. H. Schafer, Chemical Transport Reactions, Chapter
    3, Academic Press, New York 1964.
28. Paul F. Eastman and Ivan B. Cutler, J. Amer. Ceram.
    Soc., 49 (10) 526-530 (1966).
29. P. J. Wyllie, J. Amer. Ceram. Soc., 50 (1) 43-46 (1967).
30. A. J. Shaler, in Sintering and Related Phenomena, G. C.
    Kuczynski, N. A. Hooton and C. F. Gibbon, eds., Gordon
    and Breach, New York, 807 (1965).
31. T. P. Hoar and J. M. Butler, Jour. Inst. of Met., 78,
    351 (1950).

32. G. Bockstiegel, Powder Metallurgy, 10, 1971, 1962.

DISCUSSION

K. D. Reeve (Australian AEC): Have you any evidence that OH⁻ is incorporated in the MgO lattice?

Author: We have no direct evidence. However on the basis of the similar ionic radii, one does expect some OH^- to be substituted for O^{2-} in the MgO lattice.

G. H. Frischat (Tech. Univ. Clausthal): How did you check the loss of the constituents F, Cl, S etc. you added? Did you find any loss during sintering?

Author: The loss of anions during fabrication was minimized by selection of the hot-pressing technique where the closed environment of the die assembly helped to contain the additives. Also, sufficiently high as-pressed densities were obtained to prevent continuous exchange from the interior of the specimens to the environment during subsequent sintering. Standard analytical techniques on the bulk MgO specimens did show loss of these anions during sintering. Further, the levels of F and OH retained in the material were generally higher than those of Cl and S.

PROPERTY CHANGES AT INTERFACES AND FREE SURFACES

Heinrich Engelke

Institut für Werkstoffwissenschaften III
(Glas and Keramik), University of Erlangen-
Nürnberg, Erlangen, Germany

If two glasses of different chemical composition are
placed in contact and heated to sufficiently high temperatures, diffusion occurs at the contact interface and they
fuse together. One result of this diffusion is that the
stepwise change in refractive index across the interface
is replaced by a smoother transition.

The present work was a study of this diffusion to
determine not only the related diffusion constants but
also the alteration process of the refractive index profile. The latter is of interest in similar physical situations such as fiber optics. Furthermore, it was hoped
that diffusion and refractive index measurements would
yield complementary results and help to understand the
formation mechanism of the refractive index profile.

Concentration profiles for determination of the diffusion coefficients were measured using an electron probe
microanalyser. Fractional loss of sample components
under the electron beam was found to occur; this effect
and ways of reducing it will be discussed.

PREPARATION OF THE SPECIMENS

The chemical composition and physical data of the
two kinds of glasses used are shown in Table 1. Glass
B was obtained by substituting about 4 wt. % K_2O for CaO.
Extreme care was taken to obtain initially homogeneous
refractive indices for both glasses. By a special sequence
of melting, cooling and granulation followed by stirring and
multiple pouring between containers, an index variation of
less than ± 0.00003 was achieved.

Specimens were prepared by placing samples of glass
A and B on top of one another, the contact surfaces having
been polished. They were heat treated for various times
and temperatures from 70.55 days at 580° C to 14.5 hours
at 920° C. After careful cooling, small plates were cut
from the specimens by sawing perpendicular to the contact
interface. The plates were then used for diffusion profile
and refractive index measurements.

Table 1. Chemical Composition and Physical Properties of
Glasses Used

	Glass A		Glass B	
SiO_2	73.95	wt. %	73.97	wt. %
K_2O	15.42	%	19.49	%
CaO	9.94	%	5.92	%
impurities	0.30	%	0.27	%
loss on ignition (550 - 600° C)	0.25	%	not determined	
lin. therm. exp. coeff.	$83 \times 10^{-7}/°C$		$89 \times 10^{-7}/°C$	
n_D	1.51026		1.50359	
transformation point	639° C		582° C	

ELECTRON PROBE MICROANALYSIS

Experimental Procedure

Concentration profiles were measured with the electron probe microanalyser. Fig. 1 shows a recorder trace of one such measurement. The recording was produced by registering the characteristic radiation of potassium and calcium, while the sample was moved in the diffusion direction through the stationary electron beam at a rate of 20 μm/min. There was concluded to be a systematic error in this measurement, because the concentrations of potassium and calcium appear to decrease in the same direction. This cannot be the case, because in the original glasses CaO was exchanged for K_2O and this condition must persist in the sintered specimens when the distance from the diffusion zone is great enough. A comparison with the known compositions of the starting glasses shows that the concentration of calcium was correctly measured, but that the potassium concentration was systematically in error.

The source of this error is the loss of potassium in the surface layer caused by the action of the electron beam. This effect can be seen in Fig. 2. Here a homogeneous glass was investigated. The specimen was moved first with the normal velocity of 20 μm/min. and then stopped at the point indicated by the arrow. It can be seen clearly how quickly the potassium was removed from the analysed zone. This effect is so strong in the high-potassium glass that less potassium was detected in this glass than was present in the low-potassium glass. The standard methods used

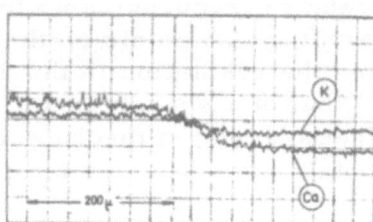

Fig. 1. Measurement of concentrations of K and Ca across a diffusion zone with a systematic error in K concentration.

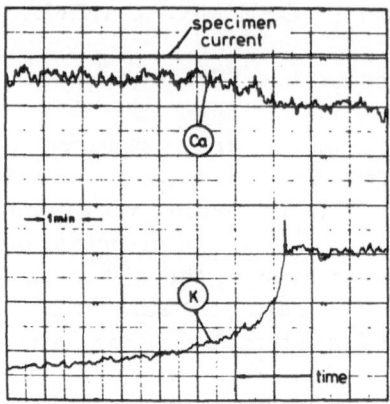

Fig. 2. Loss of potassium due to the electron beam in the
microprobe.

to reduce effects of the electron beam on the sample was
ineffective. For example, a reduced accelerating voltage
or a lower beam current would require unacceptably long
measuring times; a larger beam diameter would sacrifice
resolution.

The problem was solved by vapor plating the specimens
with a 0.4 μm thick aluminum layer instead of the usual
carbon layer and by using a special measurement technique.
A schematic diagram of this arrangement is shown in Fig.
3. Instead of remaining stationary, the beam rapidly scans
a line 320 μm long, oriented parallel to the interface between
the glasses, thus distributing the time-average intensity of
the beam over an area large enough to produce negligible
surface damage. Scanning frequencies up to 1 Hz proved too
slow; 50 Hz was adequate. All points on the line had the
same composition and the resolution in the diffusion direction corresponded to that of a sharply constrained beam.
The specimen was then translated mechanically in the diffusion direction to obtain the concentration profile.

For better counting statistics a multi-channel analyser
was used for counting. This was coupled with the movement

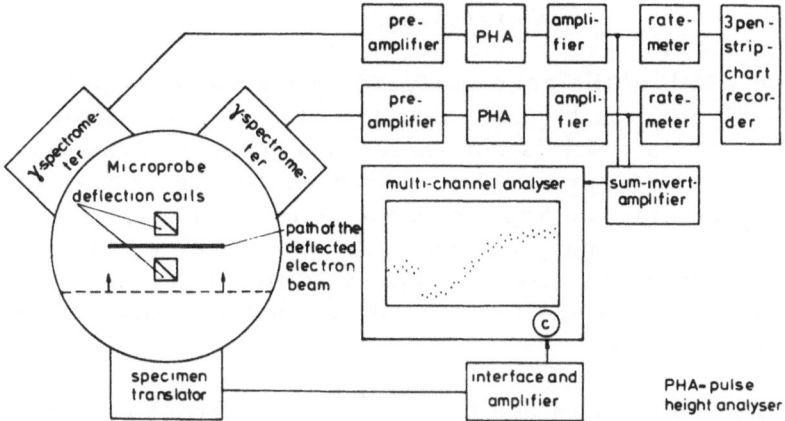

Fig. 3.   Schematic diagram of the measurement arrangement.

of the specimen and divided the distance into equal segments
and associated a counting channel with each.   In this way the
pulses of characteristic radiation of several measurements
over the same path could be summed.   Typical results are
shown in Fig. 4.   The solution of the diffusion equation with
the proper starting and boundary conditions, an error function, was fitted by computer to the data points.   From the
slope of this function at its middle one can obtain $\sqrt{Dt}$, and
thus the diffusion coefficient D.

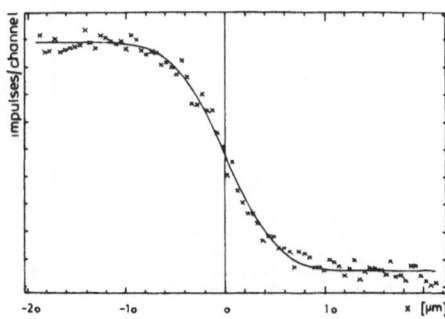

Fig. 4.   Data points (crosses) and fitted curve (solid line)
for a concentration profile in a diffusion specimen (690° C,
6. 69 days).

Corrections

The width of the zone analysed by the electron probe microanalyser was broader than the electron bream diameter[1]. If this width is not small with respect to the distance over which appreciable concentration changes occur, a correction must be applied to the concentration profiles. The width of this analysed zone can be determined by measuring a sharp concentration step; the differentiation of the measured result is the function of the analysed zone which is sought.

In Fig. 5 one can see a measurement across such a concentration step. The results may be fitted quite well by an error function, the differentiation of which yields a Gaussian curve for the profile of the analysed zone, having a half-width of about 4 μm. A simple formula for the correction of the diffusion coefficients was found[2].

$$D_{actual} \cdot t = D_{measured} \cdot t - \frac{1}{4B^2}$$

where B is the width parameter of the Gaussian curve. With this correction, diffusion coefficients a factor of ten smaller can be determined with the same diffusion time.

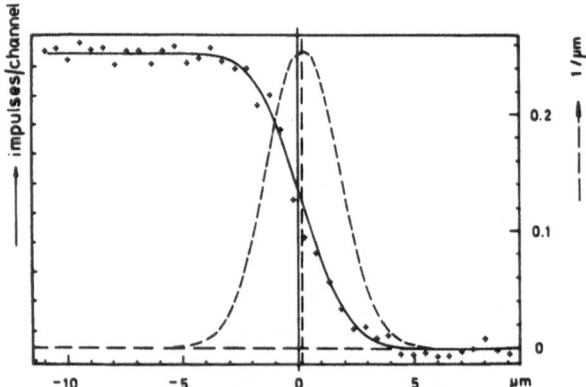

Fig. 5. Measurement for determinating the profile of the analysed zone (dashed line).

Results and Discussion

The corrected diffusion coefficients are presented in Fig. 6. The data obtained above the transformation temperatures of the glasses are fitted by a straight line. The activation energy is 51.6 kcal/mole for potassium and 79.5 kcal/mole for calcium[3].

It is noteworthy that at 650° C the diffusion coefficient of potassium is about 100 greater than that of calcium, while at about 1000° C the coefficients will be nearly equal. Therefore at the lower temperatures, Ca is incapable of producing electrical neutrality of the diffusing potassium ions. The necessary implication is that O^{2-} must diffuse in the K^+ diffusion direction and/or Si^{4+} must counterdiffuse.

REFRACTIVE INDEX MEASUREMENTS

Changes in refractive index occured within ten to about one hundred μm, which means that the interference micro-

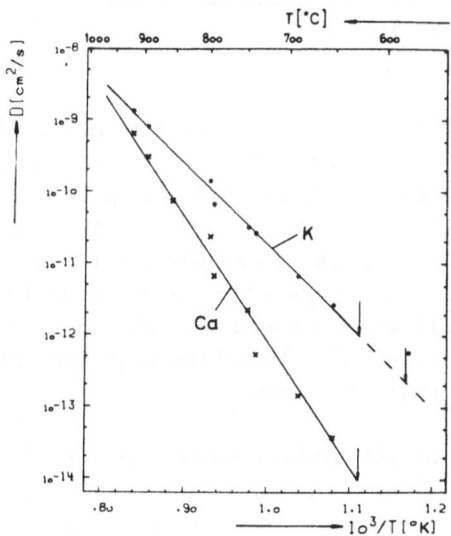

Fig. 6. Diffusion coefficients for potassium and calcium.

scope can be well applied to the measurements. Specimens
like those used for diffusion couples (in which the refractive
index changes in one direction only) can be oriented in the
interference microscope so that parallel interference lines
result, whose deflections are linearly related to changes in
refractive index.

However, it is not a simple matter to obtain good interferograms from specimens with a large refractive index
gradient, as can be seen from Fig. 7. Away from the diffusion zone there are parallel interference lines, but the
deflections in the region of interest cannot be resolved.

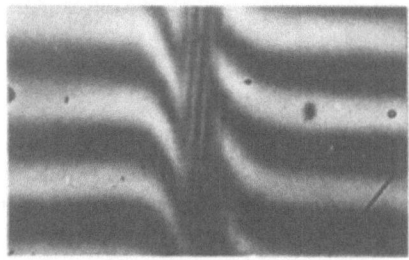

Fig. 7. Interferogram of a diffusion sample with a large
refractive index gradient.

A model calculation was made, assuming that the refractive index change had the form of an error function,
varying from 1.50 to 1.51 (Fig. 8b). Fig. 8a shows the
calculated paths of a few representative light rays as they
pass through the thickness (Y) in the region around the
interface (X = O). The rays enter normal to the surface
(Y = O) but are deflected toward the higher-index side of
the gradient. At about Y = 30 a "focal point" is formed,
beyond which the rays diverge.

Clearly one cannot make optical measurements directly on samples which incorporate such nonuniform light
paths. Good measurements require reduction of the sample
thickness and a correction for the ray deviations. A simple
formula was derived for the latter.

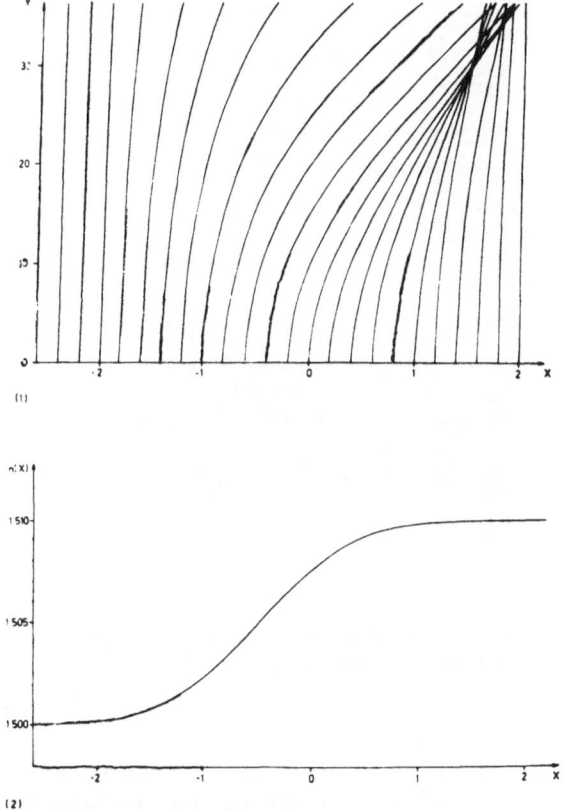

Fig. 8. (a) Beam deviations in specimens with the refractive index gradients shown in (b).

Fig. 9. shows interferograms of thin diffusion specimens oriented, so that an upward deflection corresponds
to an increase in refractive index. The specimen in Fig.
9a was heat treated at 920° C for 14.5 hours; Fig. 9b shows
the interferogram of a specimen, heat treated for 70.55
days at 580° C. The latter specimen is only 180 μm thick,
consequently more of the interface pattern is resolved except adjacent to the interface, where the interference lines
continue to show interruption. However, using experimental or mathematical techniques one can relate the individual curve portions; two portions of the same curve are
marked in Fig. 9b. If one follows the curve from left to

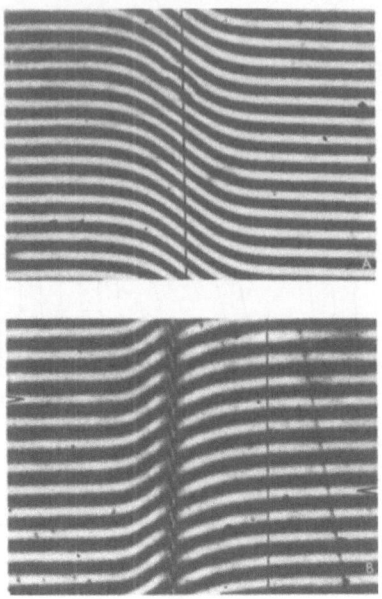

Fig. 9. Interferograms of diffusion specimens, (a) 920°C,
14.5 hours; (b) 580°C, 70.55 days.

right the refractive index at first increases to a value higher
than those present in either of the original glasses, then decreases abruptly to a value which is lower than those of the
original glasses, then rises to the value of the starting glass.

As will be explained later, the form of this refractive
index curve is easily interpreted from the diffusion behaviour
of the glass components. Fig. 10 pictures the calibrated refractive index curves for the interferograms of Fig. 9 and
two other specimens, corrected for ray deviations. In Fig.
10c the dotted line indicates that here the experimental error
of the curve is a little higher than in the solid line regions.

COMPARISONS AND DISCUSSION

For comparisons one has to know the effect of diffusion
coefficients on the refractive index curve.

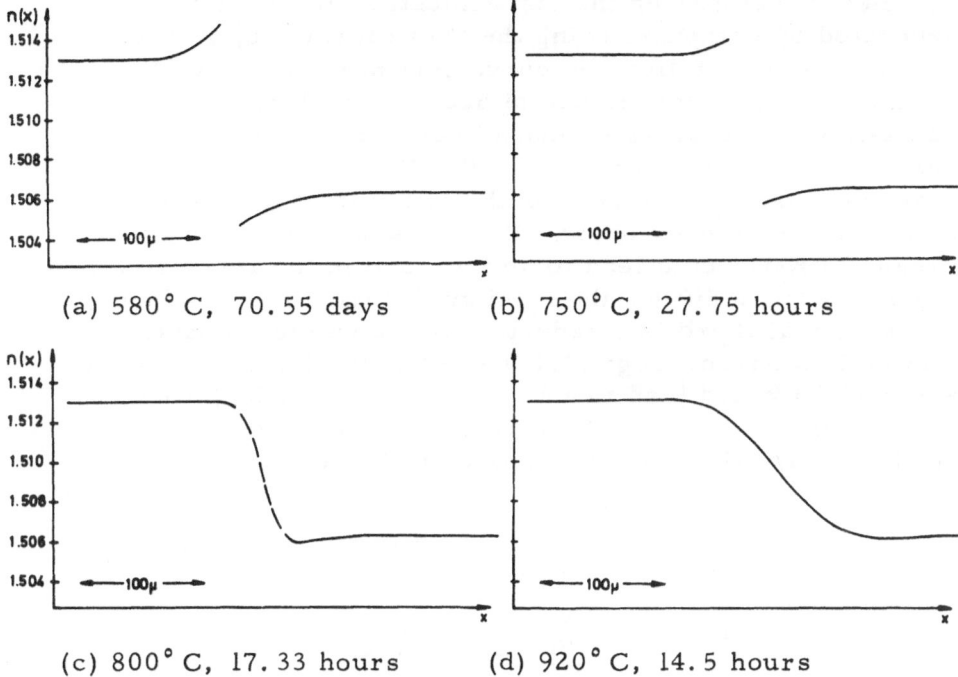

(a) 580° C, 70.55 days   (b) 750° C, 27.75 hours

(c) 800° C, 17.33 hours   (d) 920° C, 14.5 hours

Fig. 10.   Refractive index curves of diffusion specimens.

The refractive index, n, depends on the concentrations, c_i, of the glass components, i, as follows:

$$n = \Sigma f_i \cdot c_i$$

where the factors f_i can be obtained from the literature[4]. The concentrations depend on the distance x from the interface, as n does also:

$$n(x) = \Sigma f_1 \cdot c_1(x) \text{ with } c_1(x) = c_{oi} + \Delta c_i \left(1 + \Phi\left(\frac{x}{2\sqrt{D_i t}}\right)\right)$$

where c_i = concentration of component i at $x = -\infty$;
Δc_i = difference of concentrations between $x = -\infty$ and $x = \infty$; Φ = error function; D_i = diffusion coefficient of i; t = diffusion time.

Before comparing the experimental curves with curves
generated by computer using the above formulas, it would
be instructive to follow the curve generation step by step.
At the diffusion temperature of 580° C, the diffusion of
potassium was extensive, that of calcium minimal. The
relationships are shown in Fig. 11. Because of diffusion,
there is more K_2O present on the left side of the interface
center line than previously. In the glasses used, an increase in K_2O would lead to an increase in the refractive
index. If, in addition, the calcium diffusion is considered,
one sees that there is a reduction in the calcium concentration in a narrow region at the left of the interface. Such
a reduction would lead to a lower refractive index, whereby
the influence of CaO on the refractive index in these glasses
is about 2-1/2 times as great as that of K_2O.

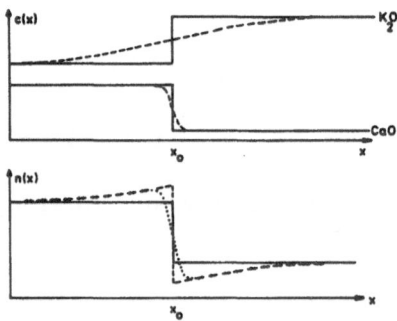

Fig. 11. Relation between diffusion of glass components and
refractive index. Solid lines are values before, dashed lines
(above) after diffusion. Dashed line below gives values calculated for K_2O only, the dotted line includes the effect of
CaO.

The computer-calculated refractive index curve is compared with the directly measured curve in Fig. 12. Although
agreement is good, it is possible to recognize differences;
the direct method shows more details, for instance the asymmetry of the curve, which arises from the concentration-
dependent diffusion coefficient of potassium. In contrast,
the curve calculated from measurements made with the

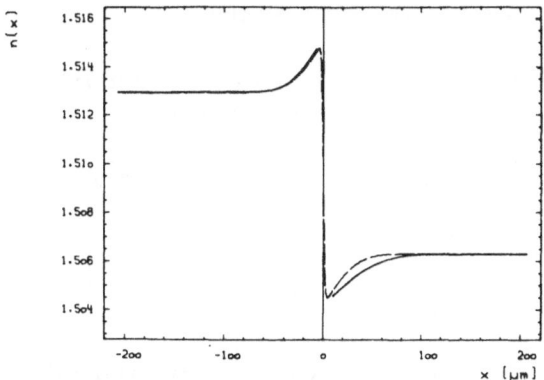

Fig. 12. Comparison between directly measured refractive index curve (solid line) and that calculated from diffusion data (dashed line).

electron probe does not show these details. It does however, provide evidence concerning the shape of the curve at its middle.

Several important points concerning the refractive index curves should be emphasized: Whether or not maxima and minima occur and how large they are, is a function of the diffusion temperature only, because the temperature determines the ratio of the diffusion coefficients to one another. The depth to which diffusion extends into the bulk glass is, on the other hand, a function of the diffusion time; the $\sqrt{t}$ - rule applies.

It is reasonable to assume that other diffusion controlled properties can be treated in the same way as refractive index. The refractive index curves may be used to determine diffusion coefficients. We have in fact done this; the results are in good agreement with those, determined from electron probe measurements[3].

The question of the elementary diffusion steps is still unanswered, that is, whether O^{2-} diffuses together with K^+ or whether Si^{4+} counterdiffuses or whether both processes play a significant role in this system.

424   H. ENGELKE

ACKNOWLEDGEMENTS

I wish to thank Prof. H.J. Oel for supporting this
project with his continual interest and encouragement,
also Drs. G. Tomandl and H. Schaeffer for many helpful
discussions, and Mrs. B. Schaeffer for performing the
wet-chemical analyses.

REFERENCES

1. J.A. Belk in: "Electron Microscopy and Microanalysis
   of Metals." Belk and Davis, Eds., Elsevier Publ. Co.,
   Amsterdam, London, New York (1968) 185.
2. H. Engelke and G. Tomandl: Eine Methode zur Ver-
   hinderung unzulässiger Probenerwärmung, in Microchim.
   Act., Wien, Suppl. V, 1973.
3. H. Engelke, Thesis, Univ. Erlangen-Nurnberg, Germany (1972).
4. P.S. Irlam et R. Rimmer: Investigation of the Mathematical Relationship between Glass Composition and
   Refractive Index. "VIIe Congres International du Verre,
   Bruxelles, 29 Juin - 3 Julliet 1965." Charleroi (1966)
   I/1, 3.1-3.20.

DISCUSSION

G.H. Frischat (Tech. Univ. Clausthal): Did you find any
evidence of phase separation in your glasses?
Author: There was no evidence from our measurements
but we did not use electron microscopy to investigate it.
W.B. Crandall (IITRI): Was there evidence of devitnifi-
cation? At these times and temperatures?
Author: Only at higher temperatures and then only if the
specimens were heat treated much longer than the times
used.
W.C. LaCourse (Alfred): How did you form the initial
interface at the low heat treatment temperatures; did you
have any problems with bubbles?
Author: The specimens for use at lower temperatures

were prepared by fusing the glass pieces together at higher
temperatures using times sufficiently short so that the diffusion of potassium was smaller than 0.2 μm. There were
no problems with bubbles at the interface; the surfaces at
the interface were plane polished to optical quality and
carefully cleaned with organic solvents before placing the
two pieces of glass together.

J. Wong (GE): From the compositions of your glasses it
seems to me that the diffusion of K^+ and Ca^{2+} were measured
in a countercurrent condition, i.e., K^+ and Ca^{2+} diffusing in
opposite directions. Did you observe any interference effect
due to this mode of diffusion?

Author: No diffusion of calcium in the concentration gradient
of potassium or vice versa was observed. However, a variation in the silicon content was observed qualitatively so that
the concentration of SiO_2, K_2O and CaO at each position
added to unity after completion of the diffusion process.

J.A. Pask (UC Berkeley): Do the concentration profiles of
K, Ca and Si obtained with the electron microprobe provide
information on the nature of the diffusion mechanisms at the
lower and higher temperatures?

Author: The variation in silicon concentration would seem,
at the first glance, to indicate that there was a counter-
diffusion of a Si^{4+} for $4K^+$. However this same variation in
silicon content would occur, if one assumed that an O^{2-} diffused with each $2K^+$ and that the oxygen occupied the space
it would normally require. The net effect of this process
would be a reduction in silicon concentration. In order to
know whether oxygen diffuses with potassium or silicon
counterdiffuses, one must use isosopic tracers.

MICROSTRUCTURE ANALYSIS FROM FRACTURE SURFACES

P. H. Crayton

New York State College of Ceramics at
Alfred University, Alfred, New York

Densification by hot pressing has been the subject of kinetics-oriented research for two decades[1]. Two mechanisms and several models have been proposed, resulting in a growing realization that several mechanisms probably operate simultaneously[2,3]. It was of interest in this study to obtain quantitative data for the proposed, previously undocumented, mechanisms of particle fracture, particle reshaping and grain growth. While grain size measurement in dense solids has been extensively treated[4], the techniques for grain shape are cumbersome at best[5] and no consideration has been given to low-density compacts. The method for simultaneous size and shape measurement developed here was based on the ellipse graticule of Fischmeister[6]. It was the objective of this work to develop this technique to provide an observational technique applicable to study of the early stages in hot pressing. The procedure was demonstrated for one sequence of experiments with alumina.

EXPERIMENTAL

A small portion of sample was stirred in a drop of oil on a microscope slide. A second slide was placed over it and rubbed back and forth to promote dispersion and deagglomeration. This mixture was thinned by adding more oil and reworking until a dispersion of proper concentration was obtained. The slides were examined and photographed using transmitted light at 200X. A similar procedure was employed to produce films for photography at 1000X on the electron microscope. Prints at various enlargements provided particle images which could be matched to graticule patterns.

A series of four figures of equal area were drawn as ellipses of axis ratios 1:1 (circle), 1:2, 1:3 and 1:4. This set was photographically reproduced on clear film at nine magnifications whose areas formed a $\sqrt{2}$ series. Thus a graticule was produced containing four shape classes and nine size classes.

Hot pressing of one-inch diameter specimens was done in graphite dies using induction heating. Pressure was applied and measured hydraulically while temperature was measured by a Pt-Pt, 10 Rh thermocouple embedded in the die. Samples were pressed at 1500°C under 352 kg/cm^2 pressure for 3 to 22 minutes to provide densities of 71 and 85%.

A 50% density specimen required a special technique owing to the extreme mechanical weakness of specimens pressed to this density by normal procedure. A powder mixture of alumina and 33 volume percent of soft glass was intimately mixed by hand. Whereas the alumina powder used (Alcoa A14) does not begin to densify below 1160°C, this mixture began to yield at 550°C, densifying up to a temperature of 840°C where the process stopped. The procedure afforded the alumina no opportunity to change character but allowed the glass to soften, providing 50% alumina density in a mechanically strong matrix. The glass was later leached from the sample by aqueous hydrofluoric acid prior to gold plating for SEM observation.

RESULTS AND DISCUSSION

Powders

The measurement of size distribution using a graticule of the Fischmeister type provides a simultaneous record of shape distribution. For the alumina powder used in this study the most frequently met particle had a projected area of 6.5 square microns. Examining the detail of this size class revealed that about half of these particles were approximated by an ellipse of axial ratio two. Other size classes showed different distributions of shape, and a trend of higher frequencies for equidimensional particles occurred at the extremes of the size classes. The summary of shape data showed that only a quarter of the particle projections were circular while nearly half were elliptical with an axis ratio of two. Volume frequencies showed similar trends in both size and shape distribution.

Calibration of the Method for Solids

The objective in developing the method was to study changes during densification by hot pressing, i. e., in solids of low density (between 50 and 90%). Direct correlation of the powder method and its application to solids thus depended on examining solid samples in which original powder grains had no opportunity to change during fabrication.

Examination of the leached fracture surfaces proved to be a problem. The surfaces were both too deep and too irregular for the optical microscope to focus an area large enough for practical observation. Replication techniques provided marginal success although the films were often stronger than the solid. Vapor-deposited gold films on the surfaces fitted them admirably for SEM viewing. These revealed (Fig. 1) a highly porous solid comprised of a jumble of particles with surface characteristics of the original powder grains.

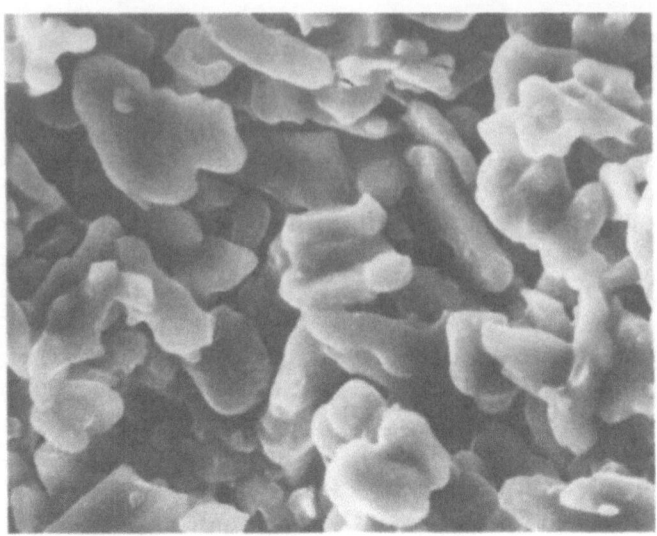

Fig. 1.   Particles in a Low Density Compact.   SEM; 1800X.

        Simultaneous size and shape measurements on these
surfaces provided the frequencies reported in Tables I and
II.   Since the raw data gave low frequencies in the smaller
size classes, some correction procedure similar to that
used for polished surfaces of dense samples must be applied.   The Correction 1 data for the number frequency
was obtained by normalizing the set of number frequencies
generated when observed frequencies were divided by the
mean sphere diameter for the respective size class
($F' = F/d$).   While the lower size class frequencies were
enhanced, the technique appeared to overcorrect.   Consideration was then given to the fact that the low density
of the piece provided a lower order of hiding of one
particle by another than is the case for high density pieces
and that the observed surface is not a flat cut but a layer
of particles and voids.   The data reported as Correction
2 are the results normalized by dividing the observed
number frequency by the respective class mean-sphere
diameter raised to a power equal to the relative density
($F' = F/d^{0.5}$).   The differences between the size distribution frequencies and this correction technique were at

Table I. Size Distribution Corrections for Low Density Solids.

Class Diam. (μ)	Powder	Frequency, F		
			Matrix	
		Raw	Correction 1	Correction 2
0.5	--	0.007	0.027	0.014
0.7	0.058	0.027	0.075	0.046
1.0	0.140	0.083	0.159	0.119
1.4	0.189	0.186	0.255	0.226
2.0	0.228	0.252	0.242	0.256
2.8	0.201	0.211	0.144	0.181
4.0	0.136	0.153	0.073	0.110
5.6	0.041	0.061	0.021	0.037
8.0	0.005	0.015	0.004	0.008
11.3	--	0.005	0.001	0.002
16.0	--	0.001	0.0001	0.0004

Correction 1: $F_i'' = \dfrac{F_i}{d \Sigma F'}$

Correction 2: $F_i'' = \dfrac{F_i}{d^D \Sigma F'}$

F: frequency, original
d: diameter
D: relative density of solid
F': corrected frequency
F'': normalized corrected frequency

the limits of difference calculated for normal observation error.

The distribution of shape class frequencies between the powder and the 50% density solid showed wide variation as did the elongation factor. The factors involved in disguising true shapes were considered to be particle shape (elongation) and density of the observed layer. A correction factor consisting of the ellipse axial ratio, R, raised to a power equal to the relative density, i.e., $S' = SR^D$, provided excellent agreement as seen in Table II. The combined results established the procedures for evaluating size and shape characteristics for particles in low relative-density solids.

Table II. Shape Distribution Corrections for Low Density
Compacts.

Axial Ratio	Powder	Compact			
		Raw	Corrected		
1:1	0.274	0.383	0.273	±	0.018
1:2	0.483	0.433	0.465		0.020
1:3	0.193	0.140	0.192		0.011
1:4	0.048	0.043	0.070		0.006
n elongation factor	2.01	1.84	2.06		0.08

Correction term: $F'' = \dfrac{FR^D}{\Sigma F'}$, where R = figure axial
ratio; D = relative density of solid; F = raw frequency;
F' = corrected frequency; and F'' = normalized corrected
frequency.

Microstructural Evidence During Early Densification

The process of densification by hot pressing has been
described as occurring in three stages, each covering a
range of relative densities: Stage I, 50-75%; II, 75-90%;
III, 90-100%. Mechanisms have been discussed in the
literature without much concrete evidence relative to the
first two stages.

The size data, presented in Table III, showed that
little or no size profile change can be detected even up to
85% densification. This rules out particle fracture as well
as grain growth as potential mechanisms of densification
of alpha alumina for the first two stages under these conditions of temperature, pressure and initial grain size.
Since the pressure application was uniaxial, surfaces
both perpendicular and parallel to the pressure axis were
viewed. Observed size differences between orientations

Table III. Corrected Size Frequency Distributions During Hot Pressing Stages I and II.

Diam. μ	Frequency				
	Powder	Matrix	D = 0.71	D = 0.85	
0.5	--	0.015	0.005	0.027	± 0.002
0.7	0.058	0.047	0.094	0.086	0.005
1.0	0.140	0.121	0.147	0.164	0.009
1.4	0.189	0.230	0.231	0.236	0.012
2.0	0.228	0.258	0.223	0.256	0.014
2.8	0.201	0.183	0.179	0.148	0.012
4.0	0.136	0.111	0.087	0.061	0.012
5.6	0.041	0.038	0.024	0.019	0.007
8.0	0.005	0.007	0.008	0.003	0.003
11.3	--	0.003	0.002	--	0.002

were found in all three solids and to about the same extent. The initial compaction step evidently had so shifted the particles that they appeared larger in the perpendicular plane. This suggests that the particles were more plate-like than ellipsoidal. Pressure compaction had forced the plates to stack with their major area perpendicular to the direction of applied pressure (and bulk movement).

The distribution of shapes presented in Table IV, shows a distinct shift. The elongation factor (mean figure axial ratio) increased as densification took place, particularly between the 71 and 85% pieces. Close examination revealed a pattern of shape change pictured in Fig. 2. The distributions of shape frequencies for the powder and compact (D = 0.50) were quite the same. As densification proceeded, an initial shift away from equidimensional particles occurred leading at first to an increase largely of "shape 2" particles and ultimately to "shape 3" particles. When the two orientations of the 85% density samples were examined, a preponderance of elongated particles was observed in the parallel plane as evidenced by the elongation

Table IV. Corrected Shape Distributions During Hot Pressing Stages I and II

Axial Ratio	Powder	Matrix	D = 0.71		
1:1	0.274	0.274	0.215	0.215 ±	0.018
1:2	0.483	0.466	0.528	0.485	0.020
1:3	0.193	0.193	0.187	0.227	0.011
1:4	0.048	0.070	0.070	0.073	0.006
n	2.01	2.07	2.11	2.16	0.08

factor of 2.25 for the parallel and 2.06 for the perpendicular sections. The evidence therefore showed that elongation of particles took place particularly in the second stage and along the pressure axis.

The use of this technique of examining the micro-structure of low density samples provided yet another avenue for evidence gathering. In the "fracture layer" both surfaces of former intergranular contact and grain interfaces with pores were readily observable. These features for the 71 and 85% density stages showed readily

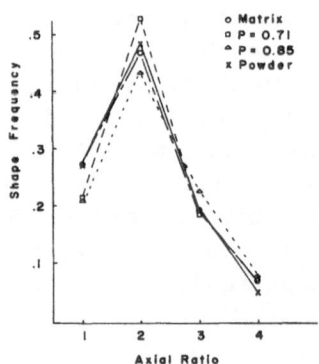

Fig. 2. Shape Frequency Profiles During Early Densification.

observable changes (Fig. 3). Between 50 and 71% density
the interfaces of grain contact had been widened suggesting plastic flow or pressure-enhanced diffusional creep.
In the Stage II sample the grain surfaces had developed
evidence of ridges which suggested basal slip as a mechanism relieving the pressure stress in the grains. Closer
study will be given to this line of evidence in future work.

SUMMARY AND CONCLUSIONS

The examination of the fracture-exposed surface
layer of low density solids by SEM has been used as a
basis for application of the simultaneous size-shape profile measurement technique using the Fischmeister graticule. A correction factor from these surface measurements for size was determined to be $F' = F/d^D$, where F
was particle frequency, d mean sphere diameter and D
relative density of the solid examined. The shape frequency

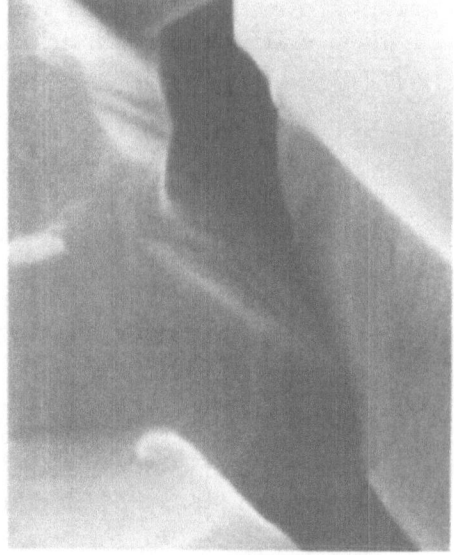

Fig. 3. Microstructural Details in SEM: (a) At the End
of Stage I, 1300X; and (b) At the End of Stage II, 26,000X.

correction was determined to be $S' = SR^D$, where S was
shape frequency, R the axial ratio of the ellipse figure
and D the relative density of the solid.

This technique was applied to the problem of microstructural changes during the first two stages of hot
pressing of a "five-micron" alumina at 1500° under 5000
psi. It was demonstrated that for these conditions neither
particle fracture nor grain growth were mechanisms of
densification. Elongation of grains along the pressure
axis was observed, particularly in the second stage.

ACKNOWLEDGMENTS

The author is deeply indebted to Dr. H. E. Exner for
suggesting the technique, for assisting in the definition of
the stages and for extensive council and encouragement;
also to the Max Planck Gesellschaft for providing working
space and financial aid at the Institut für Metallforschung
in Stuttgart-Busnau during the powder characterization
stage. Further thanks go to the College Center of the
Finger Lakes for financial aid during the adaptation of
the technique to low bulk density solids.

REFERENCES

1. L. Ramquist, Powder Met. 9, 1-25 (1966).
2. E. J. Felton, J. Am. Ceram. Soc., 44, 381 (1961).
3. R. M. Spriggs and Z. Atteraas, Proc. Third Berkeley
    International Materials Conf., June 13-16, John Wiley,
    New York, 1960.
4. H. E. Exner, Internat. Met. Rev., 17, 25-42 (1972).
5. H. H. Hausner, Planseeber., 14, 75-84 (1966).
6. H. F. Fischmeister, C. A. Blände and S. Palmquist,
    Powder Met., 82-119 (1961).

DISCUSSION

R. W. Rice (Naval Res. Lab.): The flat surfaces and terraces you observe are very suggestive of surface diffusion, forming steps whose net surface energy is a minimum. Bill Scott of the University of Washington has done a very nice study of such terracing in pores in sapphire bicrystal boundaries. Also, some investigators (e.g., Coble I believe) have observed faceting of sapphire spheres. Have you investigated surface diffusion formed terracing as a mechanism responsible for the observed steps and flats? Have you attempted to check texture in your samples to confirm possible particle rotation effects?

Author: Your comment and our interest for further work agree very well. Thank you. We will be checking particle rotation by means of x-ray analysis of orientation.

P. F. Johnson (Univ. Florida): Is it possible that the "lines," or "terraces" shown in the last slide are artifacts of earlier fracture?

Author: I don't believe so. These features appeared for the first time at 85% density and were not present at 71% density.

W. B. Crandall (IITRI): Did you see grain fracture on compaction of the Al_2O_3 at 3000 psi without heating?

Author: We did not prepare samples under pressure without heating. The 50% density samples taken to 850° C under 3000 psi did not provide evidence of fracture.

FRACTURE TOPOGRAPHY OF CERAMICS

Roy W. Rice

Naval Research Laboratory
Washington, D.C.

Fracture topography is a permanent record of the fracture process, providing valuable information on crack propagation and failure mechanisms. This paper describes mostly features observed in flexural failure at room temperature. Previous papers[1-3] are complemented by emphasizing features and techniques for finding and characterizing fracture origins to aid in use of such important, too often neglected, steps in strength studies.

TECHNIQUES OF OBSERVING FRACTURE TOPOGRAPHY

The unaided eye, hand lens and optical, scanning electron (SEM), and transmission electron (TEM) microscopes are typically used for studying fracture topography. Gross features such as crack branching are often best viewed with the unaided eye. Such viewing is rapid, versatile, allows large areas to be examined, and generally prevents details from obscuring the overall pattern. The hand lens (e.g., 10X) extends this capability, without much loss of speed or area of inspection.

The optical microscope remains one of the most useful and cost-effective tools, both in terms of initial cost as well as speed, effectiveness, and versatility of use. Specimens

are often mounted on clay or a simple ball joint (e.g., as
made by potting half a hollow plastic ball coated with vacuum
grease in a partially enclosing socket) so they can be rotated
for maximum reflectivity. Bright field reflected illumination is used, frequently with contrast methods, e.g.,
Nomarski. Objectives below 8X usually intercept insufficient light reflected at various angles from fracture irregularities. Objectives of 20X to 40X are normally the upper
limit that can be used, and these generally must be special
lenses, e.g., for universal, hot stage, or interference
microscopy to give sufficient working distance. While moving
specimens into focus, the concentrated light beam may high-
light features such as incomplete crack branching, so observation with the unaided eye from the front of the microscope
can reduce the amount of microscope viewing.

The SEM is a major advance but it is best used in conjunction with optical microscopy. Optical contrast delineates
fine topographic features (e.g., Fig. 1) such that fracture

 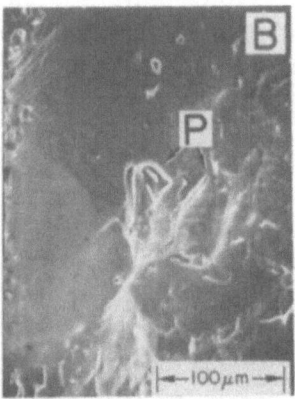

Fig. 1. Fracture origin in commercial hot pressed B_4C.
(A) Optical photo. Wallner lines, fracture steps and tail
show the origin came from an internal boundary between
usually large grains. (B) SEM of same area showing origin
at or near pores (P) at the boundary. Note the SEM depth of
focus, revealing pores, but less clearly delineating other
features.

origins often cannot be found on the SEM unless first located optically. As in most microscopy, it is important to view matching fracture halves; however these should often be mounted so the depth of focus of the SEM can be used to view the tensile and fracture surface simultaneously.

TEM is always of replicas (sometimes also used in optical microscopy). The difficulty of locating specific areas by TEM normally limits it to general characterization, e.g., grain size, pore size, degree of intergranular failure, etc. It delineates features such as structure of fine pores, and especially fracture steps, better than the SEM, but often fails to reveal larger features such as large pores or cracks, due to replica breakage at these points.

GENERAL FRACTURE FEATURES

Intergranular vs. Transgranular Failure

The degree of intergranular and transgranular fracture (Fig. 2) in polycrystalline bodies is important since it may correlate with strength and cause of failure, and because a sufficiently high degree of transgranular failure is usually necessary to identify some features. Cleavage or fracture steps, which form the well-known "river patterns" of transgranular failure can generally be followed back to the fracture origin (Fig. 13C) since they form normal to the crack front[1, 3-5]. Fracture steps are also caused by dislocations and inhomogenities such as pores, and especially failure of the fracture plane to be exactly aligned with a preferred cleavage plane, as is particularly likely on transgranular fracture due to grain-to-grain misorientation. Since overlapping cracks tend to pull together, fracture steps generally disappear, but new sources often more than balance this loss, especially in polycrystalline fracture. The density and height of fracture steps for similar grain sizes and orientations generally vary substantially, apparently being greater in materials which have a strong preference for a limited set of cleavage planes. Thus, MgO and CaO, with

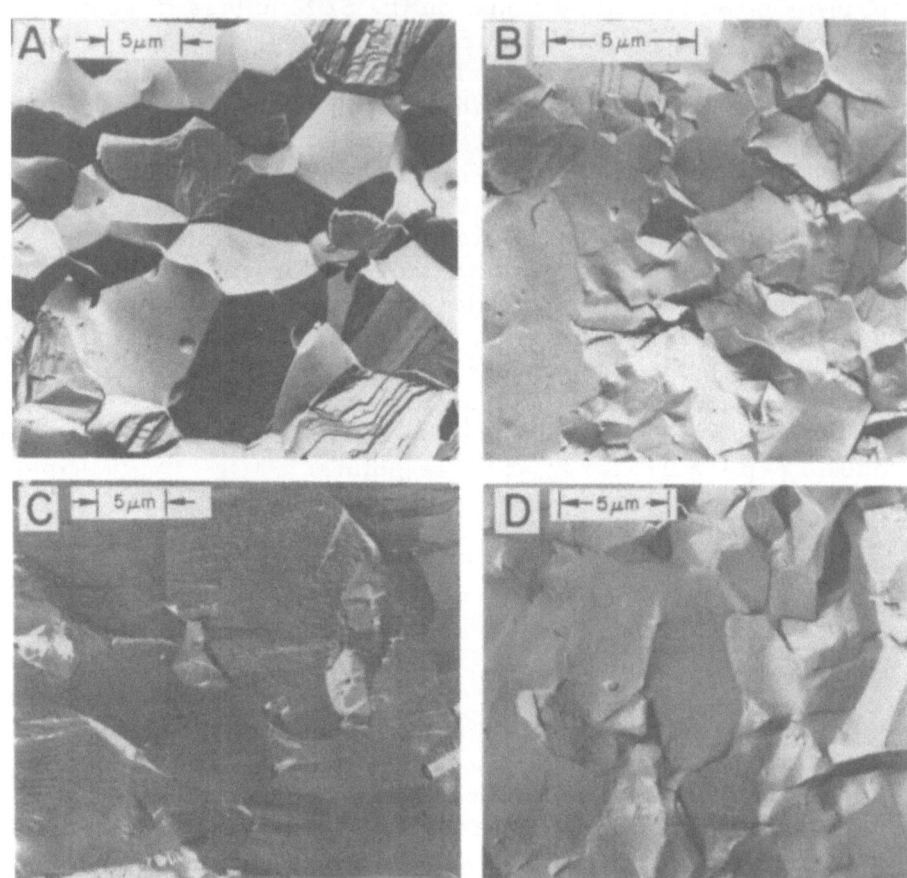

Fig. 2.   Polycrystalline fracture topography in dense bodies.
(A) MgO with mostly intergranular failure and a few small
grain boundary pores.   (B), (C) and (D) predominantly transgranular failure of $MgAl_2O_4$, ZrO_2, and Al_2O_3, respectively.   Note large frequent transgranular fracture steps on
MgO; finer "grainy" steps on ZrO_2 (crack propagating left
to right), and very few steps on $MgAl_2O_4$ and Al_2O_3.

preferred $\{100\}$ cleavage, have a substantial density of
fairly pronounced steps, and limited observations on TaC
indicate it does.   On the other hand, $Mg\,Al_2O_4$, Al_2O_3 and
B_4C have fewer, less pronounced steps, while ZrO_2 bodies
have a high density of fine, "grainy" steps (Fig. 2).

Intergranular failure generally increases with test temperature, and grain boundary impurities and porosity. Since these factors also generally lower strength, increased intergranular failure may indicate one of them as the cause of lower strength. However, caution must be used with each of these, in part due to grain size and anisotropy effects. Although specific data are limited, intergranular failure is usually associated with decreasing grain size, and hence, increasing strength. Thus, correlations of decreasing strength with decreasing intergranular failure must generally be limited to similar grain size bodies, as reported for MgO (6).

Many have felt that increasing intergranular failure with decreasing grain size is an intrinsic effect; however, there is no clear theoretical reason for this. Simple geometry does not explain it, since the total length of a fracture path around or through regular polyhedrons of the same shape is independent of their size. Finer grain sizes often have more anion impurities[6,7], and grain boundary porosity to aid intergranular failure. Observations of large amounts of transgranular failure in dense submicron ZrO_2 also question whether finer grain sizes intrinsically increase intergranular failure[8].

Elastic anisotropy (EA), thermal expansion anisotropy (TEA), and transformation-induced anisotropy (TIA) occur respectively in all noncubic, and polymorphic materials. All result in stress concentrations near the grain boundaries which can favor intergranular fracture. $PbTiO_3$ exhibits increasing spontaneous grain boundary fracture with increasing grain size on cooling below its Curie temperature[9]. Also, transgranular failure in pure Al_2O_3 increases with increasing grain size, in contrast to cubic materials such as MgO. Since internal stresses are independent of grain size, they cannot explain these effects. However, adaptation of Davidge and Green's strain-energy model of cracking around second phases[10] appears to reasonably estimate these effects. Intergranular fracture is assumed to begin when the strain energy in a grain is sufficient to supply the necessary fracture surface energy.

The strain energy varies as the cube of the grain size while
the fracture energy varies as the square, so that a grain
size critical for fracture is implied. Glassy grain boundary
phases, as in many commercial aluminas, may limit
anisotropy stresses and thus intergranular failure and then
aid it as test temperatures approach the glass softening
point. Significant contents of CaO and SiO_2 in MgO grain
boundaries do not appear to decrease transgranular failure
at lower temperatures, but they do so significantly at higher
temperatures, where high-purity MgO and CaO often have
intergranular failure only in the immediate vicinity of the
fracture origin[11, 12]. Texture, i.e., preferred orientation,
can substantially increase transgranular failure, as observed
in hot extruded MgO[11] by aligning cleavage planes and reducing anisotropy. Changes in anisotropy with temperature
may also alter fracture modes. It has also been reported
that the degree of transgranular failure may be higher on the
compression side of a flexure specimen[11].

Fracture Interaction with Twins, Domain, and Slip Bands

Twin boundaries can be preferential paths of fracture
as in the apparent cleavage features of sapphire[13]. Fracture
across twin boundaries may result in formations of fracture
steps similar to grain boundaries if there is appropriate
twin-matrix misorientation. Ferroelectric domains are
often observed to perturb transgranular failure (e.g., Fig.
3), but detailed studies apparently have not been made.
Magnetic domains apparently have little or no effect on
fracture, as expected. Screw dislocations are considered
to be an important source of fracture steps[1]. The author
has also observed fracture steps along edge slip bands that
apparently preceded (often causing fracture) in MgO and
CaO[11, 12, 14].

Crack Interaction with Pores, Inclusions, and Other Cracks

Unless there is a strongly preferred surface of fracture,
fluctuations will result in the halves of a crack, separated
during passage around a pore, completing that passage on

Fig. 3.   Transgranular fracture interaction with ferroelectric domains in PZT.

different planes.   The two halves are then joined by an
additional fracture at ~90° until they rejoin on a single
plane.   This results in a fracture "tail" on the opposite
side of the pore from which fracture approached it (Fig.
4) in ceramics and other materials.   Kerkhof and Sommer[15]

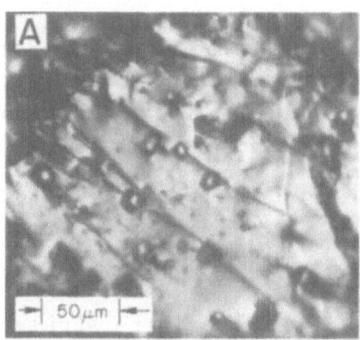 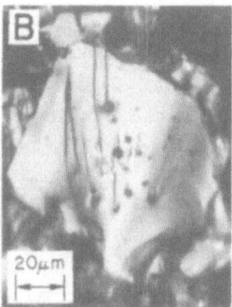

Fig. 4.   Fracture "tails" from pores within transgranularly
fractured grains of (A) $BaTiO_3$ (fracture direction: upper
left to lower right), and (B) $MgAl_2O_4$ (fracture direction:
bottom to top).

have elegantly shown that a crack accelerates locally as it approaches a hole in glass, and then is significantly retarded as it leaves the pore. Lange[16] has shown cleavage evidence for retardation of crack fronts by pores in MgO. Fracture tails can also form from cracks and inclusions, e. g., if fracture propagates around the inclusion or through it on a plane suitably inclined to the fracture plane.

Pore-fracture tails are generally observed in glasses, single crystals, and grains, even in the presence of preferred cleavage, as in MgO and NaCl. The only suitable surface to keep both halves of a crack together around a pore is a grain boundary, so grain boundary pores smaller than the grain do not yield fracture tails if the fracture stays on the boundary. However, grain boundary pores traversed by transgranular fracture generate fracture tails, as may pores at the point of transition from intergranular to transgranular failure. Intergranular failure around pores bigger than the grains also have fracture tails since different grain boundary or transgranular paths can be followed.

FRACTURE ORIGIN-RELATED TOPOGRAPHY IN GLASS AND POLYCRYSTALS

Generally, the fracture origin, an "inner mirror," an "outer mirror," and macroscopic crack branching occur as progressive stages of fracture. As shown in subsequent discussion, these features occur at characteristic levels of stress intensity:

$$K = \alpha \sigma_f \sqrt{C} = \sqrt{2EY_f} \tag{1}$$

where α = a geometrical constant, σ = failure stress, C = crack size, E = Youngs modulus, and Y_f = fracture initiation energy. Thus, if the failure stress is too low in relation to the specimen size, those features requiring values of C outside of the specimen to have sufficient stress intensity will not occur, i. e., inner mirrors, outer mirrors and macroscopic crack branching require progressively larger speci-

men size in which to occur. Initiation of test failure off-
center provides more room for part of the above features
to form.   If the stress field and the specimen are homogeneous, the crack will propagate symmetrically from
symmetric origins, and progressively approach symmetry
from origins of asymmetric shape.

Macroscopic Crack Branching

In specimens with sufficient size relative to their
failure strength, a single crack propagates part-way through
the specimen, then separates into two branches (Fig. 5).

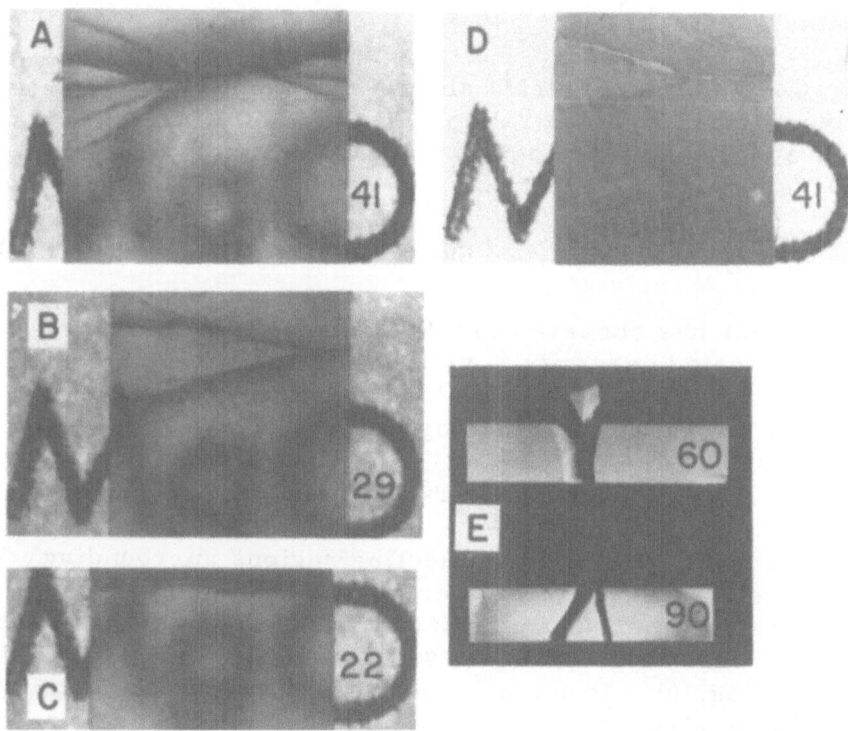

Fig. 5.   Crack Branching. (A) - (C) Dense, translucent
MgO. (D) ~1% porous MgO. (E) Dense hot pressed Al_2O_3.
Numerical values on the right of each specimen are the
failure stress in 1000 psi.

With increasing strength, and hence stored elastic energy,
these and subsequent branches may, in turn, branch, e. g.,
in sapphire and especially glasses, failing at a few hundred
thousand or more psi, fracture areas are reduced to fine
granular fragments. Higher fracture energies of polycrystals substantially reduce crack branching, e. g., the
extent of branching observed in glass failing at ~15, 000
psi requires $\geq$25, 000 psi in dense polycrystalline MgO and
Y_2O_3-stabilized ZrO_2, and ~50, 000-150, 000 psi in similar
Al_2O_3, partially stabilized ZrO_2, and Si_3N_4 bodies (Fig. 5).
Crack branching decreases with decreasing strength, and
hence with increasing grain size and porosity. Reduced
crack branching by limited porosity not significantly affecting strengths (Fig. 5) has been attributed to pores within
grains[6].

Congleton and Petch[17] showed that the distance cracks
propagated from a central hole in plates of polycrystalline
MgO and Al_2O_3 to the point of branching, R_B, varied with
failure stress, σ_f, such that

$$\sigma_f (R_B)^{1/2} = B_B = \text{a constant for a given body.} \qquad (2)$$

The author has observed crack branching in a variety of
flexure-tested ceramics consistent with Eq. 2. The observations complement Schardin's (2) earlier elegant spark
camera recording of branching in glass.

Description of Fracture Mirrors

Mirrors, flat highly reflecting regions surrounding
fracture origins, have been extensively studied in glass[18-23].
Mirrors are bounded immediately by fine ridges (mist) and
then coarser ridges (hackle) which radially point back to the
origin (Fig. 6). Previous observations have been restricted
to silicate glasses except for a note of some features in
Pyroceram (22) and some rocks (4). Recently, the author
and colleagues have extended mirror observations to non-
silicate glasses (e. g., As_2S_3 and glassy carbon, Fig. 6),
several single crystals, and many polycrystals. Kirchner
and Gruver[24] have also recently studied mirrors on some
polycrystalline bodies.

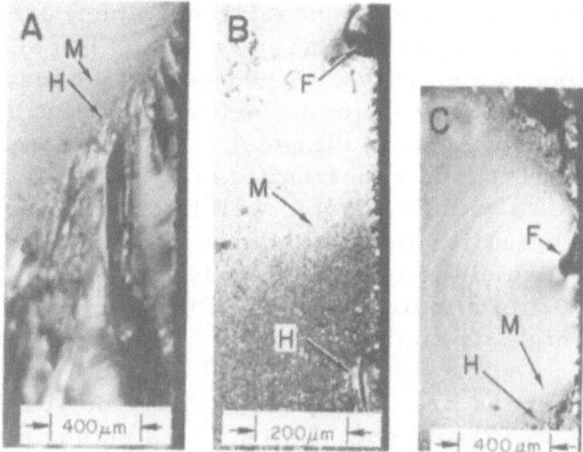

Fig. 6. Glass fracture mirrors. (A) Soda lime glass, arcs near top are parts of Wallner lines. (B) Glassy carbon: note symmetrical flaw (F) and broader area of mist with more diffuse boundary. (C) As_2S_3, complete mirror. Note Wallner lines (arcs inside mirror) and irregular upper part of flaw with associated trail of irregular mist and hackle.

Mist, taken as the boundary of the typical mirror (herein referred to as the inner mirror), is relatively closely spaced to the adjoining hackle in most glasses. Thus, little attention has been given to the boundary between the mist and hackle which outlines the area referred to here as the outer mirror. However, Johnson and Holloway[25] and the author showed that the outer mirror radius (R_o) for glasses obeys the same type of relation as the well-known inner mirror radius (R_I) relation

$$\sigma_f (R_I)^{1/2} = B_I = \text{a constant for a given body,} \qquad (3)$$

$$\sigma_f (R_o)^{1/2} = B_o = \text{a constant for a given body.} \qquad (4)$$

This distinction between inner and outer mirrors becomes important when their two boundaries become more widely separated. This occurs in glassy carbon (Fig. 6) and all polycrystalline bodies (Figs. 7, 9), possibly because of higher fracture energies, (e.g., γ for glassy carbon is about twice that of silicate glasses). An alternate possibility is that increased opportunities for crack micro-branching due to grains in polycrystals or pores in glassy carbon lead to earlier mist formation. However, the separation between the onset of hackle and branching remains small for all materials, possibly being essentially the same at high strengths.

Outer mirrors are detectable in many polycrystalline bodies (Figs. 6, 7) and obey Eq. 4. The author has observed that polycrystalline bodies with substantial trans-granular failure have a region that is analogous to the inner mirror in glass (Figs. 7, 8). It is not yet certain whether the onset of mist in glass corresponds exactly to the greater grain-to-grain fracture variation which appears to define the inner mirror on polycrystals. However, poly-crystalline inner mirrors must represent a consistent

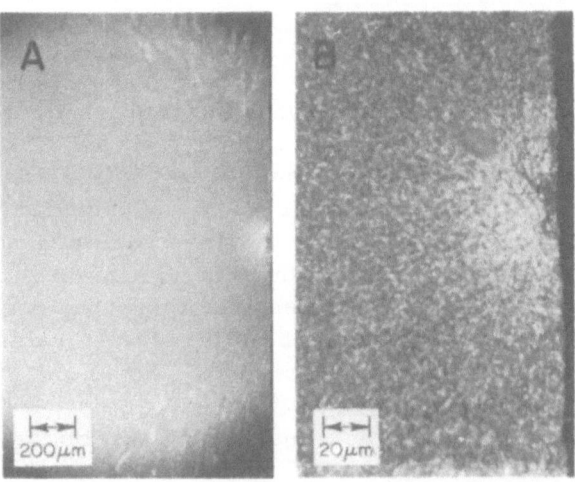

Fig. 7. Inner mirror in dense, fine grain $MgAl_2O_4$. Optical photos showing inner mirror: (A) with outer mirror (near top and bottom photo edges), and (B) at higher magnification.

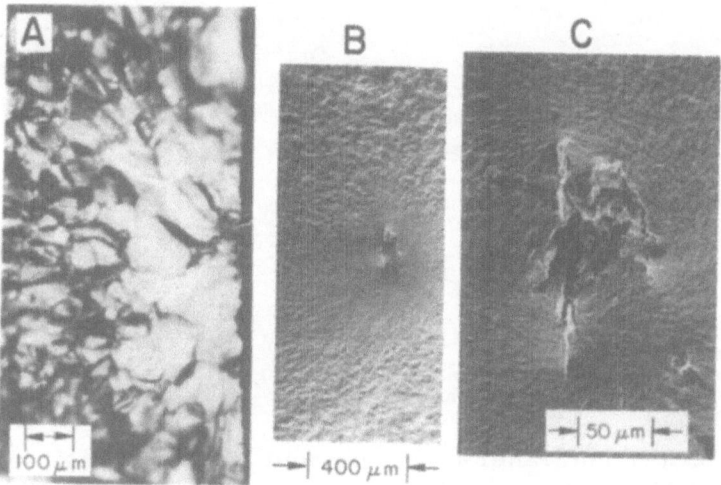

Fig. 8.   Inner mirror in ZrO_2-Y_2O_3 bodies.   (A) Optical
photo of clearer than average inner mirror in fully stabilized ZrO_2; note probable flaw origin (arrow) σ_f = 42,000
psi.   (B) SEM of partially stabilized ZrO_2 failing from a
large internal pore.   Failure stress at tensile surface
(right edge) ~ 105,000 psi; at pore ~ 85,000 psi.   (C) Higher
magnification SEM of pore in B.   Note fine "grainy" fracture steps leading into sharp areas at top and bottom of
pore, indicating fracture initiation from these areas.

physical process since they obey Eq. 3.   Further, substantial data on eight bodies representing six ceramics
ranging from PZT materials through Al_2O_3 and B_4C gave
ratios of outer to inner mirror radii averaging 5 ± 1 and,
as shown later, both radii may be related to flaw sizes.
The transgranular failure of inner mirrors appears to involve less grain-to-grain variation and fracture across
larger than random grain cross sections near the origin,
diminishing to normal fracture at or near the mirror
boundary (Fig. 8A).

   The clarity of polycrystalline mirrors and their
boundaries generally decreases with decreasing strength
and increasing porosity and grain size.   Thus, for example,

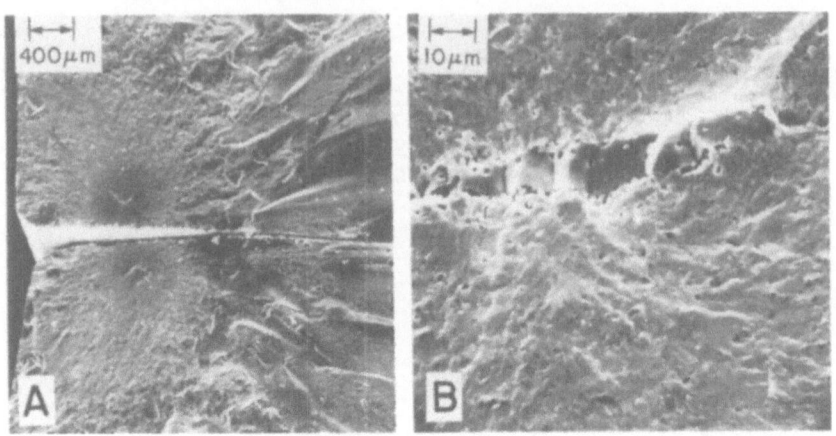

Fig. 9. Outer and inner mirrors, and fracture origin of
$BaTiO_3$. SEM of (A) matching fracture halves, and (B)
higher magnification of internal origin (upper half of A)
from large grain band and associated grain boundary pores.

lower strengths decrease available energy, so hackle become smaller, flatter, and less dense. Often, the only
apparent remmants are a few flat, flake-like features
which have a bright, diffuse lighting in optical microscopy
due to cracking which extends partway underneath and
around them. The sparsity of these flakes at low strengths
may lead to their onset being missed, and hence an over-
estimate of mirror size. Porosity and larger grains reduce detectability of mirrors by reducing strength, and
by obscuring them. Fracture tails from pores perturb
mirror flatness and obscure outlining mist or hackle as
porosity and, especially, pore size increase. Mirrors
generally have not been detected in typical polycrystalline
bodies with ~10% or more porosity, nor in typical commercial graphites. However, outer mirrors are quite clear
on Poco graphite (~1/2 μm pores), glassy carbon and unsintered Vycor (~10^{-2}μm pores), and inner mirrors are
found in the latter two (Fig. 6). Grain-to-grain fracture
variations similarly progressively perturb flatness and
obscure outlining mist or hackle as the grain size increases. However, suitable experience and technique

can often overcome this problem; e. g. , use of specimens large relative to the mirror size gives greater opportunity of following hackle ridges back to their initiation (Figs. 9, 13). Generally, the transgranular or intergranular nature of the fracture will not affect the detectability of the outer mirror. Polycrystalline inner mirrors, observed only with a predominance of transgranular fractures, are often less distinct or well defined because their flatter, more diffuse outlines are perturbed more by the effects noted above. Also, because of their closest proximity to the origin, inner mirrors more extensively reflect asymmetries of the fracture origin.

The above strength-grain size-porosity effects do limit detectability of some mirrors, mostly on weaker bodies of less interest. Mirrors are generally detectable on fractures of dense bodies having average or higher strengths for each particular material. Overall, outer mirror sizes can be determined to an accuracy of $\sim 10\%$. Inner mirror sizes can often be measured to nearly the same accuracy. The author has found good agreement between independent determinations by his colleagues and good agreement with the independent work of Kirchner and Gruver[24].

FRACTURE ORIGIN-RELATED TOPOGRAPHY IN SINGLE CRYSTALS

Single crystals generally exhibit the same fracture features as glasses and polycrystals, but with some modification and exceptions because of anisotropies in crystalline behavior. Thus, in MgO, for example, tensile stresses normal to $\{100\}$ result in fracture following this preferred plane without any crack branching. Fracture from tensile stress not normal to $\{100\}$ also usually leads to the crack still approximately following $\{100\}$, often with some branching (Fig. 10). However, the plane of initiation, and hence subsequent propagation patterns, can be surface finish-dependent. Ground MgO crystals, stressed along <110> with $\{100\}$ tensile surfaces, have fractures initiating on $\{110\}$ planes, then branching onto one or two $\{100\}$ planes

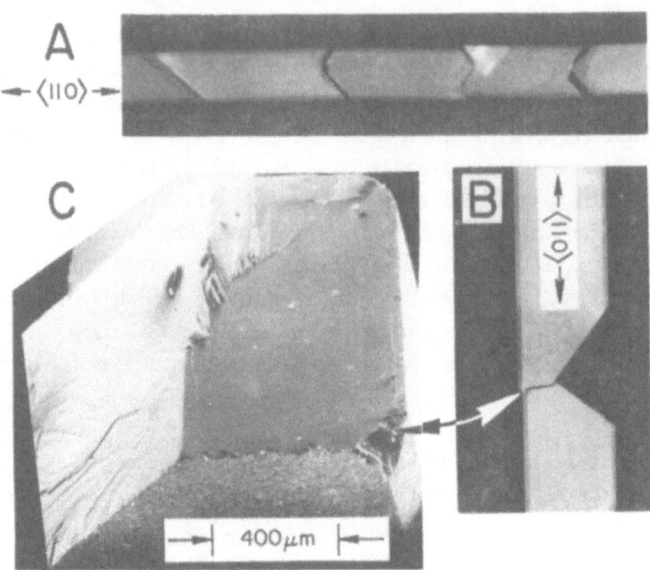

Fig. 10. Crack branching in MgO crystals. Tensile axes
as shown (A) side view of specimens with $\{110\}$ tensile
surfaces, and (B) top view of specimens with $\{100\}$ tensile
surface. Note tendency for cracks to branch onto $\{100\}$
planes. (C) Higher magnification of B showing initiation
on $\{110\}$ plane with branching onto $\{100\}$ plane toward left.
Note small approximately semi-circular flawlike feature
above chip on lower right-hand corner, and roughness of
machined surfaces.

(Fig. 10). Such crystals fracture on a $\{100\}$ without any
branching after being chemically polished. $MgAl_2O_4$ failure appears more complex. While specimens with $\{100\}$
tensile surfaces show a distinct preference for fracture
on $\{110\}$ planes, specimens with a $\{110\}$ tensile surface do
not maintain the $\{100\}$ initiating, $\{110\}$ branching condition,
and mist and hackle do not occur (Table 1). Sapphire and
rutile present still more complex behaviors not yet well
understood. However, branching onto rhombohedral planes
of sapphire (e. g. , with tension along the C axis) is fairly
common.

Table 1. FRACTURE OF SPINEL CRYSTALS

| Tensile | | Fracture | | Fracture |
Axis	Surface	Initiation Plane	Branching Plane	Mirror
$\langle 110 \rangle$	$\{100\}$	$\{110\}$	none	none
$\langle 100 \rangle$	$\{100\}$	$\{100\}$	$\{110\}$	
$\langle 100 \rangle$	$\{110\}$	$\{100\}$	none	none

Mirrors have been found on almost all fractures in fairly extensive studies of MgO, Al_2O_3, $MgAl_2O_4$, and TiO_2 crystals, as well as in preliminary observations in TiC, SiC, $BaTiO_3$ and $(NH_4)H_2PO_4$ (ADP). There are clearly pronounced crystallographic effects which are not yet completely documented or understood. However, several established observations can be given. MgO typically has tongue-shaped mirrors on extensions of mirrors with $\langle 100 \rangle$ axes (Fig. 11), with the tongue becoming more pronounced with increasing failure stress. $MgAl_2O_4$ has no mist or hackle on those fractures without crack branching (Table 1), so the whole fracture is mirror-like, but mist and hackle begin to appear, outlining expected mirror sizes with small deviations from these orientations. Other orientations result in a variety of mirrors, e. g., Fig. 12. Sapphire and rutile also show a variety of mirror shapes, again generally with exquisite symmetry, e.g., like Fig. 12 and Fig. 14. Sapphire specimens with $\langle 1\bar{1}00 \rangle$ tensile axis form mist and hackle only within $\sim \pm 30°$ of a line parallel with the "C" axis through the fracture origin on the fracture surface.

The diameter or radius (R measured along the tensile surface) of mirrors symmetric about an axis normal to the tensile surface obey constant $\sigma_f R^{1/2}$ relations, as do mirror radii taken (1) by projecting the outline of partial mirrors, (2) normal to the tongue axis near the surface, and (3) the depth of "gull wings" (Fig. 10). MgO specimens with failure initiating on $\{110\}$ planes (Fig. 10) also obey this relation, where R is the width of the $\{110\}$ portion of the fracture.

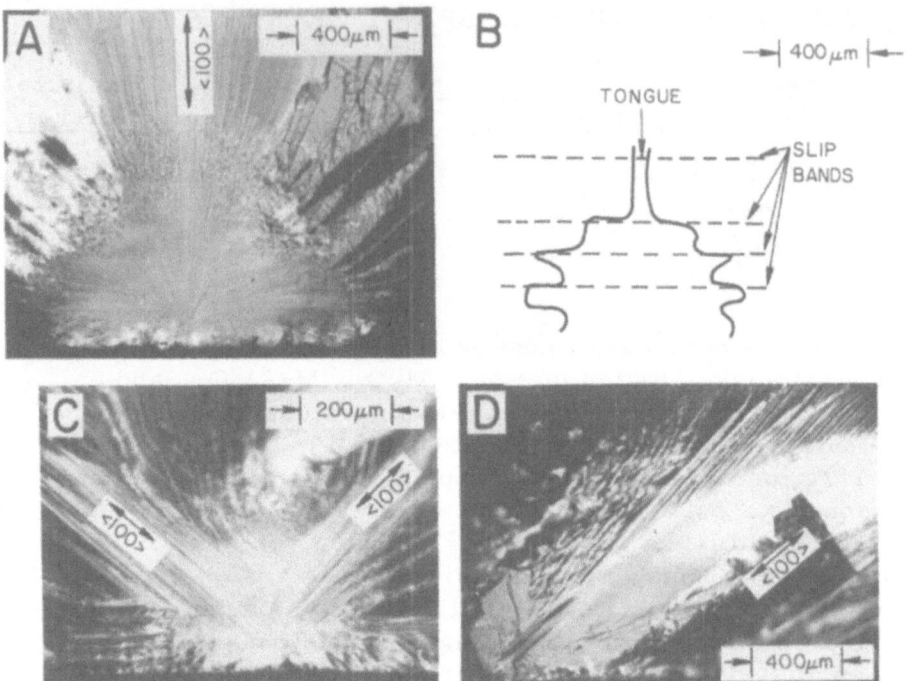

Fig. 11. Fracture mirrors on MgO crystals. Tensile axes
and surfaces are respectively: (A) <100> {100}; (B) is a
sketch of A. (C) <100> {110}; and (D) {110} <111>. Note (1)
tongue character and cleavage steps across mirror areas
to origin in all, and (2) modulation of mist and hackle by
slip bands (A and B).

ANALYSIS OF CRACK BRANCHING AND MIRROR
FORMATION

The general reason why branching and mirror formation
occur can be seen for example by considering Eq. 1 which is
identical to the Griffith equation and the strain energy, D, at
failure:

$$D = \frac{\sigma_f^2}{2E} = \alpha \frac{\gamma_f}{2C} \tag{5}$$

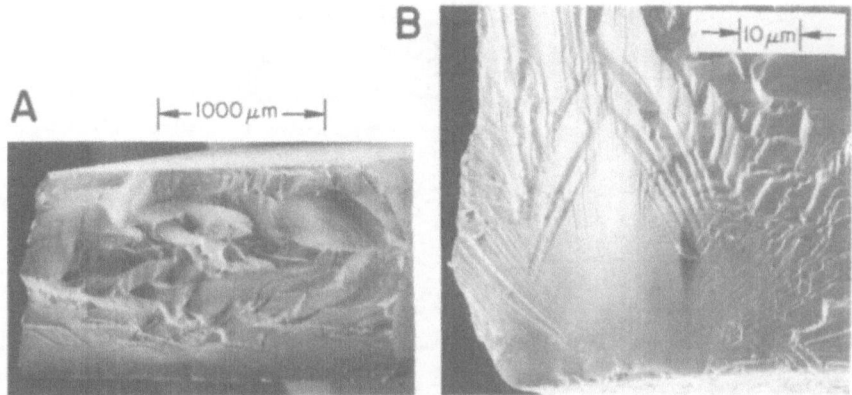

Fig. 12. Examples of $MgAl_2O_4$ mirrors. (A) "Gull-wing", and (B) "cathedral" mirrors on specimens with <111> tensile axis and {110} tensile surfaces.

A specimen failing from a flaw $C_1 = C_2/2$ still has nearly twice the strain energy needed to initiate, let alone main- tain, propagation of a crack of size C_2 as $C_1 \longrightarrow C_2$, pro- vided C_2 is small relative to the specimen dimensions. Hence, as a crack grows, it has progressively more strain energy than needed to sustain propagation, so it can pro- gressively use more energy to increase its velocity, gener- ate mist, then hackle, then branch macroscopically. Some analyses show that as crack velocity increases, the maxi- mum crack tip stress flattens and/or shifts about a line normal to the crack, thus allowing the crack to deviate and/ or branch, but other analytical models do not predict these effects[29]. Independent of these analyses, mist and hackle are typically considered to be increasing stages of incom- plete propagation of branch cracks before complete branch propagation is possible.

It is clear from comparing Eq. 1-4 that as a crack grows, mist, hackle and branching each begin to form at characteristic stress intensity levels (i.e. $B_I < B_O \leq B_B$) for each material, as is also shown by other recent invest- igators[17, 26]. Relating fracture features to stress intensity or energy[25] is not necessarily inconsistent with earlier

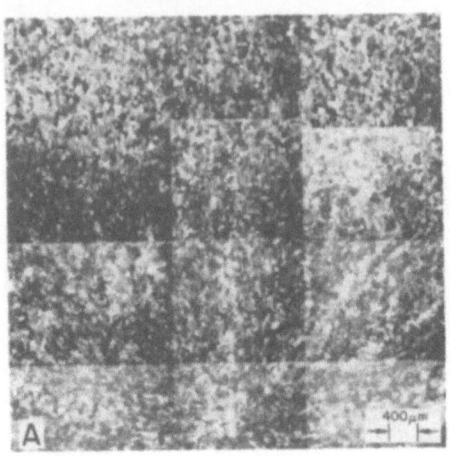

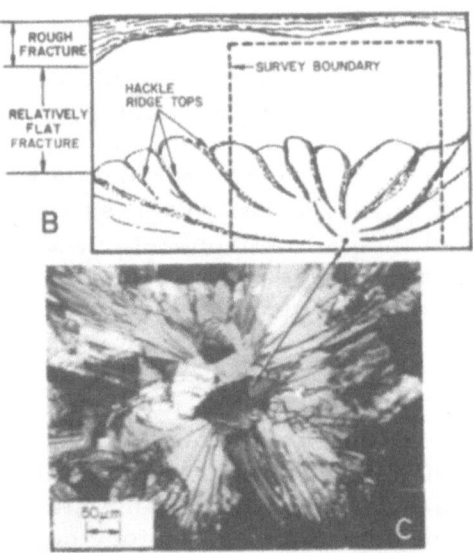

Fig. 13. Polycrystalline fracture topography of textured
MgO. (A) Composite of 12 optical photos surveying part
of a mostly transgranular fracture. (B) Sketch of complete
fracture including area covered in A. (C) Fracture origin
from grain boundary surface with surrounding transgranular fracture with fracture (cleavage) steps radiating
from boundary origin.

investigators[21, 27, 28] relating them to crack velocity, since stress intensity and crack velocity are typically related. However, stress intensity appears to be the more basic physical parameter since it is analytically relatable to energy associated with fracture processes.

Single crystal observations led to several important conclusions. First, the crystallographic dependence of the shape, or partial or complete absence of mist and hackle to define mirror dimensions shows that appropriate levels of stress intensity or velocity are a necessary but not sufficient condition for formation of mirrors or branches. This is corroborated by texturing reducing branching in polycrystalline MgO[14]. Second, mist and hackle formation represent either a different degree of the same mechanism of cleavage or fracture step formation, or a different mechanism, since such steps very frequently extend completely back across the mirror region to the origin (Fig. 11). Such steps also occur on crystal fractures having no mist and hackle, and may also extend back across mirrors of glasses or polycrystals to their origins (Fig. 8). Third, mirrors are not a demonstration of a truly brittle failure; since deformation can be associated with mirror formation. MgO crystals often undergo plastic deformation before failing and forming a mirror pattern, and resulting slip bands clearly affect mist and hackle (Fig. 11). Also, etching studies of MgO and Al_2O_3 both show an increase of dislocations at the onset of mist and in the mist regions of 10^2 and more commonly, 10^4, above background (i.e., normal mirror) densities. Finally, MgO specimens with <110> tensile axis and {100} machined tensile surfaces start to fail on {110} planes, presumably initiating along edge components of dislocations - a favored crack path. Similarly, it is noted that the preferred fracture planes of $MgAl_2O_4$ (Table 1) are always parallel with an edge component of its {111} <110> slip system.

Johnson and Holloway[25] showed that fracture mirrors in glass are distorted if the stress normal to the fracture surface varies, e.g., elongated mirrors due to flexural

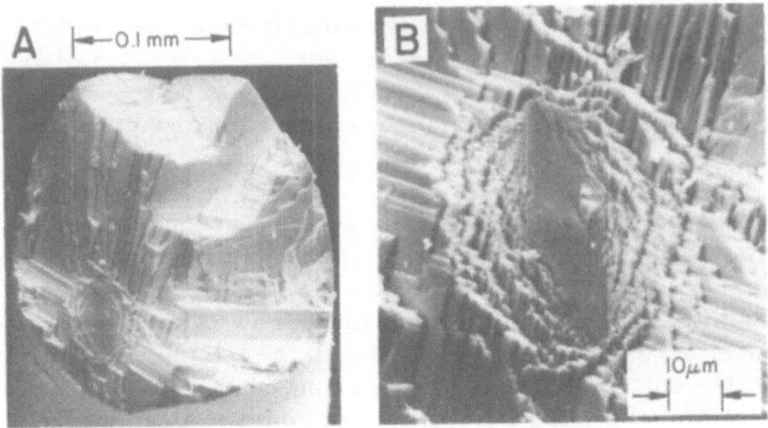

Fig. 14. Fracture of C axis sapphire fibers. (A) Lower and (B) higher magnification. Note fourfold symmetry.

stress gradients and mist and hackle do not form to continue the mirror outline beyond about halfway to the neutral
axis. The author has extensively observed this effect in
single and polycrystals. He has also found that mirror
sizes in specimens failing off the flexural test span center
correspond to the interpolated stress at the point of failure
rather than to the stress at the center of the span; again
confirming mirror determination by local stress. Correspondingly, local (e.g., residual surface) stresses affect
mirror size, e.g., as demonstrated by Kirchner and
Gruver[24]. Failure to eliminate or account for residual
stresses can be an important source of error at large
mirror sizes (i. e., low strengths).

Johnson and Holloway[25], using an energy criterion and
assuming that as $R \rightarrow R_o$ (the hackle boundary), the crack
approaches terminal velocity (V_t), derived an expression
for B_o:

$$\sigma_f R_o^{1/2} = \sqrt{\frac{8\gamma E}{2\pi - \frac{j\rho v_t^2}{E}}} = B_o \qquad (6)$$

where j = a constant, ρ = density. Congleton and Petch[17] derived from stress consideration (neglecting velocity effects) an equation for crack branching:

$$\sigma_f R_B^{1/2} = 2 \sqrt{\frac{E\gamma}{\pi}} = B_B \tag{7}$$

Solving both equations for γ, the energy associated with the onset of hackle or branching,

$$\gamma = \frac{\sigma_f^2 R}{8E} \left(2\pi - \frac{k\sigma V_t^2}{E} \right) \tag{8}$$

$$\gamma = \frac{\sigma_f^2 R\pi}{4E} = \frac{D2\pi R}{4} \tag{9}$$

If velocity or kinetic energy effects are neglected as Congleton and Petch did, the two expressions are then the same. Since as noted earlier, the onset of hackle and branching occur at similar values of R, these equations also indicate no significant difference in the energy required for the onset of hackle and branching. As shown in Eq. 9, these equations can also be expressed as a function of strain energy, using Eq. 5, showing that the branching occurs at one-fourth of the average strain energy (e. g., divided among four surfaces) times the crack circumference. Integrating D over the volume and γ over the surface area shows that normally only a small fraction of the mechanical energy is dissipated in forming the fracture. These expressions also show γ increases linearly with the crack size. This can also be seen by considering the condition, $\sigma_f R^{1/2}$ = B, and the Griffith equation. Letting R = bC (R = any radius since no special theoretical difference was assumed for the various radii) and using the same form of the Griffith equation used by Congleton and Petch, one has

$$\sigma_f = \frac{B}{\sqrt{R}} = \sqrt{\frac{2EY_f}{\pi C}} = \frac{B}{\sqrt{bC}}$$

Thus $$Y_f = \frac{\pi B^2}{2bE} \quad \text{or} \quad bY_f = \frac{\pi B^2}{2E} \qquad (10)$$

From Eq. 7 above, one then has:

$$bY_f = \frac{\pi}{2E} \left(\frac{4EY}{\pi}\right) = 2Y$$

The factor of two is due to the formation of four rather than two surfaces in crack branching, so prior to this occurrence,

$$Y = bY_f. \qquad (11)$$

It is tempting to use $Y = bY_f$ with $b \sim 10$, as found for glasses, to eliminate the discrepancy of ~3 that Kirchner and Gruver[24] got between observed $\sigma_f R_o^{1/2}$ values and those calculated using only Y_f in Eq. 7. However, this simple substitution of $b \sim 10$ does not appear to be justified for polycrystals. Though complicated by flaw shape and size variations, and probably flaw size-grain size effects, study of several bodies indicates the inner mirror-flaw size ratio averages ~6, and the outer mirror-flaw size ratio averages ~30. (True tensile testing may aid in resolving some of these uncertainties by eliminating stress gradients over flaws of large size, irregular shape, or differing locations.) It is interesting to note that the inner mirror-flaw size ratio in glassy carbon also averages ~6.

On the other hand, Rice[30] and Freiman, et al[31] have indicated that single crystal fracture energies, which are often 0.1-0.2 of polycrystalline values, may be more appropriate for calculating polycrystalline failure criteria when the flaw size is of the same order as the grain size. Rice also suggested that a crack must encompass a sufficient number of grains before typical polycrystalline fracture energies are required for propagation. Therefore, if fracture energies start from lower levels but increase with crack size due to both microstructural and velocity

effects, agreement between calculated and observed B
values could be achieved with larger outer mirror to flaw
size ratios.   Thus more study is clearly needed to clarify
the meaning of polycrystalline fracture mirrors and their
exact relation to flaws and the fracture process.

OTHER FRACTURE FEATURES AND TESTING

Other fracture features or variations of those previously
discussed can occur.   "Gull-wings" may form around pores
or inclusions in glasses or single crystals[3].   Rib marks,
which indicate the shape of the fracture front, result from a
change in the plane of fracture, due, for example, to a
temporary arrest of the crack[3, 29].   Wallner lines, attributed
to interaction of the fracture front with shock waves from
nearby cracks, form arcs within the mirror region (Fig. 6).
These arcs, which are often nearly symmetric around the
origin can be used to calculate crack velocity[21, 32].   Sommer[33]
has shown that the superposition of hydrostatic and tensile
stresses reduces mirror sizes in glasses.   He has also
shown that superposition of tensile and anti-plane shear
forces (e. g., superimposing tensile and torsional loading)
significantly changes "mirror" patterns[34].   His study of
resulting "lances," which are similar to hackle, should
also provide a better understanding of hackle formation.

The author has observed that fractures from thermal
stressing large parts with small heated zones have typical
mirror, mist and hackle, and possibly branching, near the
origin which is usually outside the heated region.   However,
hackle may often disappear further away from the origin so
much of the overall fracture is mirror-like, presumably
due to reduced stresses as indicated by Johnson and Holloway's[25] analysis of flexure behavior.   Limited evaluation
has been made of fracture surface formed at elevate temperatures.   C-axis sapphire fiber tests show some modification of the room temperature patterns (Fig. 14) as temperature rises, but the basic mirror and related markings
remain similar.   This holds true even when some macroscopic ductility is observed (e. g., at ~1800° C) again showing

that mirror features do not mean totally brittle failure.
Limited observations of C axis fibers failed in creep show
most or all of the fracture surface is generally feature-
less except for some undulation and a somewhat grainy or
mottled texture, distinctly different from short-term
fractures. Observations on high temperature fracture
of polycrystals indicate that they are similar to low temperature fracture except for increases in intergranular
failure[11,12] Again, these similarities generally persist
even when some macroscopic plastic deformation occurs.

USES OF FRACTURE MIRRORS AND BRANCHING

The single most important use of fracture features is
determining fracture origins, discussed below. Without
their aid, determining fracture origins would often be comparable to finding the proverbial needle in the haystack.
Just determining regions of fracture initiation is often useful, e. g , whether fractures originate from finished surfaces instead of edges or internal defects in finishing
studies. Presently, fracture mirrors can be used to
estimate flaw sizes or fracture energies in glasses, and
show promise for such use in polycrystalline bodies.
Strength-mirror size curves, established by normal
mechanical testing, can be used, for example, to determine
the stress at points of fracture initiation in a body in service or in testing (e. g , thermal shock failures) Also, as
discussed earlier, fracture character may distinguish creep
from short-term failures.

Locating fracture origins generally proceeds through
the list of techniques in Section II as far as necessary. In
multiple fractured specimens (e. g., hoop tensile tested
rings), the largest mirror and crack branching patterns
are used to determine the initial, i. e., lowest, stress
fracture. These patterns are used (e. g , with the unaided
eye) to determine the area of origin, often to a mm or less
if complete patterns occur. Further progess in locating
origins depends greatly on having sufficiently high levels
of transgranular failure, so (1) inner mirrors may be

identified and (2) fracture steps and tails from pores can
be followed. In large grain bodies, where mirrors are
often less detectable, transgranular fracture steps and
tails may be the only guide necessary, especially in bodies
such as MgO with distinct fracture steps (Fig. 13). In
fine grain materials, fracture steps involving several
grains (and hence granular and diffuse, especially with
intergranular failure) commonly lead back across the
inner mirror region to the origin area, and completely
back to large origins (Fig. 8). However, when origins
are a few grains or less in size, transgranular fracture
steps or tails are normally necessary to pinpoint the
fracture origin. Wallner lines can also aid in locating
origins on glasses and crystal fractures (Fig. 6)[21].

Confusion from a feature that may form on the compressive side of flexure specimens, and resemble a mirror to the unaided eye or at lower magnification, can be
avoided by keeping track of tensile and compressive surfaces. This feature, which apparently results from the
exiting of the fracture sufficiently sooner in one area, can
also often be eliminated in seeking origins by observing
fracture steps and tails, as well as other tensile-compres-
sive fracture differences (e.g., Fig. 13B). Also, unless
perturbed by inhomogenieties or concentrated loading
(e.g., as might occur in three point flexure), this feature
will occur nearly opposite the origin (i.e., shortest travel
path).

FRACTURE ORIGIN CHARACTERIZATION

Using the above technique, fracture origins can be
determined in a majority of glass specimens, many single
crystals, and dense polycrystalline bodies, especially
those with significant (e.g., 50% transgranular failure.)
In glasses, fractures originate from features whose appearance and dimensions are approximately those expected of
a "Griffith flaw" penny-shaped surface flaw. Origins may
come from a single one of these, or a few to several close
or overlapping flaws. Even on small glass specimens

(e.g., 0.1-0.2" in width), several other similar but smaller flaws can be found elsewhere across the tensile surface of flexure specimens.

Fracture origins similar to those in glasses are frequently found in spinel crystals, and on some orientations of sapphire. In sapphire with C axis tension, features of high geometrical symmetry are observed at both surface and internal origins (Fig. 14). MgO crystals with a <100> tensile axis and most surface finishes fail from sources at or near the surface with several sources showing different stages of development sometimes being observed. Similar MgO specimens with machined surfaces can have sufficiently workhardened surfaces so failure initiates internally[37] (Fig. 15). In either case, the sources are consistent with dislocation crack nucleation and different from the flaws in $MgAl_2O_4$ and Al_2O_3. On the other hand, some features similar to flaws have been found at a few of the origins of other MgO orientations (Fig. 10).

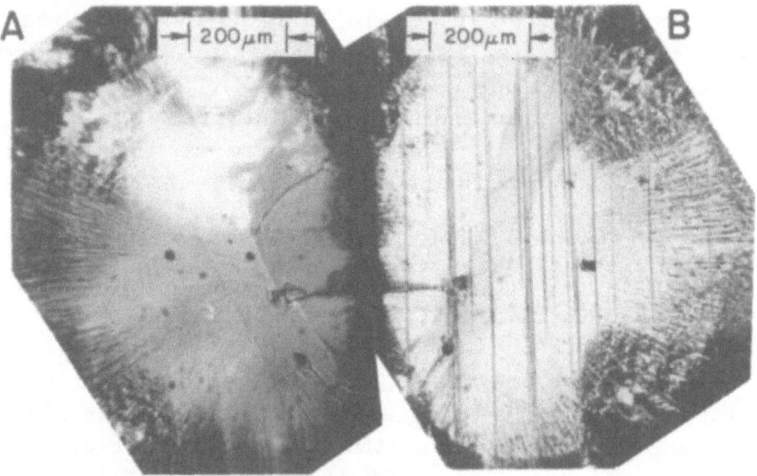

Fig. 15. Internal fracture origin of MgO crystal due to machining-induced surface work hardening. Matching fracture surfaces of specimen with <100> tensile axis and $\{100\}$ tensile surface. Right half has been etched.

Polycrystalline fracture originates from (1) pre-existing flaws, (2) cracks nucleated by microplastic flow during stressing, or (3) microplastic flow assisting the growth of flaws to critical size. Microplastic mechanisms are normally only feasible in softer materials with flaws favored in harder materials and by poorer surface finish, greater porosity, and impurities. Microplastic failure has been shown to be a cause of failure in MgO and CaO[12, 14]. Plastic deformation preceding and hence apparently leading to failure has been reported in high quality CdTe[35]. Non-zero intercepts observed, for example, on Petch plots of MgF_2, ZnSe, $BaTiO_3$, PZT, and Al_2O_3[36] may indicate failure due to microplasticity; however, Rice[30] showed that such intercepts can also result from flaw-grain size effects, and hence are not proof of microplastic failure.

Flaw origins have been shown elsewhere to be micro-structural extremes or "normal" flaws[30]. Large voids (Fig. 8), grains, or clusters of either or both (Figs. 1, 9) have been found to be important sources of failure in a wide variety of materials, e.g., large voids are a major source of failure in PZT, and voids, large grains, or both, domi-nate fracture of many Al_2O_3 bodies. "Normal" flaw origins are more common in dense bodies of relatively homogeneous grain sizes. In large (e.g., 100 μm) grain bodies, origins from flaw features within and hence smaller than the grains have been found. Origins in such bodies also occur from grain boundaries which have no associated fracture features to delineate the size, but surrounding fracture topography often indicates such flaws are no larger than the single grain. In fine grain bodies, flaw-like features, usually several-fold larger than the grain size, have been found in B_4C, Al_2O_3, $MgAl_2O_4$, MgF_2, and PZT. Recent obser-vations indicate that flaw size remains nearly constant, independent of grain size for a given material and surface finish, leading to a large reduction in the rate of strength increase with decreasing grain size when the grain size becomes smaller than the flaw size[30]. It yet remains to be unequivocably shown that the above features are truly pre-existing flaws not generated or grown by microplasticity. However, the weight of evidence indicates that they are pre-existing flaws.

SUMMARY AND CONCLUSIONS

Major fracture features related to strength and fracture initiation have been reviewed. Transgranular fracture, which can be reduced by temperature, impurities,
and anisotropies, along with fracture tails from pores,
inclusions, or other cracks, provide important guides in
locating fracture origins. Crack branching, outer mirrors
(demarked by hackle), and inner mirrors (demarked by
mist), which can further aid in locating fracture origins,
are generally identifiable and represent basic physical
processes. The occurrence or absence of these features
on single crystals, though incompletely understood, provide important clues about the formation of these features
beyond the indicated stress intensity or energy requirements. While fracture energy or flaw size estimates can
be made from these features in glasses, more work appears necessary in polycrystals. Whether exactly comparable criteria are used in identifying polycrystalline
and glass features, especially inner mirrors, must be
further established. It should also be understood why
the inner mirror boundary is further from the outer mirror in some bodies (e. g., polycrystals and glassy carbon),
but outer mirror-branching relations apparently remain
the same.

Indirect failure information obtained from fracture
mechanics or fracture features is useful, but actual failure
origins must be determined for failure analysis and material development. Use of fracture features to locate
origins from either basic mechanisms, microplastic crack
nucleation, or pre-existing flaws, is reviewed. Microstructural inhomogenieties make up a significant portion
of flaw origins, complementing cracks from machining,
and other causes.

ACKNOWLEDGEMENT

Use of some data and photos on glass and sapphire,
respectively, from J. Mecholsky and P. Bechner is grate-

fully acknowledged, as are comments on the manuscript from them and S. Freiman.

REFERENCES

1. J.R. Low, Jr., in Fracture, ed. by B.L. Averbach, D.K. Felbeck, G.T. Hahn, and D.A. Thomas, pp. 68-90, MIT Press, Cambridge, Mass., 1959.

2. H. Schardin, ibid., pp. 297-330.

3. V.D. Frechette, Proc. Brit. Ceram. Soc., No. 5, pp. 97-106, Dec. 1965.

4. F.F. Lange, Phil. Mag. 16, pp. 761-770, 1967.

5. F.F. Lange, Int. J. Fract. Mech., V(8), 287-294, 1968.

6. R.W. Rice, Proc. Brit. Ceram. Soc., No. 20, pp. 329-363, June 1972.

7. R.W. Rice, Proc. Brit. Ceram. Soc., No. 12, pp. 99-123, March 1969.

8. R.W. Rice and W.J. McDonough, in Mechanical Behavior of Materials, Vol. IV, pp. 394-403, The Soc. of Mat. Sci., Japan, 1972.

9. Y. Matsuo and H. Sasaki, J. Am. Ceram. Soc. 49(4), 229-230, 1966.

10. R.W. Davidge and T.G. Green, J. Mat. Sci. 3, 629-634, 1968.

11. R.W. Rice, J.G. Hunt, G.I. Friedman, and J.L. Sliney, "Identifying Optimum Parameters of Hot Extrusion," Final Report for NASA Contract NAS7-276, Aug. 1968.

12. R.W. Rice, J. Am. Ceram. Soc. 55(2), 90-97, 1972.

13. S.F. Pugh, Brit. J. Appl. Phys., V. 18, 129-162, 1967.

14. R.W. Rice, in Ceramic Microstructures, ed. by R.M. Fulrath and J.A. Pask, pp. 579-593, John Wiley & Sons, New York, 1968. Also see data in article by R.J. Stokes, pp. 379-405, same volume.

15. F. Kerkhof and E. Sommer, in Handbuch der Mikroskopie in der Technnik Bd IV Mikroskopie der Silikate Teil 4 Mikroskopie in der Glas-und Emailtechnik, Herausgegeben von H. Freund, pp. 173-192, Unschau Verlag, Frankfurt au Main, 1963.

16. F. F. Lange, "Theory of Dispersion Toughening of Brittle Materials," Tech. Rpt. No. 2 for Contract N00014-68-C-0323, May 1969.

17. J. Congleton and N. J. Petch, Phil. Mag. 16, 749-760, 1967.

18. P. N. Terao, J. Phys. Soc. Japan 8(4), 545-549, 1953.

19. W. C. Levengood, J. Appl. Phys. 29(2), 820-826, 1958.

20. F. Kerkhof and H. Richter, Paper No. 40 in Proc. 2nd Int. Conf. on Fract., Brighton, England, 1969.

21. E. F. Poncelet, J. Soc. Glass Tech. 42, 279T-288T, 1958.

22. E. B. Shand, J. Am. Ceram. Soc. 42(10), 474-77, 1959.

23. M. J. Kerper and T. G. Scuderi, Am. Ceram. Soc. Bull. 44(12), 953-955, 1965.

24. H. P. Kirchner and R. M. Gruver, to be published in Proc. of Symposium on Fracture Mechanics of Ceramics, Pennsylvania State University, July 1973.

25. J. W. Johnson and D. G. Holloway, Phil. Mag. 14, 731-743, 1966.

26. A. B. J. Clark and G. R. Irwin, Exp. Stress Anal. 23 (1), 321-330, June 1966.

27. S. Bateson, Phys. & Chem. Glasses 1(5), 139-142, 1960.

28. D. Haneman and E. N. Pugh, J. Appl. Phys. 34(8), 2269-2272, 1963.

29. F. A. McClintock and A. S. Argon, Mechanical Behavior of Materials, 500-501, Addison-Wesley, Reading, Mass., 1966.

30. R. W. Rice, to be published in Proc. of Symp. on Fracture Mechanics of Ceramics, Pennsylvania State University, July 1973.

31. S. W. Freiman, K. R. McKinney, and H. L. Smith, ibid.

32. E. B. Shand, J. Am. Ceram. Soc. 37(12), 559-572, 1954.

33. E. Sommer, Glastech. Ber. 40(8), 304-307, 1967.

34. E. Sommer, Eng. Fract. Mech. 1, 539-546, 1969.

35. J. C. Wurst, "Thermal, Electrical, and Physical Property Measurements of Laser Window Materials,"

Quarterly Progress Report No. 5 for Contract NoF
33615-72-C-1257, June 1973.

36. R. W. Rice, Proc. Brit. Ceram. Soc. <u>20</u>, 205-257,
June 1972.

37. R. W. Rice, J. Am. Ceram. Soc. (in press).

DISCUSSION

<u>P. J. Gielisse (Univ. Rhode Island)</u>: In recent experiments
on aluminas of controlled grain size we seemed to observe
that the variation of strength with grain size is not linear
as would be implied from your equations. Little influence
of size appeared until the grain size got down to about 25
microns when a sharp increase in strength occurred.
<u>Author</u>: The equations I presented do not predict strength-
grain size relations. Normally, strength increases with
decreasing grain size, often at a much lower rate at finer
grain sizes. However, many other variables can change
along with the average grain size, e. g., the amount, size
and spatial distribution of impurities and pores. These
can be factors even in relatively pure, dense bodies since
it is usually the local extremes of these that determine
strength. A body may be essentially pure and pore-free
and yet have an impure or porous area. Grain size distribution is also important and frequently becomes a dominant as porosity and impurity effects become more limited.
In many cases, the largest grains control strength. Since
Al_2O_3 is prone to exaggerated grain growth, especially at
higher temperatures, one can decrease the average grain
size over a considerable range (e. g., by lower processing
temperatures) before one begins to reduce the largest
grain sizes significantly. Finally, different grain-size
bodies do not necessarily respond the same to comparable
treatments. Thus, a faster cooling rate can cause more
cracking in larger grain than in smaller grain bodies, and
effects of finishing may differ. Some of these could be
factors in the effects you report. (See Ref. 29, 36.)
<u>A. Choudry (Univ. Rhode Island)</u>: The single crystal data
showed a correlation between crystallographic orientation
and crack propagation pattern and thus the cracks (prior to

fracture) are of atomic lattice origin rather than say random
flaws introduced by machining. If so, fracture strength
should be related directly to theoretical strength calculated
from various lattice parameters. Young's modulus is not a
good measure of fracture strength because introduction of
even a single microcrack in a specimen substantially reduces the strength without altering Young's modulus.
Author: The crack propagation patterns are indeed crystallographic in character. However, this does not imply that
the cracks are not random cracks. While machining-induced
cracks may generally be of one or more crystallographic
orientations, they need not be, since the results show that
if a crack starts on a less preferred plane, it will turn onto
a preferred plane. In principle, one should be able to directly calculate the theoretical and hence the actual strength
from lattice parameters and crack dimensions, but this is
not yet completely feasible. Indirectly, the Youngs modulus
is an approximate measure of the theoretical strengths
σ_T (e. g., $\sigma_T \sim E/10$), and of course E is an essential parameter in the Griffith equation. The reason a crack so
drastically reduces strength is because it localizes stress
so that the theoretical strength is reached only at the crack
tip. In fact, one of the ways of relating modulus to theoretical strength is to use the Griffith equation and let the
crack length approach lattice dimension.

THE ROLE OF SURFACE ENERGY ON THERMAL SHOCK OF CERAMIC MATERIALS

C. R. Manning, Jr. and L. D. Lineback

Department of Materials Engineering, North Carolina State University, Raleigh, North Carolina

MODEL FOR THERMAL STRESS RESISTANCE OF A TRULY ELASTIC MATERIAL CONTAINING MORE THAN ONE CRACK

The thermal shock resistance of a material is a measure of three energies associated with the material, i.e., the elastic strain energy, the energy associated with the formation of new surfaces (surface energy), and the kinetic energy associated with the formation and growth of cracks. Within limits, the work done per unit volume in stressing a material is converted to strain energy where, if Hook's Law applies:

$$SE = \sigma\epsilon/2 \tag{1}$$

or, $$SE = E\epsilon^2/2 \tag{2}$$

The limits for a truly elastic material are established when the material is stressed in tension to the level necessary to overcome the energy barrier for creation of a new surface by breaking bonds of atoms lying in a plane of fracture. This barrier is the surface energy, γ, of the material. Griffith[1] established a relationship between the surface free energy and the amount of stress, σ_g (the Griffith stress), necessary to start growth of a crack of half width c:

473

$$\sigma_g^2 = \frac{2\gamma E}{\pi c} \tag{3}$$

If the strain energy in a material which has been stressed to the Griffith stress is defined as the Griffith energy, GE, then by combining Eq. 1 and 3,

$$GE = SE = \frac{\sigma_g^2}{2E} = \frac{\gamma}{\pi c} \tag{4}$$

The Griffith energy, for any material having a given surface energy, is thus inversely proportional to the crack half width if the modulus of elasticity used in Eq. 3 is the same as that in the strain energy (Eq. 1 and 2). They are not as may be seen shortly. The elastic modulus used in the Griffith expression is that of the material locally, i.e., uncracked, and shall be designated the intrinsic modulus of elasticity, E_{int}. The modulus of elasticity used in the strain energy equations is that of the body as a whole and shall be described as the effective modulus of elasticity, E_{eff}. Eq. 4 may now be written:

$$GE = \frac{\gamma}{\pi c} \frac{E_{int}}{E_{eff}} \tag{5}$$

Walsh[2] noted that uniaxial elastic compression of rocks yielded nonlinear stress-strain behavior and hysteresis, with the slope of the stress-strain curve increasing as the stress increased. He developed the result that the measured elastic modulus of a body containing open cracks is less than that of an uncracked body. As later developed by Berry[3,4] this may be written:

$$E_{eff} = \frac{E_{int}}{(1+2\pi Nc^2)} \tag{6}$$

where N is the number of cracks through unit cross-section.

If all the cracks in a material are of the same size and they are separated from one another by a sufficient distance such that their stress fields do not interact, then the assumption can be made that the number of cracks does not

affect the Griffith stress; rather it is the crack size, the
surface energy, and the intrinsic modulus of elasticity
that affect it.   Information may now be gathered from the
two material parameters γ and E_{int} and the two physical
parameters of N and c.

The Griffith energy for a material containing cracks
may now be written:

$$GE = \gamma \left(\frac{1}{\pi c} + 2Nc\right) \tag{7}$$

Again, the work to start a crack growing is independent of
the elastic modulus and depends only upon surface energy,
crack size, and crack number.   The term in parenthesis
can be defined as the energy factor because for a given
material, i.e., a given surface energy, it determines the
Griffith energy.

Constant Griffith energy contours are given on a crack
size vs. crack number plot in Fig. 1.   Similarly, lines of
constant crack number on plots of Griffith energy vs. crack
size and lines of constant crack size on plots of Griffith
energy vs. crack number are given in Figs. 2 and 3.   Fig.
2 indicates that for a given number of cracks, as the crack
size increases from zero the Griffith energy decreases to
a minimum and then increases.   This minimum occurs when:

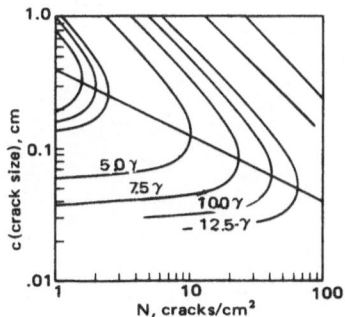

Fig. 1.   Lines of constant Griffith energy on a grid of number of cracks vs. crack size.

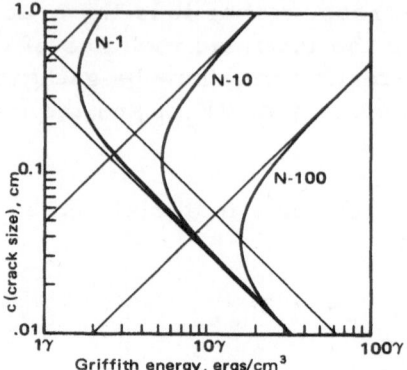

Fig. 2.   Lines of constant numbers of cracks on a grid of Griffith energy <u>vs.</u> crack size.

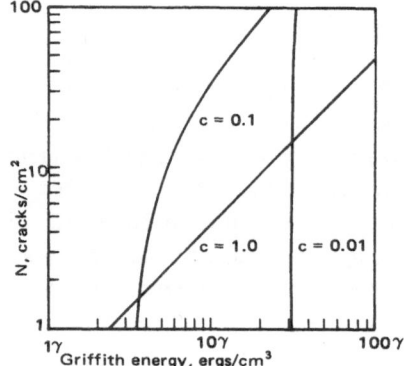

Fig. 3.   Lines of constant crack size on a grid of Griffith energy <u>vs.</u> number of cracks.

$$N = \frac{1}{2\pi c^2} \tag{8}$$

The minimum Griffith energy then becomes

$$GE_{min} = \frac{2\gamma}{\pi c} = 4\gamma Nc \tag{9}$$

or at the minimum Griffith energy the $1/(\pi c)$ term is equal to the $2Nc$ term.

It may be noted by inspection of Eq. 7 that as the crack size gets smaller, the Griffith energy approaches

$$GE = \gamma/(\pi c) \tag{10}$$

which is identical to Eq. 4, the original expression for Griffith energy without correction for the effect of open cracks on the modulus of elasticity. Similarly, as the crack size gets large, the Griffith energy approaches:

$$GE = 2\gamma Nc \tag{11}$$

The point of intersection of the lines described by Eq. 10 and 11 in Fig. 2 gives the crack size for the minimum Griffith energy and occurs at half the minimum Griffith energy. The number of cracks may also play an important role as shown in Fig. 3. For a given crack size in a given material the Griffith energy always increases with increased number of cracks. It is further observed that the rate of increase of Griffith energy is higher for the material having the larger sized cracks.

The results of increases in both crack size and number is shown in Fig. 1. Values of Griffith energy for a given material may be increased in two ways, i.e., by making the crack size very small, and by making a large number of larger cracks. In the first instance the effect of number of cracks is small as evidenced by the gentle slope of the Griffith energy contours. In the second instance, for a given Griffith energy, initially as the crack size increases, the number must also increase. However, after passing the point of minimum Griffith energy, the number gets smaller as the crack length increases.

A plot can be made of Griffith stress vs. Griffith strain, $\varepsilon_g = \sigma_g/E_{eff}$, which represents the locus of Griffith criteria for a truly elastic material containing more than one crack with cracks being independent of one another.

The slope of the straight line drawn from any point on the locus to the origin would represent the effective modulus of elasticity.

The effect of increasing the number of cracks in a given material is shown in Fig. 4. For initially small crack sizes, increases in either or both crack sizes or numbers does not significantly alter the effective modulus of elasticity. In the region above the knee of the locus, however, the Griffith stress drops rapidly with increasing crack size. At a point a minimum Griffith strain is reached when the crack size reaches:

$$c^2 = 1/(8\pi N) \tag{12}$$

This value of c is half that which produces the minimum Griffith energy. Fig. 4 indicates that the minimum Griffith strain is larger for a larger number of cracks. When the crack reaches a length twice that to produce the minimum Griffith strain, the minimum Griffith energy is reached. It can be shown through the use of Eq. 6 and 8 that at this point the effective modulus of elasticity is exactly half that of the intrinsic modulus of elasticity. That is not to say that an increase in intrinsic modulus of elasticity will increase the minimum Griffith energy, because Eq. 7 indicates that the Griffith energy is a function of only the surface energy. As the crack size increases from this point, the Griffith stress decreases more slowly. It may

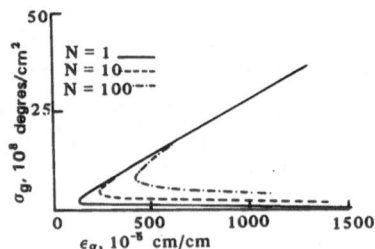

Fig. 4. The effect of number of cracks on the locus of the Griffith criteria.

also be observed that for a given Griffith stress, the
Griffith strain is larger for a larger number of cracks.
Likewise, the effective modulus of elasticity is lower and
the Griffith energy is higher.

The effect of increasing the surface energy is shown
in Fig. 5.   For a given intrinsic modulus of elasticity
and number of cracks, increases in crack sizes cause
changes in the Griffith stress and Griffith strain similar
to those in Fig. 4.   Increases in surface energy increase
the minimum Griffith strain and energy.   Similarly, for
a given Griffith stress little change in Griffith strain
occurs for a small value of crack size as a result of increases in surface energy while larger changes occur in
the Griffith strain for larger values of crack size.

Fig. 6 indicates the effect of intrinsic modulus of
elasticity on the locus of the Griffith criteria for a given
number of cracks and surface energy.   The intrinsic

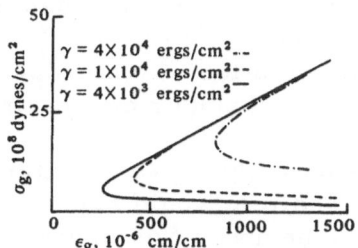

Fig. 5.   The effect of surface energy on the locus of the
Griffith criteria.

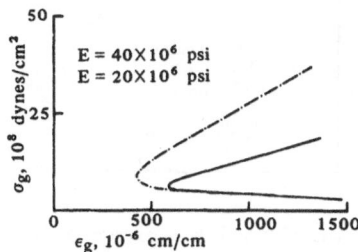

Fig. 6.   The effect of the intrinsic modulus of elasticity
on the locus of the Griffith criteria.

modulus has no effect on the Griffith energy but a decrease
does indicate an increase in the Griffith stress for a given
Griffith strain as expected from Hook's Law.

If the Griffith stress is exceeded, the cracks will begin
to grow. It is assumed that cracks of equal size will grow
in concert. The condition of constant strain will be analyzed
first because of its simplicity and similarity to materials
in a thermally stressed state. A given material containing
a number of cracks is stressed to a point $\sigma > \sigma_g$. In the
first analysis the crack number and size are such that ϵ_g
is greater than the minimum and σ_g subsequently is above
that necessary to produce a minimum as shown in Fig. 7,
indicated by oa.

If the applied stress is slightly higher than σ_g then the
area oab is more than the Griffith energy and the cracks
will grow, σ_g will decrease and the effective modulus of
elasticity will decrease. Point c may represent the condition shortly after it is in the region of instability and the
cracks must continue to grow. Berry[4] has shown that the
area acd is the kinetic energy of the cracks and that oad is
the surface energy contribution of the cracks. The area
ocb then represents the remainder of the strain energy.

The cracks grow until the locus is outside the region
of instability. This condition would appear to be met at
point e but the kinetic energy is then maximum and the
cracks continue to grow until this energy is absorbed by

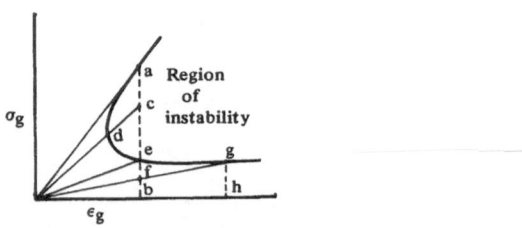

Fig. 7. Analysis of the Griffith criteria for a strong material at constant strain.

the surface energy term. This condition is met at point $\underline{f}$ with the final crack size being determined by the Griffith stress at point $\underline{g}$. Berry indicated that at this point the area $\underline{efg}$ will be the same as $\underline{ade}$.

The cracks are then subcritical. The material contains a residual strain energy ($\underline{ofb}$) but this is less than the Griffith energy ($\underline{ogh}$) and the residual stress σ_f is clearly less than the Griffith stress $\sigma_g(a)$. From the values of $\sigma_g(a)$ and $\sigma_g(g)$ it can be determined that, for a constant strain condition, initially short cracks grow to very long cracks because of the high kinetic energy associated with them. The Griffith stress and thus the fracture stress also drops significantly as does the effective modulus of elasticity.

For crack sizes and numbers above those which produce the minimum Griffith strain, the kinetic energy term is less significant. If such a material is stressed to point $\underline{a}$ in Fig. 8, lying within the region of instability, the strain energy $\underline{oab}$ again is larger than the Griffith energy and the cracks grow in size. When $\underline{c}$ is reached a small amount of kinetic energy remains and must be converted to surface energy and residual strain energy. Thus the cracks grow to $\underline{d}$ so that the kinetic energy $\underline{afc}$ equals $\underline{cde}$ as before. The loss in Griffith stress and consequently fracture stress is slight as may be noted by the positions of $\sigma_g(f)$ and $\sigma_g(e)$. The crack length has increased moderately and the effective modulus of elasticity has decreased moderately.

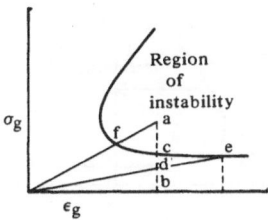

Fig. 8. Analysis of the Griffith criteria for a weak material at constant strain.

Analysis of a material containing more than one crack
at constant stress may be performed similarly. Fig. 9
gives the locus of the Griffith criteria. The material is
stressed to a point a such that the Griffith energy is only
slightly exceeded. The cracks begin to grow but the stress
level σ_a is maintained to point b. The area oabc represents
the total amount of work done on the material. The area
oae is the surface energy contribution, aeb is the kinetic
energy developed and obc is the resultant residual strain
energy at fracture. If the material fractures at b then
this strain energy will be converted to kinetic energy but
the surface energy term will remain the same. If, on the
other hand, the strain is held constant at this point the very
large amount of kinetic energy aebd must be converted to
surface energy. The Griffith stress and effective modulus
of elasticity would become very low and the Griffith strain
and Griffith energy would become very large as a result
of the very large cracks.

In reality bodies exposed to stresses large enough to
be in the region of instability are probably neither constant-
stress nor constant-strain. Constant strain is probably
more nearly correct and it will be used in the model for
thermal shock resistance to follow. As such, the thermal
stress resistances derived from this model will represent
minimum values.

A material may be exposed to mechanical stresses as
well as thermal stresses. When the sum of these two

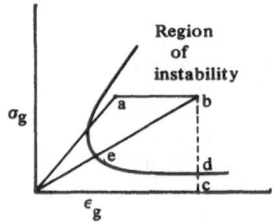

Fig. 9. Analysis of the Griffith criteria at constant stress.

stresses reaches the Griffith stress, cracks of equal size in the material begin to grow, resulting in a decrease in the Griffith stress and effective elastic modulus and an increase or decrease in the Griffith strain depending upon the Griffith criteria of the material. The Griffith energy may be increased or decreased, depending upon the size of the cracks, since the number of cracks is constant.

If it is assumed that the effect of crack size and crack number on the coefficient of thermal expansion, α, is negligible, then the thermal strain for a free body in the temperature differential. ΔT may be described as:

$$\epsilon_{ts} = \alpha \Delta T \tag{13}$$

Using Hook's Law one may also write for a body constrained in one dimension with no prestress:

$$\sigma_{ts} = E_{eff} \epsilon_{ts} = \alpha E_{eff} \Delta T \tag{14}$$

In terms of the temperature differential,

$$\Delta T = \sigma_{ts} / (\alpha E_{eff}) \tag{15}$$

Recalling Eq. 1:

$$\Delta T = 2(SE) / (\alpha \sigma_{fs}) \tag{16}$$

The maximum temperature differential to avoid crack growth:

$$\Delta T = \frac{2(GE)}{\alpha \sigma_g} = \frac{\sigma_g}{\alpha E_{eff}} = \frac{\epsilon_g}{\alpha} \tag{17}$$

$$\Delta T = \left(\frac{2\gamma}{\pi c E_{int}} \right)^{1/2} \frac{(1 + 2 Nc^2)}{\alpha} \tag{18}$$

If the criteria for thermal shock damage is that no loss in strength be incurred then Eq. 17 may be used as a measure of thermal stress resistance, while Eq. 18 may be used as a guide for material selection. The latter indicates that a high surface energy and low intrinsic elastic modulus and thermal expansion are desirable. The optimum crack size is dependent upon the number of cracks. If the crack size is small then Eq. 18 reduces to

$$\Delta T = \frac{(2\gamma E_{int}/\pi c)^{1/2}}{\alpha\, E_{int}} = \sigma/\alpha\, E_{int} \tag{19}$$

This is the classical concept of increasing thermal shock resistance by selecting a very strong material and it is, in principle, valid. In practice, many materials contain many flaws or microcracks which lower the maximum stress to which they may be exposed.

Surface energies, crack size, and crack number are very difficult to measure. Fracture stress, fracture strains, and effective moduli of elasticity and consequently Griffith energies are relatively easy to measure. Therefore, while Eq. 17 takes into account crack numbers and size it is an easier and perhaps more accurate measure of the thermal shock resistance of materials. This equation indicates that the thermal stress resistance can be measured in terms of bulk or effective material parameters. The temperature differential necessary for crack propagation in the free body is directly proportional to the ratio of the Griffith stress and the effective elastic modulus or is directly proportional to the Griffith strain. As indicated by Eq. 12 for a given number of cracks, the Griffith strain may be increased by increasing or decreasing the crack size above or below a critical value. Thus the thermal shock resistance of any material containing more than one crack decreases to a minimum and then increases again as the crack size increases. The size to produce maximum thermal stress resistance is for all materials a function of the number of cracks. Finally, the minimum Griffith strain and the minimum thermal shock resistance may be increased by increasing the surface energy and/or number of cracks and by decreasing the intrinsic elastic modulus. (Figs. 10 and 11.)

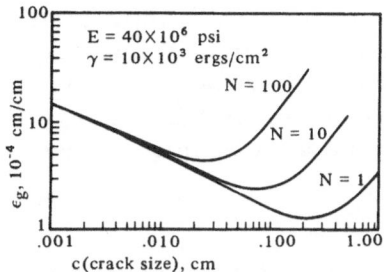

Fig. 10. Griffith strain vs. crack size for various numbers of cracks.

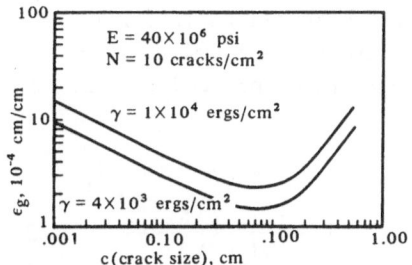

Fig. 11. Griffith strain vs. crack size for various surface energies.

If the criterion for thermal stress damage is no loss of strength when large thermal gradients are to be encountered, then a material containing cracks larger than those which produce the minimum thermal stress resistance or minimum Griffith strain must be selected. Here if the Griffith criterion is exceeded, the increase in crack size and consequently the decrease in strength is quite small. A different material having the same strength but with cracks smaller than those producing the minimum Griffith strain would fail catastrophically rather than quasi-statically because of the absorption of the kinetic energy associated with the production of very large cracks.

Determination of whether a material will suffer thermal shock damage of catastrophic proportions depends upon whether the crack size is larger or smaller than that needed to produce the minimum Griffith strain. Using Eq. 12 and Eq. 6 it can be shown that the condition of minimum Griffith strain is met when the effective elastic modulus equals eighty percent of the intrinsic modulus. Thus materials having effective moduli above that value are subject to catastrophic thermal shock failure if the Griffith energy condition for crack growth is reached as a result of large thermal differentials.

COMPATIBILITY WITH EXISTING MODELS

Hasselman[6] has recently developed a theory of thermal shock fracture initiation and propagation which is based on the intrinsic material properties and microstructure. His analysis develops along lines similar to the present model in that the total energy per unit volume at constant strain is established:

$$W_t = \frac{3(\alpha\Delta T)^2}{2(1-2\nu)} E_{int} \left[1 + \frac{16(1-\nu)Mc^3}{9(1-2\nu)} \right]^{-1} + 2 \pi Mc^2 \gamma \quad (20)$$

where M is the number of parallel cracks per unit volume. The Griffith criterion for crack instability is imposed:

$$dW_t / dc = 0 \quad (21)$$

and the following temperature differential is obtained:

$$\Delta T = \left[\frac{\pi \gamma (1-2\nu)^2}{2 E_{int} (1-\nu)\alpha^2} \right]^{1/2} \left[\frac{1 + 16(1-\nu^2)Mc^3}{9(1-2\nu)} \right]^{-1/2} \left[c \right] \quad (22)$$

The temperature differential vs. crack half length from this expression (Fig. 12) is nearly identical to that of Fig. 10 in which the Griffith strain is used instead of the temperature

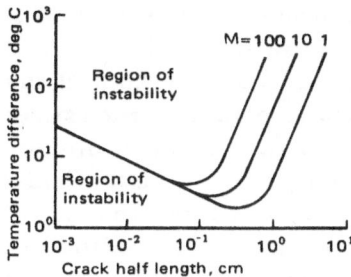

Fig. 12.   Minimum thermal strain required to initiate crack propagation as a function of crack length and crack density (after Hasselman[6]).

differential.   The temperature differential in Hasselman's expression is a function of the intrinsic materials parameters thermal expansion coefficient, elastic modulus and surface energy and the crack size.   For the present model the same plot is obtained when the temperature differential is a function of the measured thermal expansion coefficient, fracture strength, and effective elastic modulus or a function of the ratio of strain-to-failure to thermal expansion coefficient.

Comparison of Eq. 18 and 22 indicate the similarity of the two approaches if due compensation is made for the differences in strain conditions.   The product of the Poisson's ratio terms approach unity and the two expressions become nearly identical.

Earlier Hasselman[5] had suggested that the thermal shock resistance of a material containing parallel microcracks might be increased as a result of an increase in the strain-to-failure introducing microcracks.   In this analysis, however, no minimum strain was shown to occur.

From comparison of these two approaches it appears that the thermal stress resistance of elastic materials may be made on the basis of either the bulk or effective properties of the material or on the intrinsic properties of the material and the associated microstructure.   For Hassel-

man's approach, the number of cracks and their size must
be measured as well as the intrinsic properties. For the
present model, the number of cracks and their size need
not be measured, as their effect is measured indirectly by
the fracture strength and the effective modulus of elasticity.
The surface energy term is also incorporated in the fracture strength term. That is to say that the present model
treats the body as a homogeneous continuum with the effects
of its microstructure being measured in bulk parameters.
Thus the present model allows the use of existing expressions for predicting the thermal shock resistance while
information about the type of failure may be gathered by
comparison of the effective modulus of elasticity with the
intrinsic modulus of elasticity.

SUMMARY AND CONCLUSIONS

If the Griffith energy is exceeded for a material containing cracks smaller than those necessary to produce the
minimum Griffith strain, then the cracks grow rapidly,
resulting in large losses in strength or catastrophic failure.
Materials, on the other hand, containing cracks larger than
those necessary to produce the minimum Griffith strain
fail with smaller losses in strength, or quasi-statically.
Finally, the temperature differential necessary to produce
this Griffith energy in a material at constant uniaxial strain
depends only upon the following measured or effective properties of the body: fracture strength, modulus of elasticity,
and coefficient of thermal expansion. The temperature differential to cause failure depends upon the ratio of the fracture strength to effective modulus of elasticity or the strain
to failure for the truly elastic body. The temperature differential to cause failure depends upon the ratio of the strain-
to-failure to thermal expansion coefficient. Because it was
assumed that the presence of cracks had no effect on the coefficient of thermal expansion, their effect was reflected in
the strain to failure. It was suggested in the model that, as
cracks grow, the strain-to-failure initially decreases (the
material becomes less thermal shock resistant) until the
minimum is reached and that to this size the material would

fail catastrophically. With subsequent increase in crack
size the strain-to-failure increases (the material becomes
more thermal shock resistant) and the material is expected
to fail quasi-statically.

On the basis of the model, the intrinsic material properties beneficial to the thermal shock resistance of a body
are: large surface energies, small intrinsic elastic moduli,
large numbers of cracks if the crack size is large, and
small crack size. The strength of a body is not an indication of its thermal shock resistance. Rather it is the ratio
of the fracture strength to the effective elastic modulus or
the strain-to-failure that is indicative.

Regardless of the size and number of cracks of a body,
however, the surface energy as defined by Griffith is directly proportional to the thermal shock resistance. As
such, it should be a major consideration for selecting materials for application in thermal shock environments.

REFERENCES

1. A. A. Griffith. "Phenomena of Rupture and Flow in
    Solids." Phil. Trans. Royal Soc., London, 221A, (4),
    163-198, 1920.
2. J. B. Walsh. "The Effect of Cracks on the Uniaxial
    Elastic Compression of Rocks." J. Geophys. Res.
    70, (2), 399-411, 1965.
3. J. P. Berry. "Some Kinetic Considerations of the
    Griffith Criteria for Fracture: I Equations of Motion
    at Constant Force." J. Mech. Phys. Solids 8, (3)
    194-206, 1960.
4. J. P. Berry. "Some Kinetic Considerations of the
    Griffith Criteria for Fracture: II Equations of Motion
    at Constant Deformation." J. Mech. Phys. Solids 8,
    (3), 207-216, 1960.
5. D. P. H. Hasselman. "Unified Theory of Thermal Shock
    Fracture Initiation and Crack Propagation in Brittle
    Ceramics." J. Am. Cer. Soc. 52, (11), 600-604, 1969.

6. D. P. H. Hasselman. "Griffith Criterion and Thermal
    Shock Resistance of Single-phase versus Multiphase
    Brittle Ceramics." J. Amer. Ceram. Soc. 52 (5)
    288-89 (1969).

DISCUSSION

W. B. Crandall (IITRI): When you take the data from the
literature for E, what strain or strain rate will you select
for the E value? If the strain is small with a high strain
rate, this may make the data very difficult to use in your
theoretical approach.
Author: J. B. Walsh (J. Geophys. Res. 70 (2) 399-411,
1965) has indicated that the value for the modulus of elasticity must be taken by a technique employing compressive
stresses to close the cracks and minimize their effect
upon measurements. For this approach the strain rate
is not so important.
K. K. Verma (Alfred): The term used as the surface energy,
γ, quite often is not clearly defined when used in work similar to yours. Obviously this term refers to the local area
in the material where origin or propagation of a crack is
considered, and it can be expected to be different from the
values obtained at the surface by surface tension, wetting
and other such techniques.
Author: The surface energy of a material is the same for
the surface and for an open crack within the body.
A. Choudry (Univ. Rhode Island): In Fig. 11 you showed
the crack length vs. strain (Griffith); the curve goes through
a minimum at a crack length ~ 0.1 cm and then rises steeply.
The model as presented must be modified to bring the curve
down to the c-axis to prevent the paradoxical situation of
maintaining infinite crack lengths at infinite strain. Such a
modification would also force the curve to go through a
maximum which would result in sustaining large crack
lengths at 'anomalously' high strains.
Author: The model is not intended to be extended to infinity.
If one does the simple calculations, one will find that the
strain required to fracture the specimen is very small at

very large crack size. As such Mr. Choudry's suggestion would not have much value.

R. W. Rice (Naval Res. Lab.): (1) While your approach appears somewhat different than Hasselman's, many of the results are similar. Could you contrast your work with Hasselman's? (2) Following up on Bill Crandell's comment on strain rate, when one gets to extremely high heating rates, thermal stress failure can become essentially independent of mechanical parameters.

Author: Our approach allows the use of "effective" materials properties without measurement of crack size or number to determine the thermal stress, while Hasselman's approach depends upon intrinsic volume of materials properties.

SURFACE CHARGE-DEPENDENT MECHANICAL BEHAVIOR OF NON-METALS

N. H. Macmillan and A. R. C. Westwood

Martin Marietta Laboratories
Baltimore, Maryland

PHENOMENOLOGY AND MECHANISMS OF CHEMOMECHANICAL EFFECTS

The fact that chemisorbed, surface active species can significantly influence the microhardness of certain minerals was first noted by P. A. Rebinder in 1928[1]. Subsequent studies[2] established that such changes in hardness persist to depths of about 10 μm in a typical inorganic nonmetal, and that their magnitude depends not only on the environment but also on the type of bonding in the solid, the crystallographic orientation of the surface, the concentration and state of ionization of impurities in the solid, the magnitude and duration of the applied load, the temperature, the intensity of illumination incident upon the specimen, and on any potential applied to it. Recent studied have revealed useful correlation between environment-sensitive hardness and the ζ-potential of the solid, i. e., the hardness of all inorganic nonmetals and glasses examined was greatest when $\zeta \simeq 0$[2]. This "ζ-correlation," first noted for AgBr in aqueous bromide and AgI in aqueous iodide environments[3], has now been established by Westwood and coworkers for Al_2O_3 in aqueous NaOH and the n-alcohols[4]; calcite in aqueous NaOH and HNO_3[5]; quartz in buffered and unbuffered aqueous $Al(NO_3)_3$[6]; and soda-lime (s. l.) glass in aqueous KCl[7], $Th(NO_3)_4$[7] and dodecyl trimethyl ammonium bromide (DTAB) solutions, in n-alcohols[2], and in solutions of KI in isopropyl alcohol[2]. It is possible that the "ζ-correlation" is a property generic to inorganic nonmetals, both crystalline and noncrystalline. Figs. 1 and 2 provide examples of this correlation for quartz in aqueous $Al(NO_3)_3$ and glass in n-alcohols.

Since the microhardness of a crystalline solid is largely determined by the ease with which dislocations can glide in the near-surface region adjacent to and beneath the indenter, it seems reasonable that chemomechanical effects occur in crystalline solids primarily because chemisorption influences the mobility of near-surface dislocations. This was first demonstrated during studies of the etching behavior of LiF crystals cleaved in dilute aqueous solutions of organic fatty acids[10, 11]. It was noted that $\{110\}\langle 1\bar{1}0\rangle$ near-surface screw dislocations cross-slipped in response to the residual elastic stresses introduced by cleaving at a rate which was dependent on concentration (typically $\sim 10^{-6}$M) of fatty acid molecules present in the environment.

More quantitative studies of the environment-sensitivity of near-surface dislocation mobility have been undertaken for MgO, CaF_2, LiF and NaCl[12-16]. In these experiments[12, 13, 15], MgO crystals were cleaved under the test

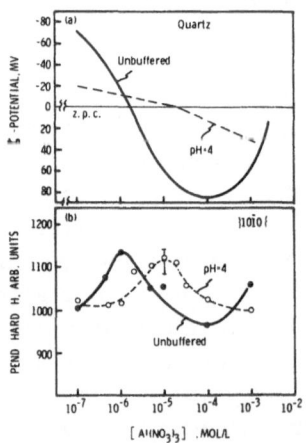

Fig. 1. (a) Variation in ζ-potential[8] and (b) pendulum hardness[6] of $\{10\bar{1}0\}$ quartz in aqueous $Al(NO_3)_3$ solutions.

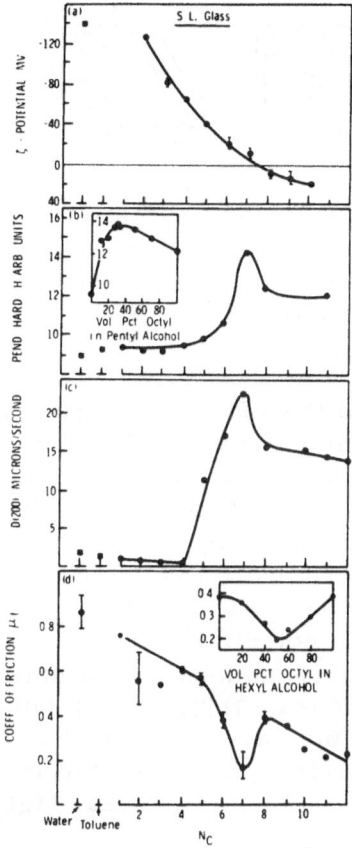

Fig. 2. Variation of (a) ζ-potential[2], (b) pendulum hardness[9], (c) rate of penetration with diamond-studded bit[9], and (d) coefficient of sliding friction[36] for glass in toluene, water and n-alcohol environments. N_C is the number of carbon atoms in the n-alcohol molecule.

environment, and the $\{100\}$ surfaces produced were indented for various times with a Vickers 136° diamond pyramid with a load of 10g. The dimensions $L_e(t)$, $L_s(t)$, $D_e(t)$ and $D_s(t)$ (Fig. 3) of the characteristic "rosette" formed about an indentation of t sec duration by edge and screw dislocations gliding on the primary $\{110\}$ $\langle 1\bar{1}0 \rangle$ slip systems were revealed by sequential etching and chemical polishing. The parameters $\Delta X(t) = X(t) - X(2)$, where $X = L_e$, L_s, D_e or D_s, were determined by optically.

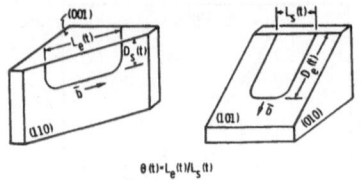

Fig. 3.   Schematic defining the extent of motion of edge dislocations, $L_e(t)$ and $D_e(t)$, and screw dislocations, $L_s(t)$ and $D_s(t)$, around an indentation of t secs duration in a $\{100\}$ cleavage plane in MgO[16].

Fig. 4 illustrates the variation of ΔL (t) with log (t-2) for MgO in several environments. In each case, edge-dislocation mobility as evidenced by the magnitude of $\Delta L(t)$ is increased relative to its value in "inert" toluene by im-mersing the crystal in some surface-active environment. Dimethyl formamide (DMF) is particularly effective. Fig. 5 illustrates the variation with pH of ζ-potential, as deter-mined from streaming potential[17], and $\Delta L_e(1000)$ and ΔL_s (1000)[15] for MgO in 10^{-2}N aqueous NaCl buffered by appro-priate additions of NaOH or HCl. $\Delta L_e(1000) > \Delta L_s(1000)$ at all values of pH, and both are least at pH $\simeq$ 12. 5 when $\zeta \simeq$ 0. Since for ionic solids $L_e(t)$ is related inversely to hard-ness[18], the data of Fig. 5 establish that MgO, like other

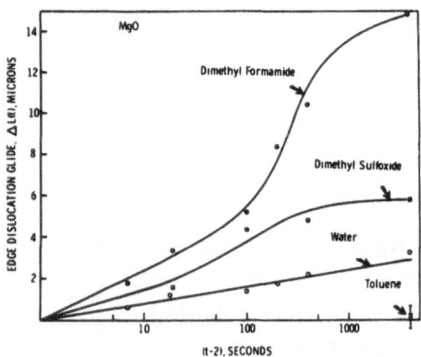

Fig. 4.   Variation of $\Delta L(t)$ with log (indentation time, t, minus 2 sec) for freshly cleaved MgO crystals exposed to various environments at room temperature.   10g load on indenter[13].

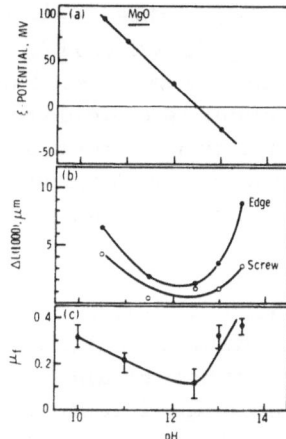

Fig. 5. Variation of (a) ζ-potential[17], (b) mobility of edge and screw dislocations[15], and (c) coefficient of friction[36] for MgO in buffered 10^{-2}N NaCl solutions. Note that both $\Delta L(1000)$ and μ_f are a minimum at $\zeta \simeq 0$.

nonmetals, is hardest at its isoelectric point and the "ζ-correlation" is shown to extend specifically to edge- and screw-dislocation mobilities.

Successive polishing and etching of adjacent pairs of 2- and 1000-sec indentations made at different values of pH revealed that $\Delta D_e(1000) > \Delta D_s(1000)$ at all pH's, and that both parameters are smaller at pH 12.5 than 13.5, consistent with the observed variation of $\Delta L_e(1000)$ and $\Delta L_s(1000)$. These studies[15] provided direct evidence of environment-sensitive motion of screw and edge dislocations in MgO at depths below the indented surface of $>10\mu$ and $>20\mu$, respectively.

The explanation for the correlation between hardness and ζ-potential remains obscure, because no detailed mechanistic understanding of the influence of environment on near-surface dislocation mobility or flow behavior exists, nor any clear understanding of the correlations between ζ-potential, surface charge and near-surface electronic structure. Our current view is as follows[15]: Chemome-

chanical effects are not thought[2,10,15] to arise as a consequence of adsorption-induced variations in the surface
free energy of the solid, as suggested by Rebinder[1].
Rather, the primary event is considered to be a chemisorption-induced change in the electrostatic potential of
the near-surface region of the solid which causes a localized redistribution of the charge carriers (electronic
and/or ionic).

For ionic or covalent crystals, the carriers are
electrons, and it is envisaged that their redistribution
changes both the band structure and the electron occupa-
tion of dislocations and point defects in the near-surface
region. Consequently, the electrostatic interactions between moving near-surface dislocations, between dislocations and point defects, and between dislocations and
the lattice are changed. Since these factors control dislocation mobility in ionic and covalent materials, the
near-surface hardness of such crystals is environment-
sensitive.

For glass, the distribution of mobile non-network
ions, e. g. , Na^+, OH^-, O^{2-}, is also likely to change in
response to an induced surface potential[2,7]. Presumably,
an excess of positive ions will accumulate near the surface
when the adsorbate donates electrons to the glass, and
negative ions will accumulate when the glass donates electrons to the adsorbate. In either case the outcome will be
an excess of non-network ions in the surficial region and,
since such ions weaken glass[19], a decrease in hardness.

The hypothesis that adsorption-induced charge carrier
redistribution is fundamental to chemomechanical effects
appears in accord with all phenomenology reported so far.
For example, dislocation mobility in ionic solids such as
MgO is markedly influenced by dislocation-extrinsic point
defect interactions[20,21], and the state of ionization of these
defects will be influenced by the surface potential in some
manner specifically dependent upon their type, concentration
and initial charge. Consequently, the chemomechanical
effects for a particular ionic crystal will be strongly depend-

ent on the concentration and electronic nature of its impurities[13]. For covalent solids such as alumina and
crystalline silicates, in which dislocation-lattice interactions dominate dislocation mobility, the hypothesis
allows for the observed lack of sensitivity of chemomechanical effects to impurity concentration[4, 6]. The glass
hypothesis predicts the observed[7] hardness maximum at
the isoelectric point, when there should be no significant
chemisorption-induced migration of (softening) non-network ions towards the surface.

That the "ζ-correlation" occurs for both crystalline
and noncrystalline inorganic nonmetals suggests the possibility of further commonalities[15]. Similarities in
chemisorption between crystals and glasses of similar
composition are not unexpected because chemisorption
is a localized process determined largely by the active
sites, and because glasses retain a network of nearest-
neighbor bonds of similar geometry and strength to those
in the crystalline form[22, 23]. If flow in glasses proceeds
via "dislocation-like" defects[24, 25], similarities may
exist in the flow mechanisms in crystals and glasses. In
view of the lack of long-range order in glasses, it is
presumably at the atomic scale that any commonalities
of mechanism must be sought. It has been conjectured[15]
that the similarities in the environment-sensitive mechanical behavior of crystals and glasses arise because
of the similar influence of chemisorption-induced changes
in near-surface electronic structure on the fundamental
unit of plastic flow, namely the formation and/or motion
of kinks along near-surface dislocations in crystals or
"dislocation-like" defects in glasses. If such mobility
is controlled by kink motion, only flow occurring within
a distance of the order of the Debye length, λ, from the
surface is likely to be environment-sensitive. But if
kink generation at the surface is the rate controlling step,
and this process is environment-sensitive, flow may be
affected at depths $> \lambda$ as the kinks penetrate into the solid
along line defects. Evidence that this may be the case for
MgO is the environment-sensitive screw and edge dislocation mobility at depths >10 μm and >20 μm, respective-

ly[15], whereas λ for this material is likely to be < 5 μm.

That the hardness maximum observed for glass in heptyl alcohol can be reproduced in binary alcohol solutions, Fig. 2b, establishes that the hardness extrema are not specific to one environment. It also reveals that such maxima are not related to impurities in the heptyl alcohol (e.g., water), or to some specific dissolution or indenterlubrication phenomenon. Such an effect can be understood, however, in terms of the relationship between ζ-potential and hardness. If one component of the binary solution imparts a negative ζ-potential to the glass and the other a positive potential, some mixture of the two components must give $\zeta \simeq 0$, and so produce a hardness maximum.

APPLICATION OF CHEMOMECHANICAL EFFECTS TO DRILLING

Workers in the USSR have for many years used environmental effects to increase the efficiency of rock drilling operations[26]. However, rocks are for the most part complex, inhomogeneous, multiphase materials. This makes it difficult to interpret environment-dependent drilling data obtained, and to optime cutting rates. Accordingly, recent studies have been concerned with the effects of chemisorption on the drilling behavior of such "model" solids as glass[9,27] and crystals of MgO and CaF_2[27,28] and then extended to alumina[4], calcite, feldspar quartz and granite[6].

A precision drill press was employed to puddle drill at a constant speed of 2200 r.p.m. and constant dead load. Rate of penetration as a function of time was recorded with an LVDT. Two bit types were employed, 2.4 mm carbide spade bits, and diamond-loaded hemispherical-ended or core bits between 3 and 6.4 mm in diameter. Spade bits blunt relatively quickly, and so give rise to a time-dependent rate of penetration. The total penetration achieved in 600 sec, D(600), was used to indicate drilling efficiency. Diamond-loaded bits, on the other hand, provide an essentially constant drilling rate after about 100 sec and so D(200), defined as the average penetration rate from 150 to 250 sec was used to assess the efficiency.

Fig. 6b shows the variation with environmental composition of D(600) for MgO crystals drilled $\langle 100 \rangle$ with a spade bit in DMSO-DMF solutions. Note the similarity of this curve to that for dislocation mobility in MgO, as indicated by ΔL (1000), Fig. 6a. Similar correlations have been observed for MgO in other environments, and

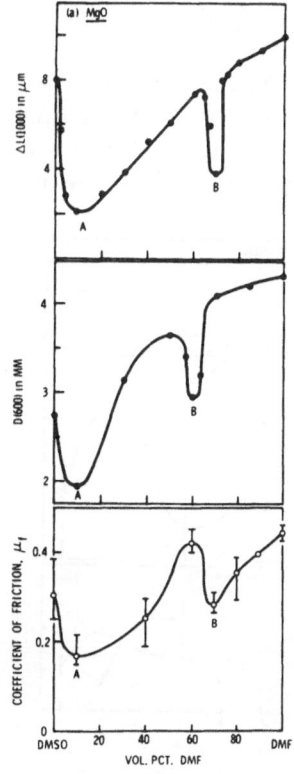

Fig. 6. Variation of (a) $\Delta L(1000)$ - a measure of near-surface edge dislocation mobility[13], (b) penetration by a carbide spade bit in 600 sec[28], and (c) the coefficient of sliding friction[36] for MgO in DMSO-DMF solutions.

for CaF_2[27, 28]. It is clear that spade bit penetration is directly related to near-surface dislocation mobility in MgO and CaF_2 and is greatest in those environments that maximize dislocation mobility and minimize hardness.

Conversely, when glass is drilled with diamond-loaded bits in n-alcohols or n-alkanes, the opposite correlation between hardness and drilling rate is observed. Fig. 2c shows that D(200) is greatest in heptyl alcohol, the environment which imparts zero ζ-potential to the glass and maximizes its hardness. Like the hardness maximum, the maximum drilling rate can be reproduced by binary mixtures of alcohols, one of which has $N_C < 7$ and the other $N_C > 7$.

Opposite effects of water with NaOH or HNO_3 on the penetration rate of calcite by carbide spade and diamond-loaded bits are shown in Fig. 7c. (The efficiency of the spade bit is defined here as D(60), the penetration in the

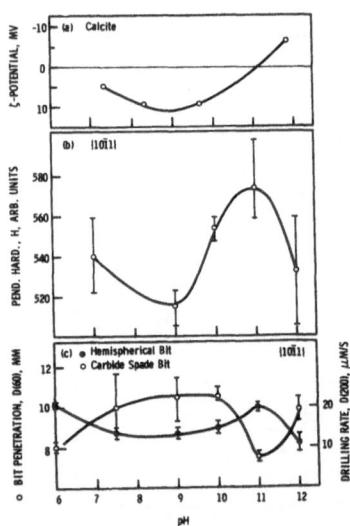

Fig. 7. Variation of (a) ζ-potential[29], (b) pendulum hardness[5], and (c) penetration by carbide spade bit in 60 sec or rate of penetration by a diamond-studded bit after 200 sec, for calcite crystals as a function of pH in buffered water environments.

first 60 sec.) Comparison with the ζ-potential[29] and
pendulum hardness[5] shows that the rate of penetration
is greatest for the diamond bit and least for the carbide
bit when $\zeta \simeq 0$ and hardness is greatest. Similar behavior has been observed when drilling MgO in n-alcohols
and glass in the n-alkanes[27]. That such opposing effects
can occur in the same environment, in the presence of
the same impurities, and with the same cooling, lubricating
and dissolution properties, demonstrates that none of the
latter factors exerts the controlling influence on drilling
rate. It is considered that such opposing influences of a
specific environment derive from the different role played
by environment-sensitive near-surface flow behavior in
the formation of chips by different types of bits[27].

For spade bits, it seems reasonable that a significant
amount of plastic deformation occurs in a "flow-zone"
ahead of the cutting edge, analogous to metal cutting[30].
Since material cannot readily escape around the edge of
the bit, strain accumulates, rapidly exhausting the limited
work-hardening capacity of materials such as MgO or
calcite. As deformation proceeds, dislocations pile up
at slip band intersections and nucleate cracks. These
grow quickly to critical size, propagate and interact to
complete chip formation. The rate of penetration of a
spade bit varies with environment in the same manner as
dislocation mobility because the dislocation motion that
comprises the essential first step in chip formation is
environment-sensitive.

In contrast, the irregularly shaped diamonds pro-
truding from a bit may be regarded as individual cutting
tools, each having a short, curved cutting edge, a large
negative rake angle, and travelling in a concentric circular groove. To the extent that environment-sensitive
dislocation motion occurs adjacent to such a tool, it is
thought to produce an outward, radial flow of material
towards the edges of the groove made by the tool, where
no further plastic strain can accumulate. Such flow is
not envisaged as the primary mechanism of chip formation.
Rather, it is postulated that chips are produced mostly by

the coalescence of cracks formed immediately behind the tool, where large tensile stresses may be expected[31] in the near-surface region just damaged by passage of the tool. Such plastic flow as does occur is presumed to lower the level of stress beneath the tool and to blunt the cracks involved in chip formation, two negative influences. In this view, any environment which facilitates dislocation mobility reduces drilling efficiency. Similar concepts may explain environment-sensitive effects in drilling of glass if some different flow mechanism is involved.

It may be concluded that the specific cutting action of the tool, the deformation characteristics of the workpiece, and the influence of environment on the near-surface flow and fracture behavior must be considered before recommending chemical optimization of any machining operation. The observation that maximum penetration rate in heptyl alcohol when drilling glass with a diamond bit may be reproduced in a binary alcohol solution implies that any liquid which imparts the same ζ-potential to the solid will provide the same drilling rate. A wide choice of cutting fluids is likely to be available in each instance, and it should be possible to formulate one that is cost effective as well as non-toxic and non-polluting. However, a given cutting fluid will be optimum only for the particular combination of solid and cutting tool for which it was developed. Any change of tool which alters the chip formation mechanism will require a different cutting fluid for optimum performance.

These concepts have been applied to diamond drilling of alumina[4] and granite[6]. Fig. 8 illustrates the variation in ζ-potential[32], pendulum hardness, and rate of penetration by a core bit[5] for Al_2O_3 crystals in water, toluene and n-alcohols. Al_2O_3 is most effectively drilled when $\zeta \simeq 0$, i.e., when its hardness is maximized. The maximum drilling rate obtained in octyl alcohol is ten times that achieved in water, and this results from a quite modest ($\sim 30\%$) increase in pendulum hardness. Drilling polycrystalline alumina[4] also can be accelerated by appropriate choice of n-alcohol environment.

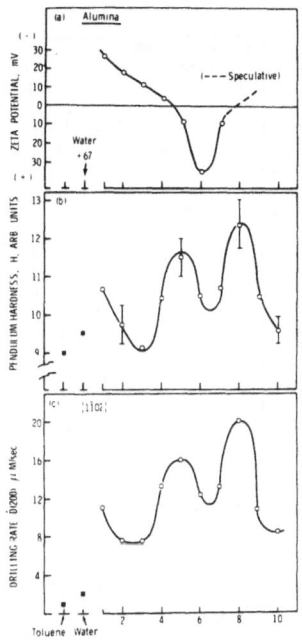

Fig. 8. (a) ζ-potential[31], (b) pendulum hardness and (c) rate of core bit drilling[4] for alumina monocrystals in toluene, water and the n-alcohols.

Fig. 9 illustrates the variation in pendulum hardness and rate of diamond drilling of quartz crystals in water, toluene and n-alcohols[6]. (Hardness and drilling data for an impure polycrystalline quartz of 1-2 mm grain diameter is almost identical.) A modest hardness increase ($\sim 30\%$ relative to the hardness in water) is sufficient to produce marked increase in the rate of diamond drilling, i.e., a 14-fold increase in undecyl alcohol.

Fig. 10 presents data on core bit rotary drilling of gray granite. The penetration rate in heptyl alcohol was twice that in water. The similarity with Fig. 9b for quartz is striking, and implies that quartz controls the drilling behavior of granite, at least in n-alcohol environments and that grain boundaries and other microstructural imperfections in the granite play no significant role in chip formation.

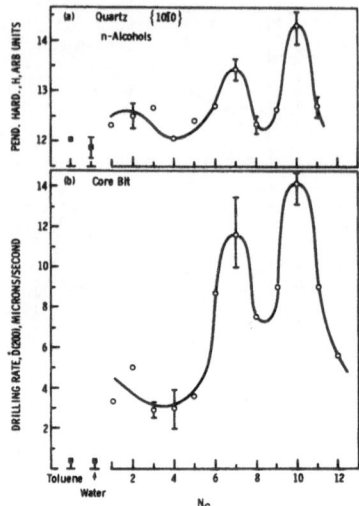

Fig. 9.   (a) Pendulum hardness and (b) rate of drilling
with diamond-loaded core bits for quartz monocrystals
in toluene, water and n-alcohol environments[4].

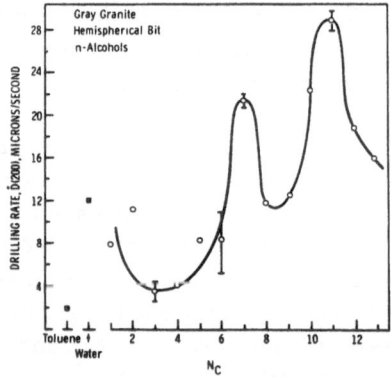

Fig. 10.   Variation in rate of drilling of gray granite in
water, toluene and n-alcohol environments.   Diamond-
loaded hemispherical-ended bits rotating at 2200 r. p. m. [4]

      Again, the drilling rate maxima for quartz and granite
in heptyl and undecyl alcohols can be reproduced in binary
solutions, implying that the drilling behavior in the n-alcohols

is probably controlled by surface charge. The fact that monocrystalline and polycrystalline quartz, granite[6], and a variety of silicate glasses[27] all show corresponding hardness and drilling rate maxima in heptyl alcohol illustrates the point that chemomechanical effects in silicates, crystalline or otherwise, are comparatively insensitive to composition.

The n-alcohols are both noxious and costly, and thus not practical drilling fluids. However, because chemomechanical effects are not specific, except with respect to surface charge, it should be possible to formulate more practical and equally effective water-based drilling fluids for drilling of granite. The large negative ζ-potential of quartz in water can be increased to zero by addition of DTAB[33]. The data for Westerly granite (Fig. 11) confirm that a more than three-fold increase in drilling rate over that in water can be obtained in 10^{-3} M DTAB solutions.

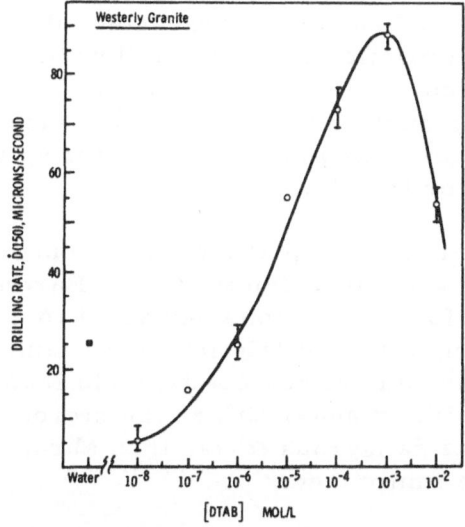

Fig. 11. Drilling behavior of Westerly granite in aqueous DTAB environments at 2200 r.p.m. Drilling rate is defined here as D(150), this being the average rate of bit penetration during the period 100-200 sec. after commencing drilling[4].

APPLICATION OF CHEMOMECHANICAL EFFECTS TO
THE CONTROL OF FRICTIONAL BEHAVIOR

The view of Bowden and Tabor[34] that friction arises in part as a result of fracturing the junctions formed by adhesion at the points of real contact (asperities) between two sliding surfaces, and in part as a result of deformation of the underlying material, is now widely accepted. These two processes may be affected in various ways by the environment in which sliding occurs. The environment may facilitate sliding by physically separating the moving surfaces (complete or hydrodynamic lubrication). Alternatively, as little as a monolayer of adsorbate may interfere markedly with adhesion of the asperities (boundary lubrication). Both effects may be expected to occur in all classes of solids. For inorganic nonmetals, the possibility exists that chemomechanical phenomena could affect frictional behavior. In this case, the coefficient of friction μ_f might be expected to reflect to some degree environmentally-induced, and surface charge-controlled, changes in near-surface dislocation mobility or flow behavior, provided that hydrodynamic lubrication does not occur, and that the sliding surfaces neither contain asperities larger than a few microns in height nor interpenetrate to depths much greater than this.

Such a possibility could provide some control over the rates of sliding of geological faults[35-37]. Earthquakes occur when shear forces developed adjacent to a fault exceed the restraining effect of friction at the fault interface. If injection of some fluid at key points could induce local sliding in a controlled manner before the stored strain energy built up to a dangerous level, then abrupt release of energy could be minimized.

To establish the feasibility of this concept, a study has been made of the frictional behavior of MgO in water, binary DMSO-DMF mixtures and buffered aqueous 10^{-2}N NaCl solutions; and of glass in water, pure n-alcohols, and a binary n-alcohol solution[36]. Specimens were prepared under the test environment, those of MgO by cleavage, and those of glass by controlled fracture using the Out-water-Gerry test geometry[38]. Two sliders were employed; a 6 mm sapphire ball (300 g load), and a standard 136° Vickers indenter (20 g load). Traverse speed was 0.5 mm/min and, for MgO, all traverses were made parallel to <100>.

Fig. 6c illustrates the variation in μ_f for sliding on MgO in bindary DMSO-DMF whose influence on near-surface dislocation mobility is shown in Fig. 6a. The same rather complex variation with environment occurs for both properties. Thus, μ_f is essentially controlled by near-surface dislocation behavior. Fig. 5c shows the variation of μ_f with pH for the same slider when MgO is immersed in buffered 10^{-2}N NaCl aqueous solutions. Comparison with the corresponding ζ-potential and dislocation mobility data in Figs. 5a and b, reveals that μ_f is largely determined by environment-sensitive dislocation motion. Moreover, μ_f is least when $\zeta \simeq 0$, so that the "ζ-correlation" appears to be relevant to frictional behavior, in the case of MgO.

Fig. 2d shows the variation in μ_f for a sapphire ball sliding on glass in water and n-alcohols, and the inset shows μ_f as a function of composition of binary solutions of octyl in hexyl alcohol. μ_f decreases more or less linearly with increasing molecular chain length, as first noted by Hardy and Doubleday[39] with a sharp local minimum in μ_f which occurs in heptyl alcohol. Fig. 2 shows that glass exhibits a zero ζ-potential and a maximum in hardness in this environment. There is a correlation between hardness, μ_f, and ζ-potential. Like the maximum in hardness, the minimum in μ_f can be reproduced by binary alcohol solutions.

It is evident that environment-sensitive near-surface flow can exert a significant, predictable, and sometimes dominant influence on the frictional behavior of inorganic nonmetals. μ_f is a minimum when $\zeta \simeq 0$. Preliminary results indicate that significant reductions (50%) in μ_f can be achieved[37].

ACKNOWLEDGEMENTS

It is a pleasure to acknowledge helpful discussions with colleagues W. M. Mularie and R. M. Latanision, and the able experimental assistance of R. D. Huntington. The financial support received from the U. S. Geological Survey, National Science Foundation and especially the U. S. Office of Naval Research is greatly appreciated.

REFERENCES

1. P. A. Rebinder, Proc. 6th Physics Conf., Moscow, 29 (1928).
2. A. R. C. Westwood and N. H. Macmillan, The Science of Hardness Testing, Am. Soc. Metals, Cleveland, Ohio, in press (1973).
3. R. W. Heins and N. Street, Soc. Pet. Eng. J., 5, 177 (1965).
4. A. R. C. Westwood, N. H. Macmillan and R. S. Kalyoncu, J. Am. Ceram. Soc., 56, 258 (1973).
5. A. R. C. Westwood and N. H. Macmillan, Rept. to U. S. Geol. Survey on Contract No. 14-08-0001-13077 (Jan. 1973).
6. A. R. C. Westwood, N. H. Macmillan and R. S. Kalyoncu, Trans. AIME (Mining), in press (1973).
7. A. R. C. Westwood and R. D. Huntington, Mech. Behavior of Materials, Soc. Matls. Sci., Japan, IV, 383 (1972).
8. A. M. Gaudin and D. W. Fuerstenau, Trans. AIME, 202, 66 (1955).
9. A. R. C. Westwood, G. H. Parr, Jr. and R. M. Latanision, Amorphous Materials, Wiley, London, 153 (1972).
10. A. R. C. Westwood, H. Opperhauser, Jr., and D. L. Goldheim, Phil. Mag., 6, 1475 (1961).
11. A. R. C. Westwood, Phil. Mag., 7, 633 (1961).
12. A. R. C. Westwood, D. L. Goldheim and R. G. Lye, Phil. Mag., 16, 505 (1967).
13. A. R. C. Westwood, D. L. Goldheim and R. G. Lye, Phil. Mag., 17, 951 (1968).
14. A. R. C. Westwood and D. L. Goldheim, J. Appl. Phys., 39, 3401 (1968).

15. N. H. Macmillan, R. D. Huntington and A. R. C. Westwood, Phil. Mag., in press (1973).

16. A. R. C. Westwood, R. D. Huntington and N. H. Macmillan, submitted for publication (1973).

17. McD. Robinson, J. A. Pask and D. W. Fuerstenau, J. Am. Ceram. Soc., 47, 516 (1964).

18. K. Inabe, K. Emoto, K. Sakamaki and N. Takeuchi, Jap. J. Appl. Phys., 11, 1743 (1972).

19. S. M. Cox, Phys. and Chem. Glasses, 10, 226 (1969).

20. W. G. Johnston, J. Appl. Phys., 33, 2716 (1962).

21. B. J. Wicks and M. H. Lewis, Phys. Stat. Solidi (a), 6, 281 (1971).

22. R. A. Huggins, Rept. of 1971 ARPA Materials Conf., Woods Hole, Mass., Univ. of Michigan, Ann Arbor, 1, 136 (1972).

23. J. H. Konnert, J. Karle and G. A. Ferguson, Science, 179, 177 (1973).

24. J. J. Gilman, J. Appl. Phys., 44, 675 (1973).

25. M. F. Ashby and J. Logan, Scripta Met., 7, 513 (1973).

26. P. A. Rebinder, L. A. Schreiner and K. F. Zhigach, Hardness Reducers in Rock Drilling, C. S. I. R. O., Melbourne (1948).

27. A. R. C. Westwood and R. M. Latanision, The Science of Ceramic Machining and Surface Finishing, N. B. S. Special Publ. No. 348, 141 (1972).

28. A. R. C. Westwood and D. L. Goldheim, J. Am. Ceram. Soc., 53, 142 (1970).

29. P. Somasundaran and G. E. Agar, J. Colloid and Interface Sci., 24, 433 (1967).

30. B. von Turkovich and G. F. Micheletti, Proc. 9th Intl. Machine Tool Design and Research Conf., Pergamon, New York, 1073 (1969).

31. G. M. Hamilton and L. E. Goodman, J. Appl. Mech., 33, 371 (1966).

32. R. A. Gortner, Trans. Faraday Soc., 36, 63 (1940).

33. W. M. Mularie, M. S. Rosenthal and A. R. C. Westwood, unpublished work (1973).

34. F. P. Bowden and D. Tabor, The Friction and Lubrication of Solids, Oxford Univ. Press, Vol. 1 (Revised) (1954), Vol. II (1964).

35. A.R.C. Westwood, RIAS Proposal No. 386 (July 1971).
36. N.H. Macmillan, R.D. Huntington and A.R.C. Westwood, submitted for publication (1973).
37. N.H. Macmillan and A.R.C. Westwood, to be published (1973).
38. A.G. Evans, J. Mat. Sci., 7, 1137 (1972).
39. W.B. Hardy and I. Doubleday, Proc. Roy. Soc., A100, 550 (1922).

DISCUSSION

J.M. Khan (LLL): You refer to dislocation effects to depths greater than 30 μm. Are there any compositional variation effects associated with deep surface influences?
Author: We have no information on this. We have studied the influence of environment on the behavior of dislocations as a function of depth only in freshly cleaved MgO crystals of ~ 99.9% purity.
L.L. Hench (Univ. Florida): Most surface-environment reactions have a significant time constant. How are the reactions responsible for maintaining the zero point of charge achieved in the very short times for which new surface is exposed during drilling experiments?
Author: The classical Langmuir adsorption model considers chemisorption as an activated process, which can be fast or slow depending upon the activation barrier height. However, recent ultra-high vacuum surface analysis experiments have established that rapid (sub-microsecond), non-activated chemisorption can occur on clean surfaces in a wide variety of adsorbate-solid systems. Possibly, therefore, similarly rapid chemisorption occurs while drilling under a liquid environment.
R. Atkin (IBM): You attributed the observed variations in diamond drilling silica and alumina to environmentally induced changes in the surface charge on the oxides. Wouldn't you also expect the mechanical properties of diamond to be environment - sensitive? Environmentally induced variations in tool wear and/or regeneration may explain the disparity in the cutting rate improvement for these two materials (minimal for Al_2O_3:diamond, but very dramatic for SiO_2:diamond).

Author: I would suppose that the mechanical behavior of diamond is also to some degree environment - sensitive. However, this cannot be the predominent environmental influence, for if it were all the solids we have cut with diamond tools would show the same dependence on environmental composition.   In fact, very different effects are seen.   For example, methyl alcohol is a relatively effective environment for cutting alumina, but detrimental when cutting glass with the same bit.   I'm surprised that you consider the ten-fold improvements in the rate of drilling of alumina achieved by cutting in octyl alcohol instead of water (Fig. 8) as "minimal".

A. Choudry (Univ. Rhode Island):   (1) The electronic transitions invoked to explain the fracture behavior could be tested independently by causing the same transitions by other methods, e.g., thermal or IR-excitation, and one should expect a marked decrease in the fracture strength under such excitation.   (2) Is there a connection between the electronic transitions model and the one based on surface charge distribution?

Author: (1) Variations in the mechanical properties of such solids as MgO, ZnO and CdTe have indeed been induced by exposure to light as well as by chemisorptive means.
(2) The two models are the same.   We have merely described the phenomena of interest in one case in terms that might be employed by a surface physicist and, in the other case, in the terminology of the physical chemist.

A. Choudry: The zeta-potential vs. pH value data as presented must first be translated into surface charge density before invoking fracture mechanisms based on electrostatic field stress.   Is there a clear mechanism which translated the ζ-potential into surface charge density for different values of pH?

Author: For an ideal polarized electrode in the absence of specific adsorption the Gouy-Chapman-Stern theory leads to a relationship between the surface charge density, and the ζ-potential of the form $\sigma = K_1 \text{Sinh} (K_2 \zeta)$, where K_1 and K_2 are constants.   This result is derived, for example, in Ch. 3 of "Double Layer and Electrode Kinetics", by P. Delahay (Interscience, 1965).

SLOW FRACTURE OF GLASS IN ALKANES AND OTHER LIQUIDS

C. L. Quackenbush and V. D. Fréchette

New York State College of Ceramics
At Alfred University, Alfred, New York

Water is generally considered to be the environmental agent most detrimental to the long-time strength of glass; consequently, it is to be expected that water will produce the highest fracture velocities for a given loading in the slow fracture range. Charles and Hillig[1] attribute this aqueous sensitivity to an autocatalized, stress-enhanced chemical corrosion of the glass network at the crack tip.

In environments in which water concentration is low, such as hard vacuum or in such liquids as the alkanes, glass also demonstrates delayed failure. This suggests that there are additional slow fracture mechanisms which are not dependent on the necessary presence of water.

MECHANISMS OF SLOW CRACK GROWTH

This study proposes that slow fracture of glass involves three independent mechanisms, with predominance dependent on local environment, velocity and glass composition. Increase in crack length intensifies the applied stress until eventually it may exceed crack tip bond strength. At this time fast fracture ensues. The proposed mechanisms by which this crack extension procedure occurs will be intro-

duced, then that mechanism which applies to alkanes will
be discussed in detail.

Chemical Corrosion

In the Charles model[2], highly strained bonds at the
crack tip are more susceptible to corrosive agents than
those along the side walls so that the crack advances maintaining its sharp tip geometry.

Surface Energy Reduction

Orowan[3] attributed static fatigue to an adsorption-induced lowering of the glass surface energy. We think of
this as meaning that it involves stress-assisted thermal
rupture of the crack tip bonds with recombination frustrated
by adsorption of the foreign species on the dangling bond
ends. The process is surely complex, with no single
physical property able to account fully for observed behavior.

Strain-Gradient Induced Diffusion of Mobile Ions

This mechanism employs strain-gradient induced diffusion of mobile ions, probably Na, to accomplish concentration of such ions at positions in advance of the crack
tip[4]. The driving forces for this preferential diffusion are
differential hydrostatic pressure and electrical polarization.
Enrichment with respect to these ions at the crack tip reduces the local Si-O bond strength allowing crack advance-
ment and apparently premature rupture.

Mechanism Predominance

In a corrosive environment chemical corrosion is
likely to be the dominant static fatigue process. Other
mechanisms operate, but their smaller effects are over-
shadowed.

In the presence of noncorrosive environments, e. g.,
alkanes or carbon tetrachloride, static fatigue is controlled

by the environment's effectiveness in reducing surface energy.

Finally, in those environments which only slightly reduce surface energy, e. g., dry nitrogen, strain-gradient diffusion dominates. Slow fracture in high vacuum (especially baked systems) is entirely attributable to this process and fast fracture is controlled by it up to limiting velocity. Chemical corrosion[2] and strain-gradient diffusion[4] have been discussed in the literature. A proposed mechanism to account for slow fracture in the presence of nonaqueous, noncorrosive environments will be discussed here.

In reviewing the lieterature on environmental effects, Hammond and Ravitz[5] observe that "in general ... environments with polar molecules have the greatest effect in reducing the fracture strength of glass". Berdennikov[6] showed that the critical stress necessary to initiate crack propagation in soda-lime glass varied in inverse proportion to the logarithm of the dielectric constant of the surrounding liquid (Fig. 1).

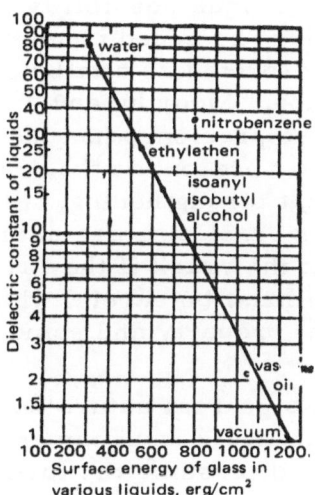

Fig. 1. Dielectric constant vs surface energy of soda-lime glass as determined by the critical stress to begin crack propagation (after Berdennikov[6]).

Such a simple relationship would indicate that the same fatigue mechanism was operating in the case of each liquid with an efficiency dictated by a single physical property, i. e. , bulk dielectric constant. Actually a more complex process, where additional factors exercise conflicting roles, would be expected. It is surprising that Berdennikov found only nitrobenzene an exception to his relationship.

In a glass fracture surface there are many atoms which must accept warped cross-bonding with neighboring surface atoms and a smaller number which must remain without complete charge satisfaction[7]. Small additions of a substance which can satisfy some of these surface bonds will lower the final surface energy and will make the original fracture surface generation process easier. Hydrogen bond formation, for instance, between the fracture surface and the liquid* surface layer, will create especially stable bonds, thus further decreasing the surface energy and the work of fracture surface formation.

Another factor which might upset a slow fracture experiment is the inadvertent introduction of a second slow fracture mechanism. One obvious possibility is the inclusion of water in the test liquid. The added chemical corrosion would then be incorrectly interpreted as attributable to the surface energy reduction scheme. Moorthy and Tooley[8] considered the opposite of the above problem when they suggested the dilution of water initially present at the crack tip (e. g. , by alcohols) and a resultant strengthening. Steric factors must also be considered, for crack tips are narrow and liquid molecules must fit into the crack if they are to reach the crack tip. Finally one must be careful when testing the trend between a bulk physical property and some characteristic of

---

*This model will be of a general nature and can apply equally to gaseous as well as liquid environments. The present discussion, however, will assume the form of the environmental agent to be liquid because in this study measurements of critical reflection angle indicated that the cracks were liquid-filled.

glass slow fracture. In the confines of a crack, reactions
will occur on the molecular level, not in bulk. This last
caution apply especially to such liquids as N-methylacetamide whose large dielectric constant is a product of long
polymer-like chains which form only in the bulk liquid .
These chains could never form within a crack for lack of
room; consequently, the molecular dielectric constant of
the liquid at the crack tip would be much lower than the
bulk value. Use of the bulk dielectric constant to characterize in-crack conditions would invite gross error.

With the above reservations in mind, the molecular
dielectric constant seems to be the most reasonable single
physical property to account for the difference in observed
surface energies and slow-fracture characteristics among
noncorrosive liquids. Consider the following argument.
Electronegativity calculations indicate that the Si-O bond
is 51% ionic, therefore partially polar. Because of this
polarity, the positive ends of naturally polar liquid molecules in the crack will be attracted to the oxygen atoms
and the negative ends to the silicons (Fig. 2). This partial
charge-neutralization decreases silicon-oxygen attraction.
The force of attachment should be a function of the molecule's inherent dipole moment and ease of polarization, as
measured by its molecular dielectric constant.

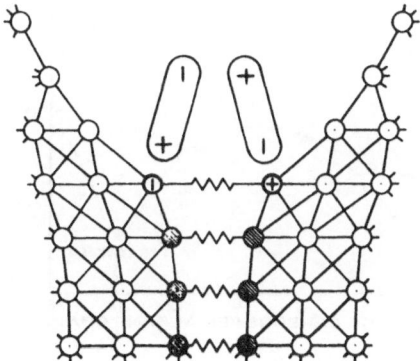

Fig. 2. "Atomic" model of the crack tip with adsorbed
polar molecules (modified after Goodier[10]).

Thermal energy will dissociate some fraction of these attached molecules and rupture some small fraction of the Si-O bonds. However, some sites will suffer bond rupture without desorption. In these instances the adsorbed molecules undergo rapid polarization to more completely satisfy the charged silicon and oxygen bond ends. As the ions vibrate back toward one another, bond re-formation is frustrated by charge neutralization and screening. Given time, the polar liquid molecules move in and allow relaxation of the structure and further Si-O separation.

EXPERIMENTAL

To study the influence of nonaqueous, noncorrosive liquids (alcohols and alkanes) on the energy necessary to propagate stable cracks in glass, Westwood et. al.[11] employed the center-loaded crack technique of Panasyuk and Kovchick[12].

One of their most interesting results is the inference that tetradecane might be at least as detrimental to the strength of glass as water (Fig. 3). It was the initial aim

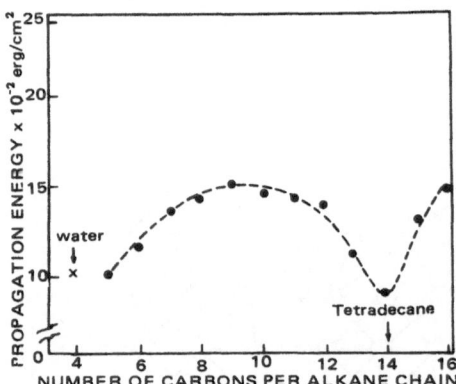

Fig. 3. Fracture energy of soda-lime glass immersed in water and also in alkanes of various chain lengths, listed according to the number of carbon atoms per alkane chain.[11]

of the present study to examine this possibly by exploring
the slow fracture vs stress intensity factor (K_I) behavior
of glass in the presence of liquids of the alkane series
pentane to pentadecane, with particular attention paid to
the reported anomalous behavior of tetradecane.

The glass samples (kindly supplied by PPG Industries)
were typical soda-lime-silica window glass of thickness
2.3 mm. Sheets 2 x 7 inches were annealed and cemented
to the strain frame between inner and outer grips of the
top and bottom cross members (Fig. 4). The crack region
was immersed in the liquid under study, which was regul-
ated to 23° C and held in a glass-walled container (not shown
in Fig. 4). Water at 23°C was circulated through top and
bottom cross beams to maintain constant dimensions in the
adjoining regions. Fixed strain was applied across the
specimen by circulating heated water through bores in the

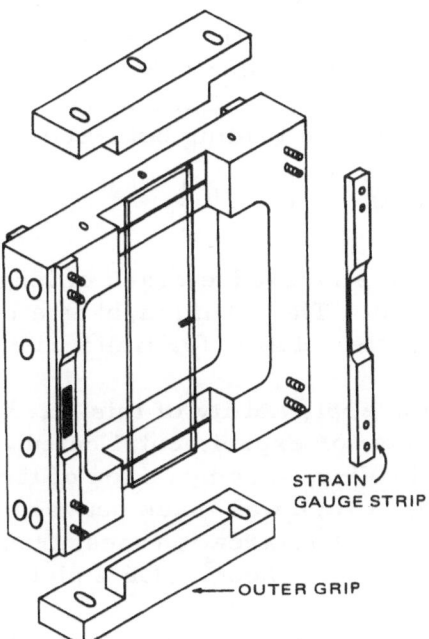

STRAIN
GAUGE STRIP

OUTER GRIP

Fig. 4. Exploded view of straining frame, specimen in
place but without liquid tank.

strain frame side arms causing thermal expansion of those members. Total elongation was monitored by a strain gauge mounted on one of four symmetrically spaced Invar strain gauge strips. Position of the crack tip was located by sighting through a 60X polarizing microscope on the stress concentration associated with the crack tip. The strain frame was mounted on the tailstock shaft of a lathe bed such that rotation of the feedscrew translated the entire frame parallel to the direction of crack propagation. The microscope, fitted with cross hairs, was mounted rigidly on the crosshead. Feedscrew rotation was coupled via a continuous-rotation, single-turn potentiometer and appropriate circuitry to read out relative position of the crack tip as a function of time on a strip chart. All crack lengths were measured relative to the sample edge position to an accuracy of better than 0.02 mm.

The data were plotted as discrete velocity vs crack length points and connected by a smooth curve. Velocity and crack length points from this representative curve were used to calculate the velocity vs K_I relationship for each sample. An exact K_I vs crack length solution is not available for plane-strain loading but Paris and Sih[15] give:

$$K_I = \sigma \, (a\pi)^{1/2} \, \left[k(a/b) \right]$$

as the solution for a single-edge crack of length $\underline{a}$ loaded under plane-stress σ. The term $k(a/b)$ is a correction to the semi-infinite plate solution for finite width 2b.

To evaluate the applicability of this fixed-stress equation as the basis for experimental calculations involving fixed-strain loading, a comparison of the stress at positions ahead of a crack was made between the Paris and Sih equation and optical retardation measurements (Sénarmont method) in an epoxy sheet*. On a plot of K_I vs crack

---

*Under conditions of fixed strain the stress distribution in a plane sheet (with a crack of length = a) is independent of material constant.

length, the Paris and Sih equation was found to lie within the scatter of the optical retardation data. Therefore, it was adopted as a reasonable approximation for the calculation of stress intensity factor.

The velocity vs K_I curves were collected according to test liquid and a single curve drawn to give the best visual representation for the trend in that particular liquid. Although the individual curve shapes are similar, there is a great deal of variation between curves, as exemplified by the hexane data of Fig. 5. One major source of this variation is the extreme difficulty in fixing the initial "zero" of applied strain. This error has the effect of displacing the entire curve along the K_I axis. One could imagine closer alignment of these curves by a gathering process involving curve translation along the K_I axis.

The representative velocity curve for each liquid is given in Fig. 6. In the present study it was not possible

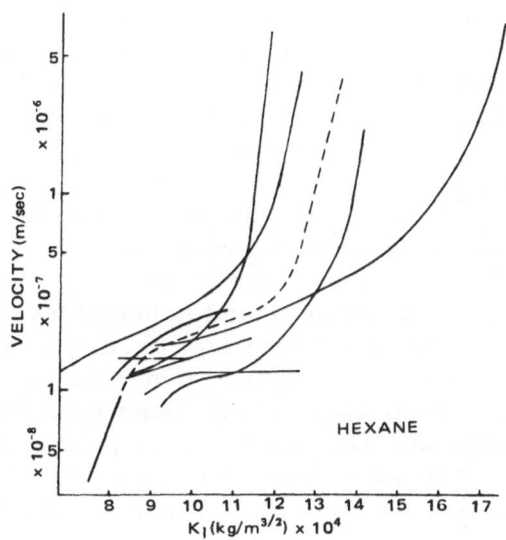

Fig. 5. Typical velocity vs K_I scatter. The solid curves are of individual samples; the dashed curve is the best visual representation of the data.

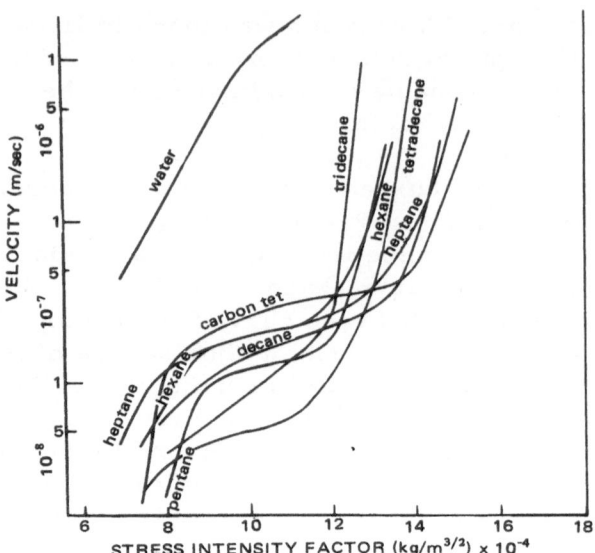

Fig. 6.   Velocity vs K_I for alkanes and water.

to discern any difference between the curves representing
the several alkanes because the scatter around each curve
was as large as the envelopes of all alkane curves.   It is
believed, however, that a chain length effect as large as
that given in Fig. 3 would be identifiable despite the
scatter.   Instead, tetradecane appeared to react in the
same way as decane.   Between the two studies the alkane-
water comparison particularly is very different.   Figure
3 shows fracture energies between alkanes and water to
be comparable for most alkanes and actually lower for
tetradecane.   This would indicate that under the same
loading conditions similar slow fracture velocities should
result.   However, the present study found slow fracture
velocities for water and alkanes to differ (according to K_I)
by between one and three orders of magnitude.   As for the
possibility of adsorption-controlled mechanical properties
of glass at the crack tip[11], Fig. 6 reveals little difference
in velocity behavior between CCl_4 and alkanes despite the
fact that negative chlorines surround the carbon atom in

CCl_4 while positive hydrogens surround it in the alkanes. Their bulk dielectric constants are similar (all near 2.0).

The shape of the crack velocity vs K_I curve has been discussed in another of our papers[16] along with a specimen thickness effect; the latter, however, is not able to fully account for the behavior of the three different sample thicknesses in the water environment. The difference in curve position between alkanes and water is attributable to the fact that water is a small, highly polar molecule (bulk dielectric constant = 85) and naturally corrosive toward glass. Consequently, it efficiently accomplishes both chemical corrosion and surface energy reduction slow-fracture mechanisms. It is concluded that alkanes support only surface energy reduction, and to a lesser extent than water because of their lower dielectric constant.

SUMMARY

Slow fracture is accounted for by three separate and independent mechanisms. Observable slow crack propagation will occur at very low stress intensity factors in the presence of corrosive environments; noncorrosive environments incorporating molecules with high dielectric constant require higher applied K_I, and very much higher K_I is needed to initiate environmentally independent slow fracture. In a noncorrosive environment where interaction with the crack tip glass network is purely electrical, the resulting slow crack propagation is predicted to be a function of the dielectric constant of the environmental molecules. This property estimates the environment molecule's force of attachment to the atoms of the polar Si-O bond at the crack tip and its ability to screen and charge-neutralize this bond. Thus, the higher the dielectric constant the more easily is the glass structure ruptured under strain and the higher the resulting velocity in the slow-fracture range.

REFERENCES

1. R. J. Charles and W. B. Hillig, "The Kinetics of Glass
   Failure by Stress Corrosion", in Symposium sur la
   résistance méchanique du verre et les moyens de
   l'améliorer, Florence, September 1961, Union
   Scientifique Continentale du Verre, Charleroi, Belgium,
   1962.

2. R. J. Charles, "Static Fatigue of Glass, I and II," J.
   Appl. Phy., 29 (11) 1549-60 (1958).

3. E. Orowan, "Fatigue of Glass Under Stress," Nature
   154 (3906) 341-43 (1944); Ceram. Abstr., 24 (1) 7 (1945).

4. C. L. Quackenbush and V. D. Fréchette, "Slow Fracture
   in Glass: I, Strain-Gradient Diffusion-Assisted Crack
   Propagation," to be published.

5. M. L. Hammond and S. F. Ravitz, "Influence of Environment on Brittle Fracture in Silica", J. Amer. Ceram.
   Soc. 46 (7) 329-32 (1963).

6. V. P. Berdennikov, Physik Z. Sowjetunion 4, 397 (1933);
   English translation in Surface Energy of Solids, trans-
   lated by V. D. Kuznetsov, Department of Scientific and
   Industrial Research, Her Majesty's Stationery Office
   (London) (1957) pp. 224-34.

7. G. Hochstrasser and J. J. Courvoisier, "Detection of
   Dangling Bonds on the Surface of Silicon and Quartz by
   Electron Spin Resonance," Helv. Phys. Acta., 39 (3)
   189-91 (1966).

8. V. K. Moorthy and F. V. Tooley, "Effect of Certain
   Organic Liquids on the Strength of Glass," J. Amer.
   Ceram. Soc., 39 (6) 215-17 (1956).

9. L. R. Dawson and W. W. Wharton, "Solutions of Some
   Alkali Halides in the Pure Liquids and in Mixtures of
   N-Methylacetamide and Dimethylformamide," Electro-
   chem. Soc., 107 (8) 710-13 (1960).

10. J. N. Goodier, "Mathematical Theory of Equilibrium
    Cracks," pp. 1-66 in Fracture, an Advanced Treatise,
    Vol II, Mathematical Fundamentals, H. Liebowitz ed.,
    Academic Press, New York (1968).

11. A. R. C. Westwood and R. D. Huntington, "Adsorption-
    Sensitive Flow and Fracture Behavior in Soda-Lime
    Glass," Third Technical Report to O. N. R., Office of

Naval Research, Contract Number N 00014-70-C-0330
NR-032-524, June 1971 Research Institute for Advanced
Studies, Martin Marietta Corp. 1450 S. Rolling Road,
Baltimore, Md. 21227.

12. V. V. Panasyuk and S. E. Kovchik, Soviet Physics-
     Doklady, 7 835 (1963).
13. G. I. Barenblatt, Adv. in Appl. Mech., 7 55 (1962).
14. V. D. Fréchette, C. L. Quackenbush and J. R. Varner,
     "A Tensile Device for the Study of Controlled Slow
     Fracture," to be published.
15. P. C. Paris and G. C. Sih, "Stress Analysis of Cracks,"
     p. 30 in Fracture Toughness Testing and its Applications,
     A. S. T. M. Spec. Technical Publication No. 381, Am.
     Soc. Testing and Materials, 1916 Race St., Philadelphia,
     Pa. (1965).
16. C. L. Quackenbush and V. D. Fréchette, "Slow Fracture
     in Glass: II, A Mechanistic Interpretation," to be pub-
     lished.

DISCUSSION

R. W. Rice (Naval Res. Lab.): Dr. Freiman of NRL has
investigated the effects of straight-chain alcohols on fracture surface energy of glasses using a constant stress
intensity test. He did observe some limited effects of
chain length on fracture energy at low velocities but later
found that the differences were attributable to small quantities of water in the alcohols.

Author: Our choice of the alkanes in preference to alcohols
to look for effects similar to Westwood's was based on their
lesser tendency to take up water. The danger is always
hard to avoid, however.

A. Choudry (Univ. Rhode Island): What order of time
intervals are involved in stress variations, e. g., would
there be complications due to dynamic elastic factors,
inertia, etc. ?

Author: Measurements extended, in general, over many
hours and in no case less than some minutes, so that
dynamic effects must be considered absent.

L. L. Hench (Univ. Florida): In regard to Westwood's comment concerning material surface energy-liquid interpretations (not submitted in writing, Ed.) it should be noted that in interpreting drilling data there is actually a three-phase system, i. e., tool-liquid-material, and analysis must therefore consider tool-liquid effects as well as material-liquid.

THE EFFECTS OF MICROSTRUCTURE ON THE FRACTURE ENERGY OF HOT PRESSED MgO

J. B. Kessler[*], J. E. Ritter, Jr.[*], and R. W. Rice[≠]

[*]Mechanical and Aerospace Engineering Dept.
University of Massachusetts, Amherst, Mass.
[≠]U.S. Naval Research Laboratory, Washington,
D.C.

The fracture energy of brittle materials can be defined as the energy absorbed during the extension of a crack over a unit area of surface formed during the fracture process[1]. Measurements of fracture energy for polycrystalline ceramic materials have resulted in a wide range of reported values for the same material. This is due to some extent to the method used for measurement[1-3]; however, more important is the strong dependence of fracture energy on such microstructural variations as grain size[4-9] and porosity[2,6].

Gutshall and Gross[8] found over a two-fold increase in fracture energy in fully dense Al_2O_3 as the grain size increased from 10 to 45 μm. This result was explained in terms of transition from intergranular to transgranular fracture with increasing grain size, the latter process requiring more energy. Clarke et al.[6] also observed that fracture energy increased with grain size in MgO that contained about 3-4% porosity. With fully dense MgO they found that fracture energy increased up to a grain size of about 100 μm, then decreased with increasing grain size to 150 μm. Also, the fracture energy of the fully dense MgO was less than that for the less dense samples. On the other

Work supported in part by the Office of Naval Research

hand, Simpson[2] observed that fracture energy in dense Al_2O_3 (1-2% porosity) decreased with increasing grain size.   This was attributed to the fact that the porosity in his samples was primarily intergranular, which promoted intergranular fracture for all grain sizes studied (up to 30 μm).   Because no transgranular fracture occurred, the increase in fracture energy with grain size reported by the above researchers was not observed.   Simpson postulated that the fracture energy decrease was caused by the larger residual stresses present in large-grained material as a result of anisotropic thermal contractions during cooling from the fabrication temperature.   Simpson also observed that when samples contained connected porosity, fracture energy was lowered, since connected porosity could provide low-energy paths for crack propagation.

The purpose of the present investigation was to evaluate the dependence of fracture energy on microstructural variations for relatively dense, hot-pressed MgO.   Grain size, amount and location of porosity and grain boundary precipitates were the features of major interest.   One difficulty in a study such as this is in measuring the effect of one microstructural variable while holding the others constant. This difficulty was overcome by having available a large number of samples fabricated under various conditions to give a wide range of microstructures.   This permitted us to measure, for example, the effect of grain size on fracture energy at constant porosity and vice versa.

EXPERIMENTAL PROCEDURE

Materials

All specimens were supplied by R. W. Rice of the Naval Research Laboratory and fabrication details are given elsewhere[10,11,12].   Briefly, the specimens were hot pressed in graphite dies from Mallinckrodt or Fisher reagent-grade MgO powders, both with and without LiF additions, in the form of discs about 1.5-in diameter by 0.1 to 0.25-in. thick. To obtain a wide range of microstructures, several specimens were cut from each hot pressed disc and then annealed in air at various temperatures using a silicon carbide resistance heater to 1650° C and a gas-fired furnace above 1650° C.

Sections of each sample were mechanically polished and etched from 5 to 30 seconds in boiling chromic acid. The mean grain size was determined by a linear intercept technique from optical micrographs for grain sizes down to about 10 microns and from scanning electron micrographs for finer grain sizes.

The porosity of the samples was determined from visual clarity. Those samples that were clear or translucent were considered "dense" and had less than 0.1% porosity. Opaque or cloudy samples were considered "porous" although their porosity did not exceed 1.0%.

In general, for a given annealing temperature the grain size in the MgO samples with LiF was larger than in specimens without additives, while porous specimens gave the finest grain size. Residual porosity from incomplete densification or clouding was generally fine and located at grain boundaries for annealing temperatures up to 1400 to 1500° C, corresponding to grain sizes of about 30-40 microns. For higher annealing temperatures, pores were generally larger and usually within the grains. The effect of annealing temperature on grain size and strength of the MgO specimens is summarized in Fig. 1.

Fracture Energy Measurement

All fracture energy measurements were made with the work-of-fracture technique devised by Nakayama[14] and Tattersall and Tappin[4]. The specimen used in this test contains a triangular reduced cross-section to promote a noncatastrophic, stable mode of failure (see Fig. 2). Fracture energy is found by measuring the work required for failure by integration of the force-displacement curve obtained from an Instron testing machine. The work done is converted to fracture energy by dividing it by twice the projected triangular area.

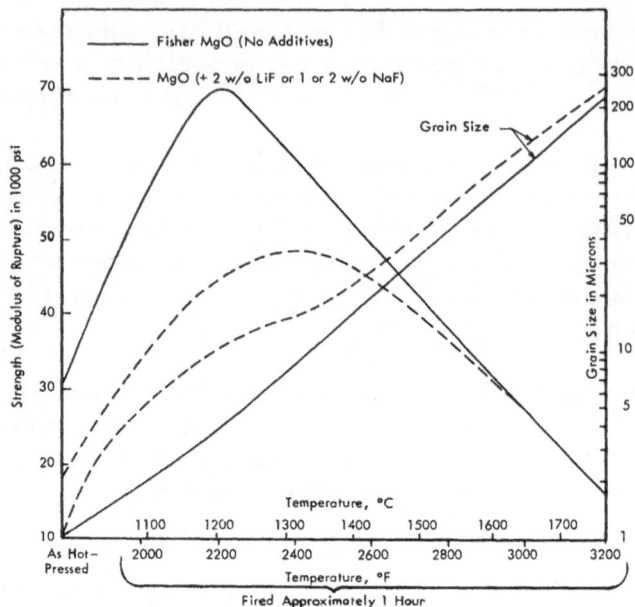

Fig. 1. MgO strength and grain size versus firing temperatures. These are only approximate curves, since results
for each hot-pressing can vary with both material and pressing conditions (after Rice[13]).

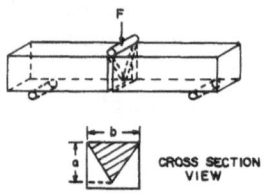

Fig. 2. Typical specimen configuration for work-of-fracture technique. The specimen is triangularly notched to
promote fracture stability.

The fracture energy samples were cut into bars about
0.1 x 0.2 x 1.0 in. and notched with a diamond wheel 0.012
in. thick. These specimens were tested in three-point
bending using an outer span of 0.625 in. at a cross-head
speed of 5.0×10^{-4} cm/min. Since stable crack propagation

in these samples was difficult to achieve, the reduced triangular cross-section for each set of samples was varied from about 60 to 20% of the original cross-section to increase the stress concentrating effect, since Simpson[2] has shown that as deeper notches are used, the proportion of stable crack propagation increases.

RESULTS

Apparent fracture energies for MgO specimens with LiF that were given an initial anneal are plotted as a function of the remaining cross sectional area in Fig. 3. This type of plot was suggested by Simpson[2] to obtain true fracture energies from the work-of-fracture test. He found that as the stress concentration increased, true fracture energies were either obtained at low reduced areas or could be estimated by extrapolation to zero reduced area. From Fig. 3 it is seen that the fracture energy decreases as the stress concentration effect increases until a minimum value, corresponding to completely stable fracture, is reached. The true fracture energy is then taken to be the average for all those tests showing completely stable fracture, which in this case corresponds to 0.98×10^4 erg/cm^2. Fracture energies of all samples in this study were

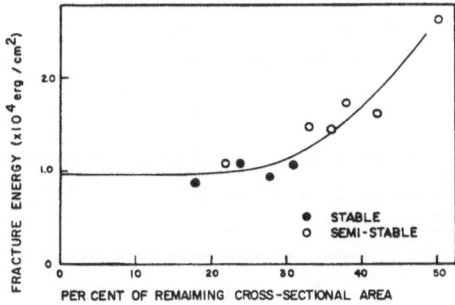

Fig. 3. Typical plot of fracture energy vs. percent of remaining cross-sectional area to find the minimum fracture energy, corresponding to stable crack propagation, for a given material. This plot is for MgO with LiF that was given an initial anneal.

determined in a similar manner except for "porous" samples with grain sizes greater than 50 microns. For these large-grain "porous" samples stable fracture was not achieved even at reduced areas of 20% and, consequently, the fracture energy could only be estimated by extrapolating to zero reduced area.

The effect of an initial anneal at 1100-1200°C for approximately 1 hr on the fracture energy of as hot-pressed MgO is given in Table I. The fracture of the as hot-pressed MgO with LiF was totally intergranular; however, after the initial anneal 10-30% of the fracture was transgranular. For as hot pressed MgO with no additives the amount of transgranular fracture decreased from about 20 to 10% on annealing and intergranular porosity became quite noticeable. The grain size of the samples with LiF was about 2 μm in the as-hot-pressed condition and about 10 μm in the initial annealed condition. For the samples without LiF the grain size of the as-hot-pressed samples was about 10 μm and the initial annealed samples about 7 μm.

The fracture energy of the "porous" samples increased with grain size as shown in Fig. 4. Included in Fig. 4 are values reported by Clarke et al.[6] for MgO with 3-4% porosity. For the samples in this study with grain sizes of 30 μm and less, the porosity was located in the grain boundaries, usually at triple points. For the larger grain size samples, the pores were found within the grains and the pore shape became spherical. Also, the percentage of transgranular

Table I. Effect of the initial anneal on the fracture energy of as hot-pressed MgO. Initial anneal was for about one hour at 1100-1200°C in air.

	Fracture Energy (x 10^4 ergs/cm^2)	
	As Hot-Pressed	Initial Anneal
MgO with LiF	0.47	0.98
MgO without additives	1.09	0.94

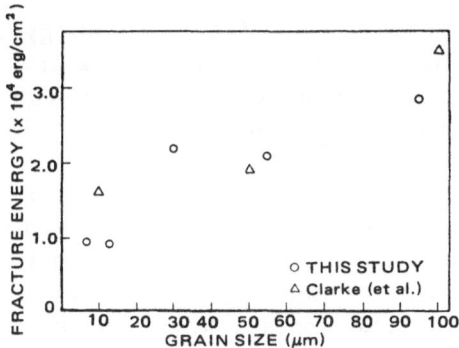

Fig. 4. Dependence of fracture energy in grain size for "porous" MgO.

fracture showed a general increase with grain size from about 10% at the 10-μm grain sizes to 100% for the 100-μm grain size samples. The fracture surface of a 100 μm grain size sample is shown in Fig. 5. This fracture was completely transgranular with numerous cleavage steps evident. The pores within grains can be seen at 1000X.

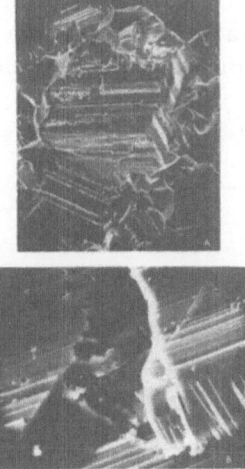

Fig. 5. Fracture surface of "porous" MgO with a grain size of 100 μm at (A) X 30 and (B) X 300.

The effect of grain size on fracture energy of dense MgO is shown in Fig. 6. Again, the values of Clarke et al.[6] are included along with the fracture energy value obtained by Evans[15] for dense MgO specimens. In comparison with the "porous" samples, the fracture energy of dense MgO is not strongly dependent on grain size, although the amount of transgranular fracture increased with grain size for the dense samples from about 20% at 10-µm grain size to 75% at 100-µm. These dense structures gave a much smoother-appearing fracture surface than "porous" samples.

DISCUSSION

Rice[11, 12], using IR, mass spectroscopy, and weight-loss measurements, has shown that hot-pressed MgO bodies, similar to the ones tested in this study, contain variable amounts of hydroxide and carbonate impurities averaging several hundred parts per million. Because of their large ionic size, these impurities are most likely to be at grain boundaries, where they could most affect fracture. In addition, hot-pressed MgO with LiF additions were found by Rice to contain substantial (of the order of 0.5%) residual fluoride, which again is most likely concentrated at the grain boundaries. Annealing progressively reduces the levels of all of these impurities, with most being lost during the initial anneal.

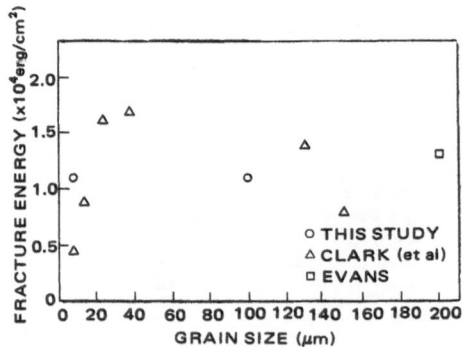

Fig. 6. Dependence of fracture energy on grain size for dense MgO.

The low value of fracture energy of as-hot-pressed MgO with LiF (and the correspondingly completely intergranular fracture) is consistent with the substantial residual fluoride content and its expected grain-boundary concentration. The marked increase in fracture energy and the partly transgranular failure of these samples after initial annealing clearly correlate with the large reduction of residual fluoride observed on annealing. On the other hand, MgO as hot-pressed without additives has a higher fracture energy than the as-hot-pressed samples with LiF and exhibits some transgranular fracture. This indicates that the hydroxide and carbonate impurities have much less effect on fracture than residual fluoride, consistent with their lower level. The lack of a clear increase in fracture energy on initial annealing of MgO without additives questions the embrittling effect attributed to these impurities[10]. However, this is still somewhat uncertain since, first, Rice[11, 12] has pointed out that annealing losses of such impurities are greatest near the annealed surface. In our fracture energy tests on deeply notched specimens, crack propagation is started well inside of the samples where the loss of these impurities may still have been quite small. Secondly, Rice has observed that porosity generally increases inside of the annealed surface toward the center of the bar. Both of these factors would limit the effect of annealing in increasing fracture energy, and, in fact, limited porosity could lead to a decrease. Some of our fracture observations, such as the appearance of intergranular porosity on annealing, with a corresponding decrease in transgranular fracture, support these suggestions.

The fracture energy of dense, annealed MgO is not strongly dependent on grain size (Fig. 6). The apparent modest initial increase in fracture energy with grain size reported by Clarke et al.[6], is probably due to residual fluoride trapped within their samples, since it is thought that their samples contained larger additions of LiF and were larger in size than the ones used in this study. Thus, it is entirely possible that not all of the fluoride had volatilized from their samples on initial annealing, resulting in the low fracture energies that they observed at the small

grain sizes (i.e., initial anneal). In contrast, the fracture energy of the porous MgO shows a definite increase
with grain size (Fig. 4). This increase can be explained
in part by the increase in transgranular fracture; however,
the dense samples also showed a similar increase in transgranular fracture even though their fracture energy remained relatively constant. Therefore, it is believed that
other factors, in particular the location of porosity, play
a more predominant role in determining the fracture
energy.

At small grain sizes the porosity is located at the
grain boundaries and the fracture is largely intergranular.
It is possible that these angular intergranular pores act
as subsidiary crack nuclei which can be triggered by the
high stress fields near the tip of the advancing crack and
thereby promote intergranular fracture. This is supported by the fact that porous MgO of 10-μm grain size had a
lower fracture energy than dense MgO of similar grain
size (0.95 x 10^4 compared to 1.1 x 10^4 ergs/cm^2). In addition, intergranular pores may be helpful in concentrating
the applied load so that less strain energy is present at
the initiation of crack propagation and, consequently,
stable fracture is facilitated. As the pores become more
rounded and located within the grains, this stress concentration would not be as pronounced and could explain why
no stable fractures were observed in the "porous" samples
with grain sizes larger than 50 μm.

At the larger grain sizes (above 30 μm) much of the
porosity is located within the grains and fracture becomes
predominantly, if not entirely, transgranular. From scanning electron micrographs such as in Fig. 5, it is seen
that in-grain porosity promotes numerous cleavage steps
as previously suggested by Rice[10] for MgO and by Forwood
and Forty[16] for NaCl. This would explain the much rougher
fracture surface of the large-grained, "porous" samples
compared to the dense samples. This increased surface
area and expected dislocation generation by pores within
grains[10] would be consistant with higher fracture energy
values (the fracture energy for "porous" MgO of 100-μm

grain size is about 3.0×10^4 ergs/cm^2 compared to 1.0×10^4 ergs/cm^2 for dense samples of the same grain size). Rice[10] has previously indicated that higher fracture energies should be observed in such "porous" MgO samples.

Fracture energy (γ_f) is generally compared with fracture strength (σ_f) with a Griffith-type relation:

$$\sigma_f = k \sqrt{\frac{E\gamma_f}{c}} \qquad (1)$$

where k is a geometric constant relating to flaw-tip geometry, E is the elastic modulus, and c is the flaw length. For given microstructures, it would be expected that k and c would be similar and that fracture energy differences would be reflected in the strength measurements. On comparing the fracture energies given in Table I and Figs. 4 and 6 with the fracture strengths in Fig. 1, it is seen that the Griffith-type relation can only in part explain the results. For example, the ratio of the fracture strengths of the as-hot-pressed samples with and without LiF is 0.57, which is in good agreement with the value of 0.66 predicted from fracture energy values. On the other hand, fracture strength decreases with grain size; however, the fracture energy of the porous samples increases with grain size and that of the dense samples shows no strong dependence. Thus, one must conclude either that k and c vary in some undetermined manner with microstructure or that under certain circumstances fracture energy and strength are not comparable through a Griffith-type relation. It is proposed herein that the latter possibility is the more likely.

In the measurement of fracture energy by the work-of-fracture technique the crack propagates in a stable manner which allows for the possibility of the microstructure interacting with the crack front in a number of energy absorbing mechanisms. Fracture strength measurements are found from crack initiation at the largest stress concentration, and unstable crack propagation ensues. Thus, in fracture strength measurements it would be expected that the effect of microstructure would be more related to crack initiation than to propagation. Extensive evidence shows that the strength of MgO bodies such as those in this study is controlled by dislocation crack nucleation[10]; and hence, strength of MgO will not necessarily be consistent with fracture energy which relates to crack propagation. The lack of a clear decrease in fracture energy with increasing grain size (i.e. decreasing strength) is consistent with the strength of MgO being controlled by crack nucleation and fracture energy by crack propagation. On the other hand, strength and fracture energy will correlate when parameters lead to the same trend for both crack nucleation and propagation. For example, weakening of grain boundaries by additives facilitates both crack nucleation and propagation; strength and fracture energy then correlate well. Similarly, Rice[10] has indicated that pores within grains may effectively work harden grains and inhibit dislocation crack nucleation. Since previous work, as well as this study, indicates that fracture energies are higher for specimens with pores within grains, a correlation again exists between strength and fracture energy, but apparently owing to different effects.

CONCLUSIONS

1. Grain boundary precipitates in the as hot-pressed MgO with LiF causes intergranular fracture and low fracture energies.
2. For dense MgO with essentially no porosity, fracture energy did not appear to depend on grain size up to 200 μm.
3. For "porous" MgO with up to 1.0% porosity, fracture energy increases with grain size up to 100 μm. This behavior is thought to be caused by a shift in the location of porosity from intergranular at small grain sizes to within grains at large grain sizes.
4. Intergranular porosity encourages intergranular fracture; however, it has little, if any, effect on fracture energy.
5. In-grain porosity produces a very rough-appearing fracture surface with numerous cleavage steps and increases fracture energy.

6. When microstructure relates to crack propagation and initiation in different ways, fracture energy (as measured by the work-of-fracture technique) and strength cannot be compared on the basis of a Griffith-type relationship.

REFERENCES

1. J. A. Coppola and R. C. Bradt, "Measurement of the Fracture Surface Energy of SiC," J. Am. Ceram. Soc., 55, 455-60 (1972).
2. L. A. Simpson, "Effect of Microstructure on Measurements of Fracture Energy of Al_2O_3," J. Am. Ceram. Soc., 56, 7-11 (1973).
3. R. W. Davidge and G. Tappin, "The Effective Surface Energy of Brittle Materials," J. Mat. Sci., 3, 165-73 (1968).
4. H. G. Tattersall and G. Tappin, "The Work of Fracture and Its Measurement in Metals, Ceramics, and Other Materials," J. Mat. Sci., 1, 296-301 (1966).
5. G. D. Swanson and G. E. Gross, "Physical Parameters Affecting Fracture Strength and Fracture," Midwest Research Institute Report, #N00019-69-C-0161, January 1970.
6. F. J. P. Clarke, H. G. Tattersall and G. Tappin, "Toughness of Ceramics and Their Work of Fracture," Proc. Brit. Ceram. Soc., 6, 163-72 (1966).
7. D. B. Binns and P. Popper, "Mechanical Properties of Some Commercial Alumina Ceramics," Proc. Brit. Ceram. Soc., 6, 71-82 (1966).
8. P. L. Gutshall and G. E. Gross, "Observations and Mechanisms of Fracture in Polycrystalline Alumina," Eng. Fract. Mech., 1, 463-71 (1969).
9. G. D. Swanson and G. E. Gross, "Factor Analysis of Fracture Toughness Test Parameters for Al_2O_3," J. Am. Ceram. Soc., 54, 382-84 (1971).
10. R. W. Rice, "Strength and Fracture of Hot-Pressed MgO," Proc. Brit. Ceram. Soc., 20, 329-63 (1972).
11. R. W. Rice, "Fabrication of Dense MgO," U. S. Naval Research Lab., Report 7334, Washington, D. C., November 16, 1971.

12. R.W. Rice, "Characterization of Hot-Pressed MgO,"
     U.S. Naval Research Lab., Report 7335, Washington,
     D.C., November 16, 1971.
13. R.W. Rice, "Strength and Fracture of Dense MgO,"
     in Ceramic Microstructures, Their Analysis, Sig-
     nificance, and Production, (R. Fulrath and J. Pask,
     editors), John Wiley & Sons, N.Y., pp. 579-593 (1968).
14. Junn Nakayama, "Direct Measurement of Fracture
     Energies of Brittle Heterogeneous Materials," J. Am.
     Ceram. Soc., 48, 583-87 (1965).
15. A.G. Evans, "Energies for Crack Propagation in Poly-
     crystalline MgO," Phil. Mag., 22, 841-52 (1970).
16. C.T. Forwood and A.J. Forty, "The Interaction of
     Cleavage Cracks with Inhomogenieties in Sodium
     Chloride Crystals," Phil. Mag., 11, 1067-82 (1965).

DISCUSSION

A. Choudry (Univ. Rhode Island): If one does not assume
Griffith's equation, then the calculation of fracture energy
from experimental stress and strain would have to take
into account the equation of state to sort out what part of
imput strain energy caused diverse changes, e.g., fracture, temperature excursion, kinetic energy of fracture
fragments, loss due to internal friction, etc., unless one
just assumes that at the moment of fracture the entire
strain energy is converted into fracture which is not possible even from simple thermodynamic arguments. Are
such factors accounted for in your work?
Author: In the WOF technique used, fracture energy is
calculated by measuring the amount of work that a mechanical testing machine has to do to break a specimen in a
slow manner so that the crack propagates stably and no
kinetic energy is given to the resultant fracture pieces,
and dividing this work by the nominal surface area of both
fracture faces, taking no account of fine-scale irregularities.
No assumptions are made regarding Griffith's equation. We
do assume that for stable crack propagation the entire strain
energy is converted into fracture.

J. Reed (Alfred): Our results on fracture strength of yttria-stabilized zirconia parallel your results for MgO; the location of pores, i. e., in grain or at grain boundary is more critical than a few tenths percent porosity in specimens about 99. 5% of theoretical density. How was grain size determined? It would seem that for plots of fracture strength vs. grain size, the grain size at fracture origin is requisite. And for WOF - grain size plots another definition of grain size would seem appropriate. Are extremes in the grain size distribution controlling behavior?

Author: The grain size reported is an average. For fracture energy vs. grain size plots, I feel that the average grain size is appropriate since fracture energy is determined from a crack propagating through a cross-section of grain sizes. For fracture strength plots, I agree that the grain size at the fracture origin is more appropriate. For a more complete discussion of this, see "Fractographics Identification of Strength-Controlling Flaws and Microstructure," R. W. Rice, Symposium on Fracture Mechanics of Ceramics, Pennsylvania State University, July 1973, to be published by Plenum Press.

L. L. Hench (Univ. Florida): The Kapadia and Leipold grain boundary mobility data in these proceedings appear to have similar grain size and porosity dependencies as shown in your work. Does this suggest that a grain boundary sliding contribution is operative in your fracture energy measurements or that variations in porosity distributions may be influencing their data?

Author: We saw no evidence from scanning electron micrographs of any grain boundary sliding in our fracture energy measurements. ·

K. K. Verma (Alfred): You mentioned that pores inhibit dislocation motion. It is true, but a couple of these things can still happen, e. g., a dislocation pile-up adjacent to the pore, dislocations terminating as the pore adding to its size, dislocation motion by climb, glide, etc. Inhibited dislocations themselves can cause appreciable amount of strain energy in the local area around a flaw. If this energy is not released by one of the above mentioned phenomena, then would it not aid in fracture of the material?

Author: I agree that pores can interact with dislocations in a number of ways; however, I feel that the major contribution of in-grain pores to fracture energy is in promoting cleavage steps.

J.A. Pask (UC Berkeley): Can the "dense" and "porous" specimens be further differentiated on the basis of presence or absence of LiF originally added as a fabrication aid?

Author: Fracture energy values after annealing were essentially independent of the original LiF additions. Rice[10] similarly observed that strengths after annealing were not dependent on the LiF additions.